Tectonics and Geophysics of Continental Rifts

# NATO ADVANCED STUDY INSTITUTES SERIES

*Proceedings of the Advanced Study Institute Programme, which aims at the dissemination of advanced knowledge and the formation of contacts among scientists from different countries*

The series is published by an international board of publishers in conjunction with NATO Scientific Affairs Division

| | |
|---|---|
| A Life Sciences<br>B Physics | Plenum Publishing Corporation<br>London and New York |
| C Mathematical and Physical Sciences | D. Reidel Publishing Company<br>Dordrecht and Boston |
| D Behavioral and Social Sciences | Sijthoff International Publishing Company<br>Leiden |
| E Applied Sciences | Noordhoff International Publishing<br>Leiden |

Series C – Mathematical and Physical Sciences

*Volume 37 – Tectonics and Geophysics of Continental Rifts*

# Tectonics and Geophysics of Continental Rifts

*Volume Two of the Proceedings of the NATO Advanced Study Institute Paleorift Systems with Emphasis on the Permian Oslo Rift, held in Oslo, Norway, July 27–August 5, 1977*

edited by

I. B. RAMBERG
*Department of Geology, University of Oslo, Oslo, Norway*

and

E.-R. NEUMANN
*Mineralogical-Geological Museum, University of Oslo, Oslo, Norway*

**D. Reidel Publishing Company**
Dordrecht : Holland / Boston : U.S.A. / London : England

**Published in cooperation with NATO Scientific Affairs Division**

**Library of Congress Cataloging in Publication Data**

NATO Advanced Study Institute Paleorift Systems, with Emphasis on the Permian Oslo Rift, Norway, 1977.
Tectonics and Geophysics of Continental Rifts.

(Proceedings of the NATO Advanced Study Institute Paleorift Systems with Emphasis on the Permian Oslo Rift, held in Oslo, Norway, July 27–August 5, 1977: v. 2)
(NATO Advanced Study Institutes Series: Series C, Mathematical and Physical Sciences; v. 37)
Bibliography: p. Includes Index.
1. Rifts (Geology) – Congresses. 2. Geology, Structural – Congresses. 3. Geophysics – Congresses. 4. Rifts (Geology) – Norway – Oslo Region – Congresses. 5. Geology, Stratigraphic – Permian – Congresses. I. Neumann, Else-Ragnhild, 1948– II. Ramberg, Ivar B. III. Title IV. Series: NATO Advanced Study Institutes Series: Series C, Mathematical and Physical Sciences: v. 37.
QE606.N37 1977, vol. 2 551.1'3s (551.1'3) 78–16085
ISBN-13: 978-94-009-9808-7 e-ISBN-13: 978-94-009-9806-3
DOI: 10.1007/978-94-009-9806-3

Published by D. Reidel Publishing Company
P.O. Box 17, Dordrecht, Holland

Sold and distributed in the U.S.A., Canada, and Mexico
by D. Reidel Publishing Company, Inc.
Lincoln Building, 160 Old Derby Street, Hingham, Mass. 02043, U.S.A.

CONTENTS

CONTENTS OF
VOL. I PETROLOGY AND GEOCHEMISTRY OF CONTINENTAL RIFTS

CONTENTS

# INTRODUCTION

## PREFACE

The NATO Advanced Study Institute "Paleorift Systems with Emphasis on the Permian Oslo Rift" was held at Sundvollen near Oslo, Norway, 26. July - 5. August, 1977. The meeting included 6 field trips to various parts of the Oslo Region. 70 official participants and 16 observers from 14 countries attended the meeting.

The majority of the invited lectures, short research papers and progress reports presented at the meeting are published in two volumes, of which this is volume No. II. A table of contents for both volumes is included herein. The field trip guide is being published in the Norwegian Geological Survey Series, Vol. 327 (1978).

We are especially pleased to acknowledge the efforts of the many authors of contributed papers, and the able assistance of secretaries, proof-readers and other staffmembers without whose help these volumes would not have been possible.

Oslo, 20. February 1978.

Ivar B. Ramberg                                        Else-Ragnhild Neumann

Organizing Committee members:

| | |
|---|---|
| O. Eldholm<br>G. Grønlie<br>J. Naterstad<br>I.B. Ramberg (chairman) | Department of Geology, University of Oslo |
| J.A. Dons<br>B.T. Larsen (secretary)<br>E.-R. Neumann (secretary) | Mineralogical-Geological Museum, University of Oslo |
| K.S. Heier (chairman)<br>S. Huseby<br>B. Sundvoll | Norwegian Geological Survey |
| M.A. Sellevoll | Seismological Observatory, University of Bergen |
| K. Storetvedt | Geophysics Institute, University of Bergen |
| P.M. Ihlen<br>Chr. Oftedahl<br>F.M. Vokes | Geological Institute, University of Trondheim |
| E.S. Husebye | NTNF/NORSAR, 2007 Kjeller |

This volume is Scientific Report No. 40 of the Geodynamics Project.

The Geodynamics Project is an international programme of research on the dynamics and dynamic history of the Earth with emphasis on deep-seated foundations of geological phenomena. This includes investigations related to movements and deformations, past and present, of the lithosphere, and all relevant properties of the Earth's interior and especially any evidence for motions at depth. The programme is an inter-disciplinary one, coordinated by the Inter-Union Commission on Geodynamics (I.C.G.) established by the I.C.S.U. at the request of I.U.C.G. and I.U.G.S., with rules providing for the active participation of all interested I.C.S.U. Unions and Committees.

# PALEORIFT SYSTEMS WITH EMPHASIS ON THE PERMIAN OSLO RIFT OPENING ADDRESS

K.S. Heier

University of Oslo, or
Geological Survey of Norway,
7001 Trondheim, Norway

It is with considerable honor, pleasure and pride that I welcome you to this Study Institute. As students and lecturers from 14 countries we are together here with one common interest; the study of world rift systems. After the 1974 NATO sponsored meeting "Iceland and midocean ridges", we were unofficially approached by members of working group 4 (WG 4) of the Inter-Union Commission on Geodynamics (ICG) to consider undertaking a similar conference on Paleorift systems.

World rift systems are today considered by earth scientists as providing significant clues to the development of the earth's crust and to relations between continental and oceanic crust, as well as between crust and mantle. World rift systems are also of interest as possible source regions of mineralization in the crust. This makes an opportune moment to review our present knowledge of the Permian Oslo Rift and discuss it in the context of world rift systems in general.

In 1810 the German geologist Leopold von Buch first described parts of what we know today as the Oslo region. The first map and regional description, Das Christiania Übergangsterritorium, was published by B.M. Keilhau in 1838. During the period 1855-1880 Th. Kjerulf produced several maps and gave correct interpretations of several tectonic features of the region. He also gave the correct interpretation of the relationships between the volcanics and the underlying Cambro-Silurian rocks. However, the Oslo region became well known from the works of W.C. Brøgger during more than 50 years. I refer here to his series "Die Eruptivgesteine des Kristiania- (Oslo-) gebietes" in seven volumes. The first one in 1894, and the last one in 1934. In 1943 a new series "Studies on

*I. B. Ramberg and E.-R. Neumann (eds.), Tectonics and Geophysics of Continental Rifts*, xiii-xv.

the igneous rock complex of the Oslo region" was initiated by Olaf Holtedahl. Contributions to this series were all published in "Skrifter utgitt av Det Norske Videnskaps-Akademi i Oslo" and today consist of 24 numbers.

The most recently published thorough discussion, is by Ivar Ramberg "Gravity interpretaion of the Oslo Graben and associated rocks" which was published as a monograph in NGU Bull. 325 during 1976. Also the final report on the Norwegian geotraverse project in 1977 which was the Norwegian contribution to the IGP contains previously unpublished data on the Oslo rift.

The age of the Oslo rocks was long uncertain. The commonly held belief was that they were of Devonian age and relatively closely associated in time with the Cambro-Silurian rocks. Early age determinations by Fajans on thorite from the syenite pegmatites in the Langesundfjord indicated a much younger age of 225 m.y. On a student excursion led by Professor Olaf Holtedahl in 1931 fossils were found in a sediment sequence closely associated with the lava series dating the sediments as lower Permian. More recent age determinations on the plutonic rocks indicates that major emplacement occurred around 270 m.y. With improved analytical techniques we decided not only to determine the relative ages of the different rocks, but also approach the problem of how long the Oslo rift was active. This we consider a fundamental problem and our preliminary results will be presented at this meeting. In the initial stages of this study we received considerable support and assistance from the Isotope laboratory at the Geology Department, University of Oxford.

Of course the Oslo province and its structures has long been known, and it is considered one of the "classical" provinces in the world. It is also conveniently located in the most densely populated parts of Norway, and one would be justified in assuming that this is an area where truly all geological observations worth while have already been made. This is far from the case. We are still frustrated by our lack of knowledge. However, the last thirty years have improved conditions for work in the region. We now have excellent aerial photographs, topographic maps on the scale of 1:5 000, and last but not least, a good system of roads, constructed for the purpose of forestry. Thus we feel that now, and the near future is when we will make a major advance in our understanding of the area. NATO, through their Advanced Study Institute Programme, is the major single contributor to this meeting, and without whose assistance the meeting could not be arranged. The Universities of Oslo, Bergen and Trondheim, and the Geological Survey of Norway also contributed generously both directly and indirectly. Financial assistance was also received from:

The Norwegian Research Council for Science and Humanities (NAVF)
Amoco Norway Oil
Den norske stats oljeselskap A/S
Elkem-Spigerverket A/S
Esso Exploration and Production Norway Inc.
Norsk Agip A/S
Norsk Geofysisk Forening
Norsk Hydro A/S
Phillips Petroleum Company Norway
Saga Petroleum A/S & Co.
Sydvaranger A/S

This assistance is most heartly acknowledged.

The organizing committee of 16 persons has relied heavily on the efforts of Professor I.B. Ramberg, Mr. B.T. Larsen and Mrs. E.-R. Neumann. For clerical assistance we are indebted to Mrs. Inger Holm who will also act as secretary during the meeting. Though it is difficult to single out individuals I am certain the organizing committee agrees in that the credit for the success of this meeting largely should be given to these four individuals.

Finally it is my hope that this meeting will contribute to the understanding of Paleorifts in general, and to an increased knowledge of the Oslo rift in particular. I hope the proceedings of the meeting will turn into a classical reference work on the subject, and that the guides to our excursions to be published as a special NGU-Bulletin will benefit a large number of students to come.

# EVOLUTION OF CONTINENTAL RIFT SYSTEMS IN THE LIGHT OF PLATE TECTONICS

Kevin Burke

Department of Geological Sciences, State University of New York at Albany, Albany NY 12222 USA

ABSTRACT. Rifts occur at all stages of the Wilson Cycle of opening and closing of oceans, but those most likely to be preserved form either at continental breakup or continental collision. All preserved rifts are "failed rifts" because they have not "succeeded" in developing into oceans. Hundreds of ancient rifts ranging up to 2.6 Ga in age have been recognized. Axial mafic dike systems have been located in both ancient and active rifts and activation of the basalt-eclogite transition in these systems may be an effective way of promoting the repeated subsidence that characterizes many rift complexes, such as those of the North Sea. Ancient rifts striking into fold belts are known as aulacogens (after Shatski). Recognition (after Wilson) that many fold belts mark places where oceans have closed has made it possible to interpret numerous aulacogens as rifts formed at continental rupture and to interpret their history in plate-tectonic terms. In many ways the best known of rift systems are those formed at the opening of the Atlantic Ocean because they have been subject to intense petroleum exploration. Eleven classes of rifts with diverse histories can be distinguished among the rifts presently facing Atlantic-type continental margins and aulacogen histories are even more diverse.

## 1. RIFT SYSTEMS AND PLATE TECTONICS

Rifts are places where the entire thickness of the lithosphere has ruptured under tension. Generally, rifts occur where the lithosphere is abnormally thin and most active rifts today are in oceans or marginal basins. This was presumably also the case in the past. Continental rifts are much less abundant than oceanic rifts, but unlike rifts within oceans they stand a chance of being preserved in the geological record.

*I. B. Ramberg and E.-R. Neumann (eds.), Tectonics and Geophysics of Continental Rifts,* 1-9.

It is now ten years since widespread awareness among geophysicists followed Wilson's recognition of the plate structure of the lithosphere [1, 2]. From the point of view of geologists and geophysicists who work on continents, the most significant application of our new understanding of lithospheric structure has been Wilson's further suggestion [3] that it is appropriate to regard the history of the earth as recording a series of complex, interwoven, cycles of ocean opening and closing. In 1974, John Dewey and I proposed [4] that these cycles be called Wilson Cycles because Tuzo Wilson was the first to realize that such cycles are inevitable on a plate-structured earth.

It may seem paradoxical that plate tectonics, which emphasizes horizontal motion of parts of the lithosphere with respect to each other at rates of tens of mm per year, has proved critical in the understanding of rifts in which horizontal extension is small and typically takes place at speeds more than an order of magnitude slower, but recognition of the Wilson Cycle has permitted, for the first time, appreciation of the different tectonic environments in which intra-continental rifts may occur. It is now clear that rifts form in a variety of tectonic environments related to different stages in the Wilson Cycle. At all stages, although intra-continental rifts may develop to form oceans, many rifts fail to develop and (regarding oceans as successes) these have been called "failed rifts" or "failed rift arms". It is these "failed rifts" preserved within continents that can be compared to the intra-continental rifts that are active today.

We cannot be altogether sure to the extent to which such comparisons are justified. We cannot know, for example, whether the active Ethiopian or Baikal rifts will continue developing to become sites on which oceans form or whether they will fail to develop and remain preserved within the continents. However, a recent review [5] has shown that failed rifts associated with the opening of the Atlantic ocean greatly exceed in total length the length of ocean margin formed by continental rupture. (This abundance of failed rifts around the Atlantic is illustrated in Fig. 1). If the rift system in East Africa propagates and continental rupture eventually ensues, we can infer by analogy with the opening of the Atlantic that at least half the presently active rifts in Africa will be left as failed intra-continental rifts. For this reason comparison of active and failed rifts does appear justified.

Although rifts develop at most stages of the Wilson Cycle, preserved rifts within continents are mainly either those resembling the active rifts of Africa today or rifts like the Rhine and Lake Baikal graben, and it is appropriate to consider these types of rifts in relation to the Wilson Cycle.

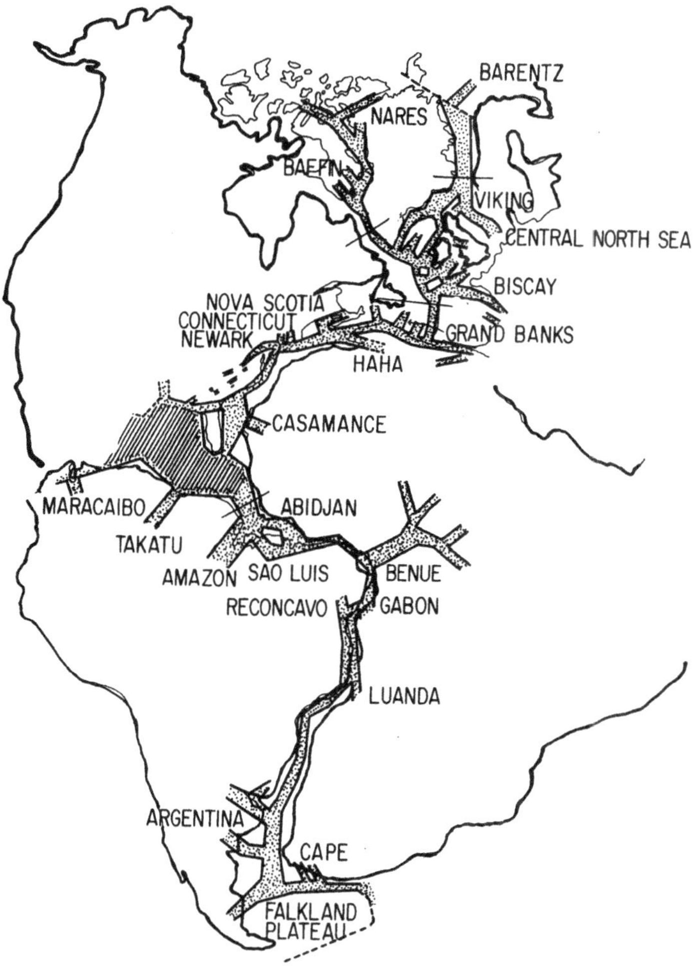


Fig. 1. Failed rifts associated with the opening Atlantic. Since failed rifts are numerous and exceed in total length the length of continental margin formed at rupture, it is inferred that there is probability that at least half the presently active rifts in Africa remaining as failed rifts if at sometime in the future that continent ruptures along the East African rift system.

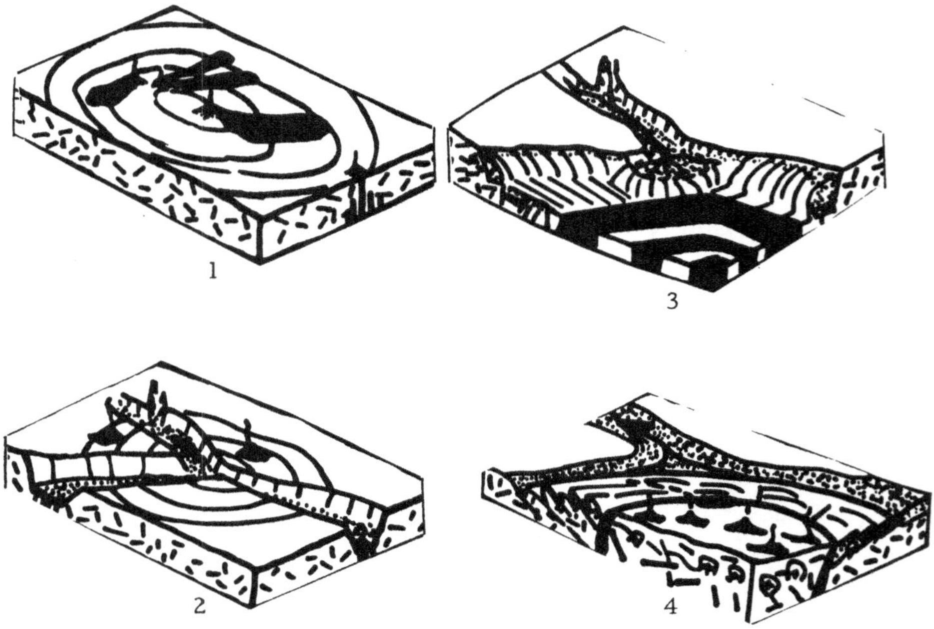


Fig. 2. African type rifts commonly form over a swell (1) where the lithosphere is thinned. A 3-armed pattern often develops (2) and two of these rifts may become a plate boundary, or boundaries, leading to ocean formation (3). A delta generally progrades down the failed rift arm. At the close of a Wilson Cycle continents collide (4), the former site of the ocean is marked by a suture (black line), sediments of the continental margin wedge shown in (3) are thrust back onto the continent from which they were eroded and eroded again to flow in the reverse direction along the failed rift, now an aulacogen. On the opposite side of the suture volcanoes cap a Tibetan-style plateau on top of a continent thickened at collision.

## 2. AFRICAN TYPE RIFTS

Modern African rifts have been active since the beginning of the Neogene and this development coincides with Africa's coming to rest over a mantle convection pattern [6,7]. Fig. 2 shows the sequence of development of an African rift from an uplift over thinned lithosphere through development of a triple-rift system to continental rupture [8]. With further development an ocean opens and a Wilson Cycle culminating in a continental collision follows. At continental collision the structure that started as an African-style rift becomes an aulacogen [9,10]. The pattern of rifting and the evolutionary sequence are generally more complex than the idealized pattern depicted in Fig. 2, some idea of the complexity being given in Fig. 1. I have suggested elsewhere [10] that eleven classes of failed rifts extending into continents may be distinguished, the classification reflecting the varied structural and stratigraphic styles of the failed rifts.

Much effort has been expended on studying the petrology and geochemistry of the igneous rocks of African type rifts. Basaltic rocks predominate and alkaline basalts are markedly more abundant than tholeiitic, suggesting equilibration of melt with mantle at depths nearer 100 km than 10 km. The great variety of rift igneous rocks can be readily explained as products of the two processes of fractional crystallization and partial melting. A most significant observation from the point of view of students of rift structure is that rift igneous rocks seldom show signs of reaction with continental crust. This is consistent with structural observations indicating that the common form of igneous intrusion in rift valleys is the axial dike, one or more of which underlies the rift floor. The cross-cutting form of these large igneous bodies makes country-rock reaction unlikely and where exposed, as in the Great Dike, little reaction is seen.

## 3. RHINE GRABEN TYPE RIFTS

Rhine graben [11] and the Lake Baikal rift [12] appear to have started to develop at times of continental collision in the Alpine and Himalayan mountains and this kind of rift (an impactogen) can be considered a product of continental impact. Well-documented fossil examples of these kinds of rift are relatively few, but Burke and Sawkins [13] have suggested that the Lower and Late Devonian rifting events of Central and Southern Europe may have been produced in this way by the Erian and Acadian collisions of the Caledonian-Appalachian fold belt [14], and Sengor has suggested (in conversation) that the Permian rifts of Northern Europe (including the Oslo rift) were produced in the same way as a result of Hercynian collision. Although the geometric relations between aulacogens and fold belts are similar to those between impactogens and fold belts (both types of rift run into a fold belt, usually at high angles), rift stratigraphy may permit the two kinds to be distinguished. Thick sediments in an aulacogen date back to continental rupture, but those in an impactogen date only from the time of continental collision. This kind of analysis has shown that the Ottawa-Bonnechere graben is an aulacogen as Schatski [15] suggested, rather than an impactogen [10].

## 4. RIFTS AT OTHER STAGES OF THE WILSON CYCLE

These rifts are apparently less often preserved than others because they are liable to be tectonized beyond recognition. Numerous rifts in the Mediterranean that have been formed during Neogene stages of ocean closure - for example, those in Western Turkey [16] and those related to the rotation of Corsica and Sardinia away from Europe can be expected to be highly tectonized if Africa and Europe continue to converge, and it is not surprising that this type of rift is seldom preserved.

The active rifts of the Rio Grande and the Basin and Range province in the Southwestern United States and Mexico are apparently products of tension related to the imperfection of the rotation of the North American plate past the Pacific plate along the San Andreas fault. Comparable features have not yet been recognized in the fossil record, but this may also be because subsequent convergence is likely to destroy this kind of rift structure.

Rifts like the Pliocene to Recent Thakkola, high in the Himalayas [18] have been produced at collision and might be thought unlikely to survive erosion, but the Lower Devonian rifts of Western Norway, of which the Solund basin [19] is the best described, seem very similar to the Thakkola. The Norwegian rifts date from immediately after the Erian Caledonide collision [14].

Marginal basins develop from Andean arcs by rifting along the volcanic crest where the lithosphere is thin. The Seoul-Wonson graben [10, fig. 4] is a rift formed in association with the opening of the Japan Sea and its origin appears to date that feature as of low Miocene age.

## 5. RIFT REACTIVATION

The structure of the entire thickness of the lithosphere is modified by rifting commonly for a distance of hundreds of km along strike and this vertical discontinuity in lithospheric structure is liable to be picked up whenever, for example, horizontal deviatoric stresses are applied to the lithosphere. This is one reason why rifts are so prone to reactivation - see for example [20]. The Oslo graben appears to have been first active in latest Precambrian times when it localized an unusually thick sparagmite sequence and suffered the Fen alkaline intrusions. This event appears related to the opening of Iapetus [10] and the Permian event, perhaps related to Hercynian collision, appears to have been localized by the older structure.

A characteristic style of rift reactivation is rapid subsidence and sediment accumulation, and this is nowhere better seen than in the North Sea rifts [21]. It has been suggested that activation of the basalt-eclogite transition in axial dikes underlying reactivated rifts is a possible way of inducing this type of subsidence [10].

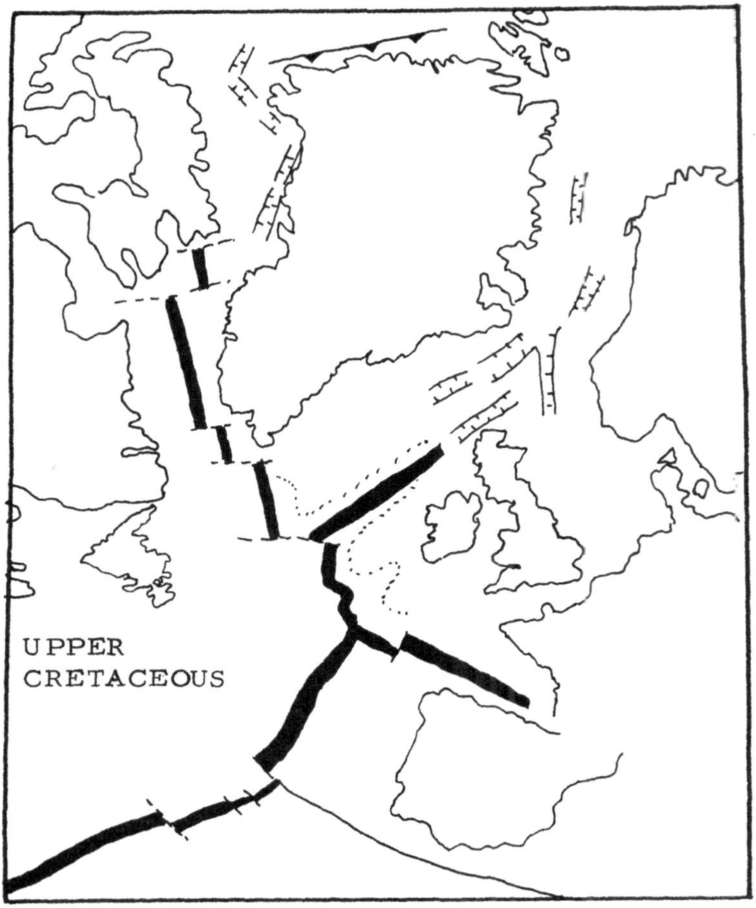


Fig. 3. Sketch based on [25] showing how ridges in the Rockall trough and Labrador Sea during the Upper Cretaceous passed into rifts along strike closer to the poles describing the motion of North America wrt Greenland and Greenland wrt Europe. Note the contemporary convergence in Ellesmere Land.

## 6. VARIATION IN RIFTING WITH TIME

Rifts are general associates of the Wilson Cycle and it has been argued that evidence of operation of the cycle can be seen in rocks of all ages [22] although the style was perhaps slightly different in the Archean [23] where no failed rifts have yet been recognized. The oldest preserved failed rift structure is the Great Dike of Rhodesia (2.5 Ga) which exposes the axial dike of a rift-valley system and terminates in fold belts, with suture zones making places where oceans have closed, in the Zambezi and Limpopo valleys. Precambrian rifts show no systematic differences from later rifts in structure, igneous rocks, or sediment fill.

Sawkins [24] has pointed out that rifts, apparently of African type, were common at about 2.1 Ga ago over much of the world and has discussed what the significance of such a concentration might be.

## 7. CONCLUSION

Rift studies are beginning to profit from the integrated approach that is clear in the program of the Oslo Institute. Studies of active and fossil rifts are complementary: in the former we can see rift processes happening and in the latter we can study sequences of rift events over tens and even hundreds of millions of years. For example, it is now clear that rifting, rather like that in East Africa today, preceded the opening of the Central Atlantic for a period of about 40 m.y. and a similar rift episode preceded the opening of the South Atlantic for about 25 m.y. [5]. Since the present episode in East Africa has been going on for about 25 m.y., it seems that it may well be a precursor to continental rupture.

Nowhere can rift evolution be better studied than in Northwest Europe where many rifts contain thick sediment piles with producing oil fields. For example, Fig. 3 is a sketch [25] illustrating conditions in the North Atlantic in the Upper Cretaceous. Spreading ridges lay in the Labrador Sea, in the Rockall trough, between Iberia and the Grand Banks and in the Bay of Biscay. Poles describing the motion between North America/Greenland and Greenland/Europe, across the first two ridges, lie near the northern coast of Greenland and contemporary rifts lie roughly along strike of the Labrador and Rockall spreading centers in Baffin Bay and the Norwegian Sea. Spreading rates across the rifts are too slow to be meaningful and it looks as though there was at this time a transition from spreading to rifting in this part of the North Atlantic. (Convergence has been shown in Ellesmere Land in the sketch as this helps in summing the motions, but the timing of folding on the ground is not well constrained.)

With the wealth of data available for offshore Northwest Europe it should be possible to work out something of the way in which the ridge to rift transition took place at this time and also about how it gave way to spreading in the Eocene.

## REFERENCES

1. J.T. Wilson, Nature, 207, 343, 1965.
2. J.T. Wilson, EOS, 58, 366, 1977.
3. J.T. Wilson, Proc. Am. Philos. Soc., 112, 309, 1968.
4. J.F. Dewey and K. Burke, Geology, 2, 57, 1974.
5. K. Burke, Tectonophysics, 36, 96, 1976.
6. R. Thiessen, K. Burke and W.S.F. Kidd, EOS, 58, 503, 1977.
7. K. Burke, Tectonophysics, 36, 197, 1976.
8. K. Burke and A.J. Whiteman, Uplift, Rifting and the Breakup of Africa, in: Implications of Continental Drift to the Earth Sciences, ed. by D.H. Tarling and S.K. Runcorn, Acadmic London, 1973.
9. K. Burke and J.F. Dewey, J. Geol., 81, 406, 1973.
10. K. Burke, Ann. Rev. Earth Planet. Sci., 5, 371, 1977.
11. A.M.C. Sengor, Geology, 4, 779, 1976.
12. P. Molnar and P. Tapponnier, Science, 189, 419, 1975.
13. K. Burke and F.J. Sawkins, in press.
14. J.F. Dewey and W.S.F. Kidd, Geology, 2, 543, 1974.
15. N. Schatski, Vergleichende Tektonik alter Tafeln, Akademie, Berlin, 1961.
16. J.F. Dewey and A.M.C. Sengor, in press.
17. J.F. Dewey, W.C. Pitman, W.B.F. Ryan and J. Bonnin, Geol. Soc. Am. Bull., 84, 3137, 1973.
18. A. Gansser, Geology of the Himalayas, Wiley, London, 1964.
19. R. Steele, Tectonophysics, 36, 207, 1976.
20. R.B. McConnell, Geol. Rundsch., 63, 389, 1974.
21. A.W. Woodland, ed., Petroleum and the Continental Shelf of Northwest Europe, Wiley, London, 1975.
22. K. Burke, J.F. Dewey and W.S.F. Kidd, Tectonophysics, 40, 69, 1977.
23. K. Burke, J.F. Dewey and W.S.F. Kidd, Dominance of Horizontal Movements, Arc, and Microcontinental Collision, in: The Early History of the Earth, ed. by B.F. Windley, Wiley, London, 1976.
24. F.J. Sawkins, Geology, 4, 427, 1976.
25. J.D. Sclater, S. Hellinger and C. Tapscott, EOS, 58, 367, 1977.

# EVOLUTION OF THE EARTH'S CRUST*

Stephan Mueller

Institut für Geophysik,
Eidgenössische Technische Hochschule
(Swiss Federal Institute of Technology),
CH-8093 Zurich, Switzerland

ABSTRACT. The transformation of continental to oceanic crust is part of a cyclic evolutionary process which seems to be episodic in nature. In the initial phase of rifting the crust-mantle system is thermally altered and a graben is formed at the surface. Necking of the sialic low-velocity zone in the upper crust is the governing mechanism at the beginning of crustal attenuation. Magmatic intrusions originating in an anomalous mantle under the rift modify the intermediate and lower crustal layers. As this process continues a "pseudo-oceanic" crust is generated, whose top part is presumably of intermediate chemical composition. During the final stage of "oceanisation" the intermediate crust vanishes completely and the lower crustal layer is severely altered by magmatic injections from mantle sources. New oceanic crust is produced in this "ridge" phase and moves away from the spreading centre to the subduction zones where it is consumed. Parts of it will return to the upper crust in the form of igneous intrusions thus contributing to the formation of new continental crust.

## 1. INTRODUCTION

The earth's crust is a relatively thin shell whose structure and composition undergo changes both in space and time, primarily as a result of dynamic processes in the upper mantle and of crustal interaction in the zones of contact between different lithospheric

---

* Contribution No. 189, ETH-Geophysics, Zurich, Switzerland

*I. B. Ramberg and E.-R. Neumann (eds.), Tectonics and Geophysics of Continental Rifts, 11-28.*

plates. There are profound differences in origin, thickness and age of the crust under the oceans and in the continents.

It is now generally accepted that the present distribution of continents has resulted from the break-up and joining together of previous distributions of the continental masses. These dynamic processes have probably been active for quite some time, possibly ever since the earth first acquired a crust. Through the concept of sea-floor spreading and the theory of plate tectonics the earth's main structural features as created during the last 200 million years can be readily understood. But how can we tell if sea-floor spreading, involving plates of lithosphere that drift about the surface of the earth, has operated throughout geological time - possibly in a cyclic fashion [55]? The best place to look for evidence is obviously on the continents, whose crust is much older than the present oceanic crust. In the continental crust clues can be found that reveal a long and complex geological history containing relics of more or less "successful episodes of oceanisation".

Crustal attenuation in continents may come about by several tectonic processes, such as subcrustal erosion (magmatic stoping), igneous injection, uplift and surface erosion, among others. When the crustal thickness decreases, isostatic subsidence will occur aided by taphrogenic rifting, and sedimentary basins are formed. If the continental crust is thinned even more, it may become so thin and perforated by magmatic impregnation that it cannot be easily distinguished anymore from new oceanic crust. In the final stage of this evolutionary chain we thus arrive at a structure which will eventually correspond to a mid-ocean ridge system. At these locations new oceanic crust is generated by upwelling material from the mantle. The aging oceanic crust thickens as it moves away from the ridge crest to the subduction zones where it is consumed, but part of it returns to the upper crust in the form of igneous intrusions and thus contributes to the formation of new continental crust. At this point the orderly cycle of crustal evolution is complete. This concept provides the explanation for many geophysical observations, such as the pattern of seismicity, the distribution of volcanic belts and heat flow provinces.

## 2. INTRA-PLATE STRUCTURE AND INITATION OF RIFTING

In general, the earth's crust shows great diversity both laterally and vertically. Deep seismic sounding results have demonstrated a block-type structure of the continental crust with continuous layering extending only over limited ranges. Nevertheless, consistent structural features can be derived from purely geophysical observations [39].

In Fig. 4 (left margin) a characteristic P velocity-depth function is depicted which may serve to illustrate the more or less typical features of the crust under a continent. Beneath a near-surface low-velocity sedimentary layer (4.0 km/sec) usually a zone of positive velocity gradient is found in the top part of the crystalline basement (5.9 km/sec). It is underlain by the sialic low-velocity zone in the upper crust at a depth of about 8 to 10 km [30]. The intermediate (6.3 km/sec; hatching in Fig. 4) and lower (6.7 km/sec) layers of the crust are separated by the Conrad (C) discontinuity, with the possibility of a slight velocity inversion immediately above it, i.e. at depths of around 20 km [13]. An abrupt velocity jump (from 6.7 to 8.2 km/sec) occurs at the Mohorovičić (M) discontinuity. There are indications of a velocity anisotropy in the uppermost mantle [3,4] as well as of the presence of a high-velocity layer (~8.5 km/sec) at depths of around 50 km [14].

Prior to the initial phase of rifting an increase in heat flow from the mantle will cause the geotherms to rise, which in turn will lead to metamorphism in the lower crust. A distinct lowering of the P velocity in the lowermost part of the crust immediately above the crust-mantle boundary will result as indicated in Fig. 1 b. In this thermally altered continental crust the classical Conrad (C) discontinuity is now interpreted as a thin - possibly laminated - transition zone, which is only a few kilometers thick and which is separated from the Mohorovičić (M) discontinuity by a depth range of lower velocity [39]. Deep seismic reflection observations (see e.g. [10]) suggest that the crust-mantle transition must be relatively sharp with a pronounced lamination at the top of the uppermost mantle in order to account for the high-frequency character of the near-vertical reflections.

A crustal model of this same type was first proposed in 1969 for the Rhinegraben area in Central Europe [32]. It is depicted by the solid line in Fig. 1 a and differs from the "standard model" in Fig. 1 b by the very pronounced low-velocity zone in the middle part of the crust. This feature as well as the second inversion in the lower crust seem to be typical for the initial taphrogenic phase in the rifting process of a continental plate. In the depth interval between 19 and 25 km a lamellar structure is indicated from seismic reflection measurements in the Rhinegraben [11,35]. It is probably due to basic intrusions into the lower crust under the rift system (see also later discussion in conjunction with Fig. 2).

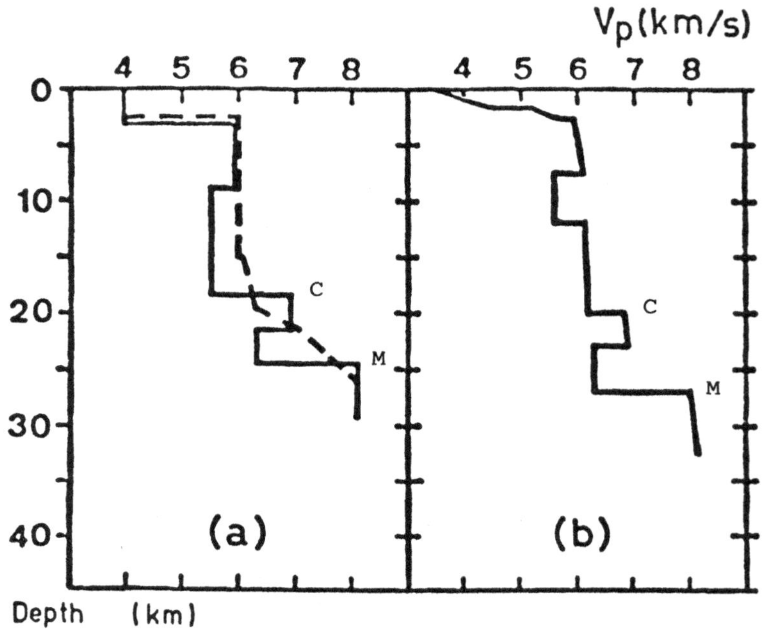


Fig. 1. Typical velocity-depth functions for tectonically active continental areas: (a) Upper Rhinegraben [32,35]; dashed line represents the average for the southern part of the graben [12]. (b) Franco-Swiss Folded Jura Mountains [38]. C = Conrad Discontinuity; M = Mohorovičić Discontinuity.

## 3. MECHANISM OF RIFTING

Thinning of the earth's crust as illustrated in Figs. 1 1 and 4 seems to be a general feature of graben formation. Crustal necking resulting from horizontal tensile stress was proposed as early as 1933 [8]. This mechanism has been revived and was applied to the Baikal rift zone some years ago [2]. In a somewhat modified form emphasizing the prominent role of the sialic low-velocity channel in the upper part of the crust this concept has been invoked to explain the mechanism of taphrogenesis in the Rhinegraben area [17]. The lowered P velocity in the upper or middle crust (see Fig. 1) is associated with a corresponding lowering of density [31,33], thus introducing an inherent gravitational instability into the crust. This zone of velocity and density inversion is presumably also a zone of weakness [38,39] and may, therefore, act as a source of tectonic movements, particularly in the early stages of graben formation when this crustal layer attains a relatively large thickness (c.f. Fig. 4).

If the sialic low-velocity channel is identified with a semicontinuous zone of granitic intrusions into the upper crust [31,33], its viscosity will be one or two orders of magnitude lower than that of the crystalline basement above or that of the lower crustal layer beneath it [44]. When the earth's crust is stretched as indicated in Fig. 2, plastic flow will occur and the layer with the lowest viscosity, i.e. the sialic low-velocity channel, will tend to decrease in thickness as long as the applied stress does not exceed the tensile strength. Continued stretching will concentrate the stresses in the more viscous brittle basement layer above. When the breaking strength is reached and exceeded, a fault will develop which may cut through the entire basement layer (see e.g. [10]). After fracturing, the creep rate in the low-viscosity layer will increase, aided by the gravitational instability, and a neck will be produced somewhere under the rift system, as for instance in the Kraichgau [32] and Kaiserstuhl [1] regions of the Rhinegraben Rift System. In this context it should be noted that the depths of earthquake foci in the Rhinegraben proper with one exception do not exceed 12 km thus placing the hypocentres close to the zone of contact of the crystalline basement and the sialic low-velocity channel [33,35,37].

Once the complete graben is developed the fractured central wedge block will sink until the gap between the uplifted basement flanks has been closed [17]. An arch or dome spanning 150 to 200 km will thus be formed by the two basement shoulders and the

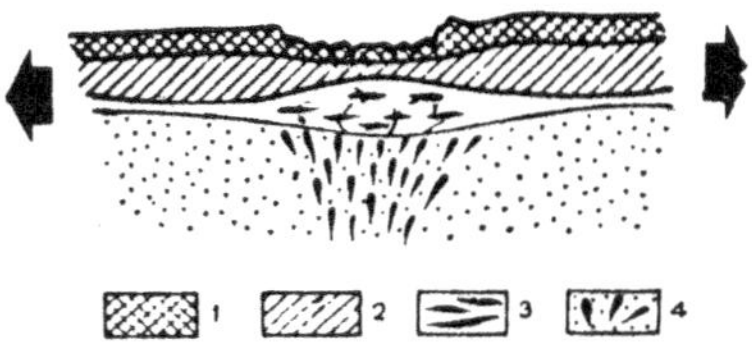


Fig. 2. Model of rifting by plastic deformation of the sialic low-velocity zone in the middle part of the crust. The sedimentary cover is omitted for simplicity.
1 = Crystalline basement with high viscosity
2 = Sialic low-velocity channel with low viscosity (Necking as the last stage of taphrogenesis must not necessarily occur under the graben axis, but could be also shifted to the flanks)
3 = Intermediate and lower crustal layer with basaltic intrusions
4 = Anomalous mantle ("cushion") impregnated by low-density magmatic material

graben segment. Underneath this arch the lower crust will be partially released from the weight of the overburden. This reduction of the lithostatic pressure will facilitate the intrusion of basaltic magma from the underlying mantle [54] as indicated in Fig. 2. By means of this process the thickness of the lower crustal layer may be considerably increased as has indeed been observed in the Rhinegraben [35]. Deep seismic reflection profiling in the northern part of the Rio Grande Rift [40] indicates the presence of intrusive bodies in the lower crust and a swelling of the lower crust under the rift. Micro-earthquake surveys in an adjoining area outlined a magma body in the lower crust, situated below a sharp discontinuity [48] which would correspond to the lower boundary of the sialic low-velocity channel in Fig. 2. It can, therefore, be surmised that the intermediate and lower crust under continental rift systems is characterized by a "lamination" of high- and low-velocity material which produces the typical striate pattern observed in deep seismic reflection profiling [11].

The sialic low-velocity channel above might be due to large syenitic and granitic batholiths of lower density and P velocity rising towards the surface by means of piece-meal magmatic stoping.

As in the Oslo Graben they could have been derived from an inferred monzo-dioritic magma chamber fed by a deeper "gabbroic source" [45] located at a level corresponding to the intermediate and lower crustal layer in Fig. 2.

As mentioned above the pressure reduction caused by the crustal arching will involve partial melting in the upper mantle [54]. The rising light magma will impregnate the uppermost mantle thus eventually elevating the crust-mantle boundary (cf. Fig. 1 a and Fig. 4). With this is associated a regional up- al uplift of the graben area and its flanks [46]. As the updoming continues a cushion-shaped body of "anomalous" mantle material will replace the original crust-mantle boundary [43]. The sialic low-velocity channel will be squeezed off to the sides and the graben wedge cannot sink deeper into this low-viscosity layer. Any further increase of crustal tension will attenuate the crust even more. The growing gaps can no longer remain closed and volcanic activity will strongly increase in the region of the graben, as is, for instance, found in the East African Rift System. Eventually the attenuated crust (of the graben type) will break up and develop into a continental margin (see Fig. 4). In this phase the entire crustal column will be lifted up by 5 to 10 km thus forming a "pseudo-oceanic" crust of intermediate composition. As this process continues the sialic part of the crust will disappear first, the intermediate crustal layer will thin out more and more, but the lower crust will increase in thickness considerably.

## 4. TRANSITION FROM CONTINENTAL TO OCEANIC STRUCTURE

It is possible to study the various phases of this acute rifting at present in eastern Ethiopia. Three different rift systems with their associated volcanic activity merge there forming the triple junction of the Afar Depression (see Fig. 3) - a rather unique feature on earth. There, in a relatively limited area, it is possible to find different stages of rifting in continental and oceanic structures. In fact, most of our knowledge about the development from a mature continental rift (e.g. the East African Rift System) to an embryonic oceanic crust (e.g. the central trough of the Red Sea or the Gulf of Aden) comes from this region.

Extensive deep seismic sounding campaigns were carried out over the past few years in the Gulf of Aden [15,25,26], the Red Sea [9,15], the Kenya Rift [20,21], the TFAI (Djibouti) area [27,47] and the Afar Depression [6]. In Fig. 3 a schematic

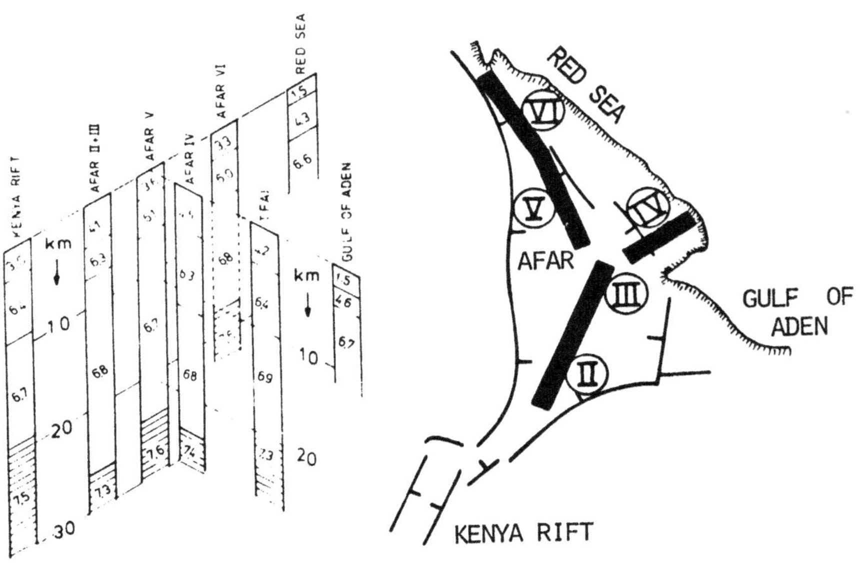


Fig. 3. Schematic fence diagram of crustal models (left) for the Afar Depression and the adjacent rift systems of the Gregory Rift in Kenya, the Red Sea and the Gulf of Aden [6]. Numbers are P velocities in km/sec. The position of the seismic refraction profiles in Afar is indicated on the right.
AFAR IV = southern portion of the Danakil Horst
TFAI = Djibouti Region, including the Gulf of Tadjura [27,47].
Horizontal hatching designates "anomalous" mantle.

summary of crustal models for the triangular Afar Depression and its adjacent rift systems [6] is presented. It can be seen that the top of the anomalous mantle is markedly depressed by 7 to 9 km under the Afar. While in the embryonic oceanic structures of the Red Sea and the Gulf of Aden the sequence of oceanic layers 2 (4.3 to 4.6 km/sec) and 3 (6.6 to 6.7 km/sec) is clearly defined, a completely different structure is found for the entire Afar (including the Danakil Horst), the TFAI (Djibouti) region and the northern end of the Kenya Rift. It is characterized by an uplifted "intermediate" continental layer with compressional velocities between 6.0 and 6.4 km/sec above a lower crustal layer with P velocities of 6.7 to 6.9 km/sec which in continents is usually interpreted as being of "gabbroic" composition. The intermediate layer is much thinner in central Afar as compared to the rift and horst regions around the depression and disappears as the "oceanic" structures of the Red Sea and of the Gulf of Aden are approached.

The existence of the Afar Depression cannot be understood if this structure is considered to be an isolated segment of the world rift system [42]. It can be demonstrated that the Arabian plate in the northeast must have rotated counterclockwise thereby creating the rift systems which ultimately led to the opening of the Red Sea and of the Gulf of Aden [19]. In the southwest the African plate has been weakened by graben faulting forming the Main Ethiopian Rift and the Gregory Rift in Kenya. An upward directed transport of mantle material caused the break-up of the crust and the dome-like upwarping of the adjacent plateaus. In this still continuing process the northern part of the Afar Depression (Profiles V and VI in Fig. 3) and the Gulf of Tadjura [47] have reached an advanced stage in the transition from a continental rift to an oceanic ridge system with its associated spreading of newly created crust.

The presence of a 3 to 4 km thick layer with P velocities of 6.0 to 6.1 km/sec in those two regions attests to the fact that the upper crust under Afar differs markedly from a "normal" oceanic crust. It is most similar to an oceanic structure in northern and central Afar, but the similarity is less striking in the southern part of the depression. Voluminous dyke injection must have altered the originally continental crust such that it appears as an attenuated "intermediate" crust. The wide variety of rock types and of volcanological forms indicates complex differentiation and transformation processes within the crust which are coupled with progressive crustal attenuation [23]. Alkaline granite intrusions subsequently followed by basaltic dykes are reported from several localities which must be related to intra-crustal processes during the rift development [24]. In a thorough discussion of the nature of the Afar crust [5] it has been stated that the nature of the basalts found is compatible with an "oceanic origin", but that rhyolites seem to be more widespread than in the ocean floor. It has further been suggested that these rhyolites must have been formed mainly by fractionation of basalts in shallow magma chambers at the intersection of different fault zones. The alternative suggestion advanced here is that these silicic rocks are the last traces of the "vanishing" intermediate crustal layer under rift zones, at a stage just prior to a complete break-up of the continental plate.

As the thinning of the crust continues the intermediate crustal layer will disappear totally and the lower crustal layer will move closer to the surface (through the stages of the Red Sea trough and the central Gulf of Aden). The structure finally approaches that of a mid-ocean ridge system (see Fig. 4), or of a spreading centre, at which new oceanic crust is being produced. Once this phase has been reached the evolutionary process of transforming continental crust into oceanic crust is complete.

## 5. CRUSTAL EVOLUTION

In Fig. 4 the sequence of structural changes in crustal evolution is summarized. Starting with a normal crust-mantle system in a tectonically active area of a continent, as e.g. the Molasse Basin in the northern foreland of the Alps [13,14], it is seen that under a sedimentary cover the crystalline basement is found at a depth of 2 to 3 km. It is underlain by the sialic low-velocity zone, which most likely consists of mushroom-shaped granitic intrusions into the upper crust [31,33,39]. The middle crust is characterized by intermediate P velocities, with the possible indication of a second zone of velocity inversion (not shown in Fig. 4) at depths between 16 and 22 km [13], which could possibly be related to the existence of a sub-volcanic storage chamber of differentiated monzo-dioritic material [45]. This reservoir has probably been fed by a gabbroic source at deeper crustal levels above the M-discontinuity. The transition from the lower crust to the uppermost mantle is relatively abrupt and is characterized by a pronounced velocity anisotropy immediately below the crust-mantle boundary [3].

Three factors seem to be essential for a break-up of a continental plate; namely, (1) the presence of an intermediate crustal layer with a P velocity of 6.3 to 6.4 km/sec in the middle part of the crust, (2) the existence of a gravitational instability in the upper crust, i.e. the presence of the sialic low-velocity zone, and (3) the large-scale thermal activation and updoming of the crust-mantle boundary with the associated formation of an "anomalous" uppermost mantle (asthenospheric diapir).

At the beginning of rifting the lower and intermediate crustal layers become thinner [22], while the sialic low-velocity zone (LVZ) increases in thickness, probably through a continued differentiation at intermediate levels in the crust. At the same time a crustal thinning of about 20 per cent is observed, as e.g. in the southern Rhinegraben [12]. An even more pronounced anisotropy of $P_n$ velocities is required for continental rift structures -

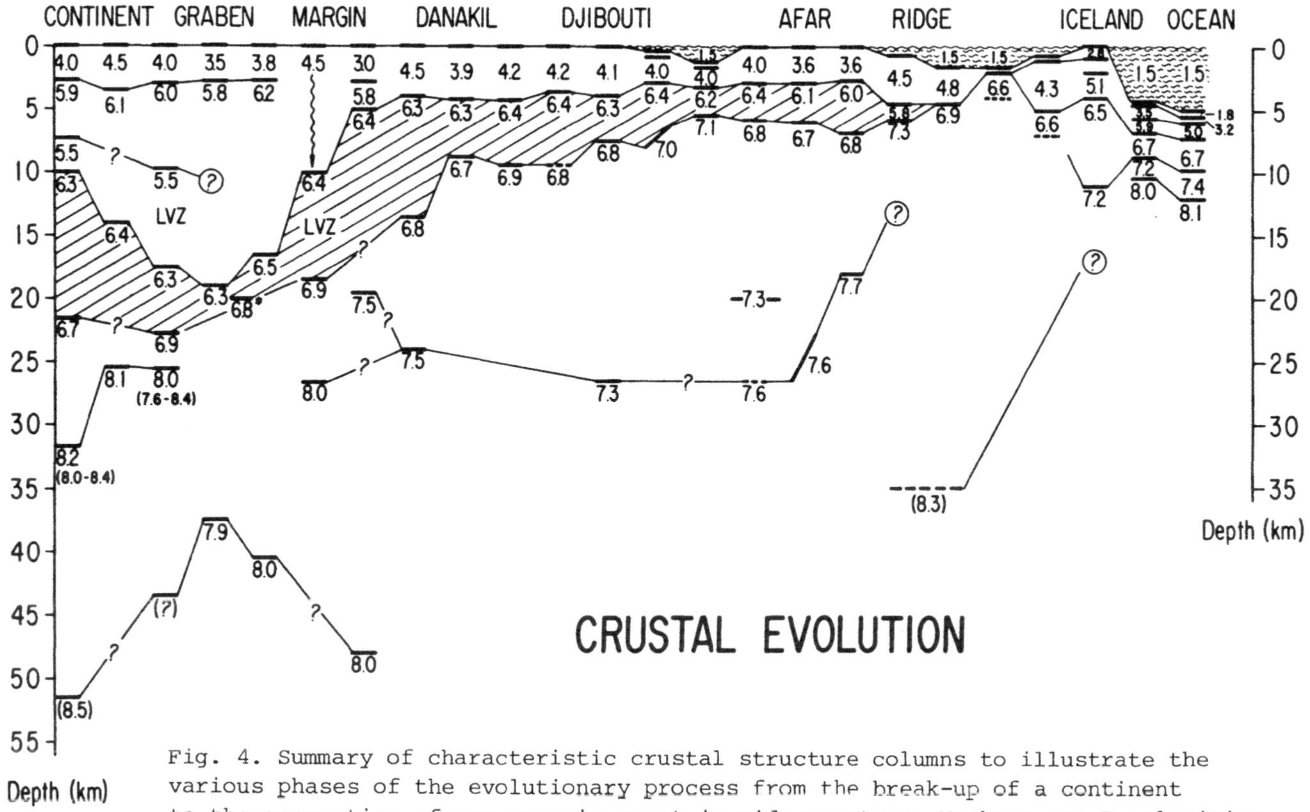


Fig. 4. Summary of characteristic crustal structure columns to illustrate the various phases of the evolutionary process from the break-up of a continent to the generation of new oceanic crust in ridge systems. Numbers are P velocities in km/sec. Hatching signifies the "intermediatc" crustal layer.

e.g. in the Rhinegraben [4] - in order to satisfy the regional observations. The high apparent velocities measured can only be explained if a preferred orientation of an olivine-bearing uppermost mantle in the present stress field is assumed [18]. In the more advanced graben stage of the East African rift system also a strong azimuth dependence of the slowness anomalies of teleseismic earthquake signals was found which was attributed to the shape of the upper surface of the anomalous mantle zone [28]. As far as the type of crustal structure in graben areas is concerned, a similar situation is encountered beneath the flanks of the Gregory Rift in Kenya [21,28], of the Rio Grande Rift in New Mexico [7,52] and under the Ethiopian Highland [6].

The transition from the "graben" to the uplifted "margin" phase of evolution is characterized by a gradual decrease in the thickness of the sialic layer of the upper crust. At the same time a sizeable thickening of the intermediate layer (6.4 km/sec), which - as, for example, in Portugal [34] - may be interrupted by a pronounced low-velocity zone (LVZ). It is tempting to speculate that this zone is identical with the monzo-dioritic magma chamber postulated to exist at intermediate crustal depths [45]. In fact, the crustal characteristics of the Oslo region [53] would put this rift into the "margin" category, somewhere between southwestern Portugal [34] and the northern part of the Kenya Rift [20,21]. This type of "structure" would probably also correspond to the "magnetic quiet zone" of continental margins.

Attenuated continental crust is also present under the Danakil Horst (see profile AFAR IV in Fig. 3). Although the available refraction seismic data [6] only refer to the southern extension of this structure, they correlate well with the results obtained in the neighbouring TFAI (Djibouti) region [27,47] as can be seen in Fig. 3. The depth to the crust-mantle boundary there ranges from 20 to 25 km, and the second layer of presumably intermediate composition has a thickness of about 6 to 9 km, i.e. it is twice as thick as in the central and northern portions of the Afar Depression, in the Gulf of Tadjura and at the western end of the Gulf of Aden.

As the intermediate crustal layer starts to decrease in thickness the lower crustal layer will grow proportionately (see Fig. 4). A new type of crust is formed, which is "pseudo-oceanic" in nature and which occupies most of the Afar Depression. It differs from true oceanic crust by its excessive thickness (15 to 20 km) of the lower crustal layer corresponding to the oceanic layer 3. In this comparison care must, of course, be exercised to correct the P velocities for the appropriate pressure and temperature conditions. Following the arguments presented for the Rhinegraben [17] it can be estimated that the velocity decrease caused by the depressurization and the temperature increase in the Afar-Djibouti crust is of the order of 1 per cent which is well within the experimental scatter. Definitely the most "mature" portion of the Afar is the axial zone in the central and northern part of the depression (profiles V and VI in Fig. 3). There the structure approaches more and more that of an active spreading ridge, although the crustal thickness still exceeds considerably that of a "pure" mid-ocean ridge. The velocity-depth distribution is also quite different from that found in the axial zone or on the flanks of mid-ocean ridge systems [51].

The most recent structural model of the Mid-Atlantic Ridge [16] has not yet been included in Fig. 4. From detailed seismic refraction work it can be shown that at and near the ridge axis the crust is about 7.5 km thick, under a water layer whose depth ranges from 1.5 to 2.5 km. The formation of the oceanic layers 3A and 3B seems to take place within a few kilometers of the axis. A normal upper mantle P velocity of 8.1 km/sec has been measured only 10 km off the axis on either side, while under the axial trough an anomalous mantle with a P velocity of 7.6 km/sec was found at a depth which corresponds to a zone of high P wave attenuation [50]. Shear waves propagate across the ridge axis, thus precluding the existence of any sizeable magma chamber within the crust. The new ridge structure is very similar to the model of rifting shown in Fig. 2, but now the lower crustal layer (3B) filled with mantle-derived intrusions has a P-velocity of 7.2 km/sec, and the neck caused by plastic deformation appears in the oceanic layer 3A, whose P velocity is markedly lowered by the high temperature postulated for that depth range [49].

For quite some time it has been maintained that Iceland is an "outcrop" of the Mid-Atlantic Ridge System. If so, there should be a crustal structure of the same type, but a comparison of the gross structures in Fig. 4 shows only a very crude similarity. It is interesting to note that there is an almost perfect correspondence between the average crustal structure of Iceland [41] and the new structural model for the flanks bordering the axial rift zone of the Mid-Atlantic Ridge [16]. There are two differences, however, worth mentioning, namely that the depth to the "oceanic" layer 3A under Iceland is rather variable, but the average thickness of 4 to 5 km is close to the "standard ridge" model. Furthermore, layer 3B (7.2 km/sec) is found under Iceland at a depth of 8 to 16 km; its thickness so far is unknown.

In the discussion of the Afar Depression it has been emphasized that this structure bears no direct similarity to an active mid-ocean ridge system. For the same reasons it is, therefore, not surprising that there is also no correspondence between the crust under Iceland and Afar [36]. In spite of this fact over the past few years several authors [19,27,29,47] have attempted to compare these two structures with the contention that both are underlain by new "oceanic" or "pseudo-oceanic" crust, possibly similar to that of Iceland. It has been pointed out above that the crust under Afar still contains significant amounts of material of intermediate chemical composition, while in Iceland this intermediate layer is completely absent. The obvious contrast in crustal structure can be understood if the different history of evolution is taken into account [36]. It is clear now that the Afar crust came about by gradual attenuation of continental crust, while Iceland must have been created by sea-floor spreading over an already active mid-ocean rift. The conditions in the upper mantle of an altered continent and in an oceanic rift system also must be different (see Fig. 4), and there is no reason to believe that even if the temperature and the thermal gradients are of the same order everything else (as e.g. physical state, petrology, volcanism) should be directly comparable.

## 6. SUMMARY AND CONCLUSIONS

In summary it can be stated that the transformation of continental to oceanic crust is an intricate evolutionary process which is not continuous, but rather seems to happen in a pulsating or episodic manner. The initial phase of break-up of continents is primarily governed by the formation of a neck in the sialic low-velocity, low-density and low-viscosity zone of the upper crust. Arching of the intra-crustal boundaries and swelling-up of the intermediate and lower crustal layers in a rift system are corollaries of this process.

In the subsequent phase of crustal attenuation the intermediate layer with P velocities of 6.3 to 6.4 km/sec plays the important role. As this process continues a "pseudo-oceanic" crust is formed, in which the intermediate layer is still present (a few kilometers thick), but the major feature of the process now is the rather extended lower crustal layer - corresponding to the oceanic layer 3 - which may attain a thickness of up to 20 km. During the final stage of "oceanisation" the intermediate layer will vanish completely and the lower crustal layer will be severely altered by magmatic intrusions fed from sources in an anomalous mantle underneath. New oceanic crust is produced in this "ridge" phase and will move away from the ridge axis to the subduction zones where it is consumed. Parts of the former oceanic crust will return to the upper crust in the form of igneous intrusions thus contributing to the formation of new continental crust. At this stage the cycle of crustal evolution is complete.

## ACKNOWLEDGMENTS

The author is indebted to his colleagues at the Institute für Geophysik der ETH Zürich for many stimulating discussions on the subject of this paper. Particular thanks go to W. Lowrie and J. Ansorge for reading the manuscript and to Mrs. E. Hirzel for painstakingly preparing the "camera-ready" typescript.

## REFERENCES

1. J. Ansorge, D. Emter, K. Fuchs, J.P. Lauer, St. Mueller and E. Peterschmitt, Structure of the Crust and Upper Mantle in the Rift System around the Rhinegraben, in: Graben Problems, ed. by J.H. Illies and St. Mueller, Schweizerbart-Verlag, Stuttgart, 1970.
2. M.E. Artemjev and E.V. Artyushkov, J. Geophys. Res., 76, 1197, 1971.
3. D. Bamford, Z. Geophys., 39, 907, 1973.
4. D. Bamford, An Updated Time-Term Interpretation of $P_n$-Data from Quarry Blasts and Explosions in Western Germany, in: Explosion Seismology in Central Europe, ed. by P. Giese, C. Prodehl and A. Stein, Springer-Verlag, Berlin-Heidelberg-New York, 1976.
5. F. Barberi and J. Varet, Nature of the Afar Crust: A Discussion, in: Afar Depression of Ethiopia, ed. by A. Pilger and A. Rösler, Schweizerbart-Verlag, Stuttgart, 1975.
6. H. Berckhemer, B. Baier, H. Bartelsen, A. Behle, H. Burkhardt, H. Gebrande, J. Makris, H. Menzel, H. Miller and R. Vees, Deep Seismic Soundings in the Afar Region and on the Highland of Ethiopia, in: Afar Depression of Ethiopia, ed. by A. Pilger and A. Rösler, Schweizerbart-Verlag, Stuttgart, 1975.
7. J. Bridwell, New Mexico Geol. Soc. Guidebook, 27, 283, 1976.
8. W.H. Bucher, Deformation of the Earth's Crust, Princeton University Press, Princeton (N.J.), 518 pp., 1933.
9. D. Davies and C. Tramontini, Phil. Trans. Roy. Soc. London, 267-A, 181, 1970.
10. G. Dohr, Erdöl und Kohle, 10, 278, 1957.

11. G. Dohr, Reflexionsseismische Messungen im Oberrheingraben mit digitaler Aufzeichnungstechnik und Bearbeitung, in: Graben Problems, ed. by J.H. Illies and St. Mueller, Schweizerbart-Verlag, Stuttgart, 1970.
12. J.B. Edel, K. Fuchs, C. Gelbke and C. Prodehl, Z. Geophys., 41, 333, 1975.
13. D. Emter, Ergebnisse seismischer Untersuchungen der Erdkruste und des obersten Erdmantels in Südwestdeutschland, Ph.D. Thesis, Univ. of Stuttgart, 108 pp., 1971.
14. D. Emter, Results from Southwestern Germany, in: Explosion Seismology in Central Europe, ed. by P. Giese, C. Prodehl and A. Stein, Springer-Verlag, Berlin-Heidelberg-New York, 1976.
15. J.D. Fairhead, Tectonophysics, 20, 261, 1973.
16. C.M.R. Fowler, Geophys. J. Roy. Astron. Soc., 47, 459, 1976.
17. K. Fuchs, Geophysical Contributions to Taphrogenesis, in: Approaches to Taphrogenesis, ed. by J.H. Illies and K. Fuchs, Schweizerbart-Verlag, Stuttgart, 1974.
18. K. Fuchs, Geol. Rundsch. 64, 700, 1975.
19. R.W. Girdler and P. Styles, The Relevance of Magnetic Anomalies over the Southern Red Sea and Gulf of Aden to Afar, in: Afar between Continental and Oceanic Rifting, ed. by A. Pilger and A. Rösler, Schweizerbart-Verlag, Stuttgart, 1976.
20. D.H. Griffiths, R.F. King, M.A. Khan and D.J. Blundell, Nature Phys. Sci., 229, 66, 1971.
21. D.H. Griffiths, Tectonophysics, 15, 151, 1972.
22. A. Hirn and G. Perrier, Deep Seismic Sounding in the Limagne Graben, in: Approaches to Taphrogenesis, ed. by J.H. Illies and K. Fuchs, Schweizerbart-Verlag, Stuttgart, 1974.
23. M.O.C. Kürsten, Stratigraphic Units of Northern Afar, in: Afar Depression of Ethiopia, ed. by A. Pilger and A. Rösler, Schweizerbart-Verlag, Stuttgart, 1975.
24. M.O.C. Kürsten, Tectonic Inventory of the Danakil Depression, in: Afar Depression of Ethiopia, ed. by A. Pilger and A. Rösler, Schweizerbart-Verlag, Stuttgart, 1975.
25. A.S. Laughton and C. Tramontini, Tectonophysics, 8, 359, 1969.
26. A.S. Laughton, R.B. Whitmarsh and M.P. Jones, Phil. Trans. Roy. Soc. London, 267-A, 227, 1970.
27. J.C. Lepine, J.C. Ruegg and L. Steinmetz, Tectonophysics, 15, 59, 1972.
28. R.E. Long, The Deep Structure of the East African Rift and its Relation to Afar, in: Afar between Continental and Oceanic Rifting, ed. by A. Pilger and A. Rösler, Schweizerbart-Verlag, Stuttgart, 1976.
29. J. Makris, Afar and Iceland - a Geophysical Comparison, in: Afar Depression of Ethiopia, ed. by A. Pilger and A. Rösler, Schweizerbart-Verlag, Stuttgart, 1975.
30. St. Mueller and M. Landisman, Geophys. J. Roy. Astron. Soc., 10, 525, 1966.

31. St. Mueller, Low-Velocity Layers within the Earth's Crust and Mantle, 10th Gen. Ass. Europ. Seism. Comm., Leningrad, Sept. 3-11, 1968.
32. St. Mueller, E. Peterschmitt, K. Fuchs and J. Ansorge, Tectonophysics, 8, 529, 1969.
33. St. Mueller, Geophysical Aspects of Graben Formation in Continental Rift Systems, in: Graben Problems, ed. by J.H. Illies and St. Mueller, Schweizerbart-Verlag, Stuttgart 1970.
34. St. Mueller, C. Prodehl, A.S. Mendes and V. Sousa Moreira, Tectonophysics, 20, 307, 1973.
35. St. Mueller, E. Peterschmitt, K. Fuchs, D. Emter and J. Ansorge, Tectonophysics, 20, 381, 1973.
36. St. Mueller, The Evolution of Rift Systems, Int. Afar-Symposium, Bad Bergzabern (F.R.G.), April 1-6, 1974.
37. St. Mueller and L. Rybach, Crustal Dynamics in the Central Part of the Rhinegraben, in: Approaches to Taphrogenesis, ed. by J.H. Illies and K. Fuchs, Schweizerbart-Verlag, Stuttgart, 1974.
38. St. Mueller, R. Egloff and J. Ansorge, Schweiz. Min. Petr. Mitt., 56, 685, 1976.
39. St. Mueller, A New Model of the Continental Crust, in: The Earth's Crust (Its Nature and Physical Properties), ed. by J.G. Heacock, American Geophysical Union, Washington (D.C.), Monograph 20 (in press), 1977.
40. J. Oliver and S. Kaufman, Geotimes, 21, 20, 1976.
41. G. Pálmason, Soc. Sci. Islandica Rit, 40, 188 pp., 1971.
42. A. Pilger and A. Rösler (eds.), Afar between Continental and Oceanic Rifting, Schweizerbart-Verlag, Stuttgart, 216 pp., 1976.
43. N.N. Puzyrev, M.M. Mandelbaum, S.V. Krylov, B.P. Mishenkin, G.V. Krupskaya and G.V. Petrick, Tectonophysics, 20, 85, 1973.
44. H. Ramberg, J. Geophys. Res., 77, 877, 1972.
45. I.B. Ramberg, Norg. Geol. Unders., 325, 1, 1976.
46. Rhinegraben Research Group for Explosion Seismology, The 1972 Seismic Refraction Experiment in the Rhinegraben - First Results, in: Approaches to Taphrogenesis, ed. by J.H. Illies and K. Fuchs, Schweizerbart-Verlag, Stuttgart, 1974.
47. J.C. Ruegg, Main Results about the Crustal and Upper Mantle Structure of the Djibouti Region (T.F.A.I.), in: Afar Depression of Ethiopia, ed. by A. Pilger and A. Rösler, Schweizerbart-Verlag, Stuttgart, 1975.
48. A.R. Sanford, Ö. Alptekin and T.R. Toppozada, Seism. Soc. Am. Bull., 63, 2021, 1973.
49. N.H. Sleep, J. Geophys. Res., 80, 4037, 1975.
50. L. Steinmetz, A. Hirn, M. Sapin, R. Whitmarsh and V. Moreira, Soc. Géol. France, Bull., 18, 931, 1976.
51. M. Talwani, X. Le Pichon and M. Ewing, J. Geophys. Res., 70, 341, 1965.
52. T.R. Toppozada and A.R. Sanford, Seism. Soc. Am. Bull., 66, 877, 1976.

53. J. Tryti and M.A. Sellevoll, Pure Appl. Geophys., **115**, 1061, 1977.
54. W.J.M. van der Linden, How much Continent under the Ocean?, Marine Geophys. Res., **2** (in press), 1977.
55. J.T. Wilson, Nature, **211**, 676, 1966.

# THE ORIGIN AND EVOLUTION OF RIFTS AND RIFT VALLEY STRUCTURES: A MECHANISTIC INTERPRETATION

S. Bhattacharji and H. Koide

Department of Geological and Geophysical Sciences, Princeton University, Princeton, N.J. 08540, U.S.A.[x], and Geological Survey of Japan, Tokyo, Japan

ABSTRACT. The tectonic effects of asthenospheric upwell and wedging of mantle-derived material on the elastic crustal plate and probable patterns of development of ridge-rift valley structures in oceanic and continental crusts are analysed by theoretical and experimental studies.

## 1. INTRODUCTION

High gravity, magnetic, seismic and heat flow anomalies, and the frequent association of volcanism and volcanic centres with recent and major ridges-rift centres on oceanic and continental crusts in time and space, indicate an intimate relationship between magmatism and the process(es) of rifting and rift valley formation in crustal rocks. In recent years, such geophysical anomalies have also been interpreted by many workers [1-4] as indication of the presence of igneous body(ies) at depth along the ridge and the axial valley.

Magmas or crustal mush at equilibrium at depths between 50 and 100 km can generate excess magma pressure (in the order of 0.3 to 2 kb) within the lithosperic plate [5-7]. Initiation of mantle diapirism by gravitational instability [8] and its continued forceful uprise with excess magma pressure from upper mantle thermal anomalies and from various depths of the upper mantle or lower lithosphere induces thermo-mechanical stresses on the lithospheric plate. Here we summarize our study on the effects of

x Present address: Department of Geology, Brooklyn College of CUNY, Brooklyn, New York, 11210 U.S.A.

*I. B. Ramberg and E.-R. Neumann (eds.), Tectonics and Geophysics of Continental Rifts, 29-37.*

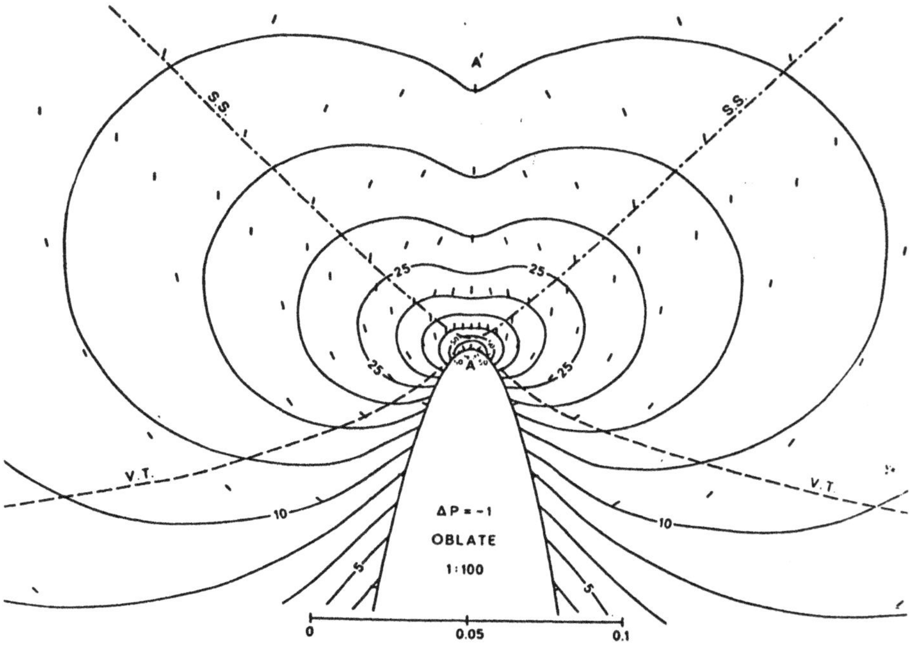


Fig. 1. Contours of maximum tensile stress (in kilobars) around the tip of a thin oblate upwell of aspect ratio 1:100 due to an excess pressure over lithostatic stress of 1 kb (corresponding to a depth of 10 to 30 km from the surface). Tensile stress is indicated as positive. Only the contours around the apex are shown. All normal stresses are three-dimensionally tensile within the surface (VT) although absolute tensile stress actually occurs close to the tip of the upwell. Short straight lines are probable directions of extension fractures (directions of maximum compressive stress). The unit of scale is arbitrary. Abbreviations: SS, shear surface; VT, vertical tension. Normal faulting along SS would initiate ridge-rift valley structure. Note the depressed contours at the axial region (AA') where maximum subsidence or central rift valley should occur.

stresses transmitted to the elastic crustal plate due to asthenospheric upwell or magmatic wedging of various aspect ratios and evolution of ridge-basin and ridge-rift valley structures with increasing stress.

## 2. ANALYSIS AND BASIC DATA

We used a mathematical solution of three-dimensional theory of elasticity [6-7] to obtain stress distribution around igneous intrusions or protrusions of different sizes and shapes (determined by aspect ratios) and probable development of structures and fractures in an infinite, elastic, homogeneous rock body. Partial melting of the upper mantle and development of isolated oblate magma bodies of increasing aspect ratios over rising currents and hot spots were assumed [9-11]. We analysed stress distribution (s) around and at the tips of igneous bodies of aspect ratios 1:1.4 to 1:1000. Details are presented elsewhere [6, 7, 9]. The analyses show that the elastic stress concentration is greatest at the tips (Fig. 1) and increases with increasing aspect ratio for both oblate and prolate igneous bodies [6, 7, 9]. The stress concentration around an oblate body, however, is much higher than around a prolate one of same aspect ratio. The effect of changing Poisson's ratio from 0.25 to 0.4 (i.e., elastic to almost viscous) of the lithosperic plate above a hot spot was found to be negligible on the magnitude of stress concentration as compared to aspect ratio (shape factor). The area of maximum stress concentration is also the likely location for initiation of fractures and their further development into faults and ridge-rift valley structures, although other important controlling factors include pre-existing lineaments, anisotropies and their orientation to magmatic stress axis. This is important to the nature of transmitted stress on the crustal plate and probable directions of extension and shear fractures, and rift propagation due to progressive wedging or enlarging magma body(ies) from different depths of the asthenosphere, time duration of stationarity of the plate over the hot spot and increasing volume of magma [9].

## 3. RESULTS AND DISCUSSION OF THE EVOLUTION OF RIDGE-RIFT VALLEY STRUCTURES

Analyses show that wedging of an igneous mass or upwell into the elastic lithospheric plate produces wide areas of compression at the sides at depth but causes horizontal extension immediately above, in a broad, funnel-shaped zone (VT in Fig. 1) over a deep-seated mantle mass (Fig. 2). Horizontal maximum compressions have been measured outside active rift zones in Iceland [12], in the Rhine graben rift belt [13], and also in volcanic belts of Kurile-Kamchatka [5]. Dykes and volcanic centres perpendicular to main rifts and parallel anticlines between main rift zones in Iceland [14], and the formation of more recent secondary transverse rifts (Kavirondo, Speke Gulf, etc.) on the flank of the main Gregory rift of the East African rift system may also be related to horizontal compression due to deep-seated wedging of mantle-derived material along the main rift zone (Fig. 3). Geophysical evidence indicates the presence of mantle material along the main Gregory rift [15], the Red Sea axial trough [16], and in Rhine graben [13].

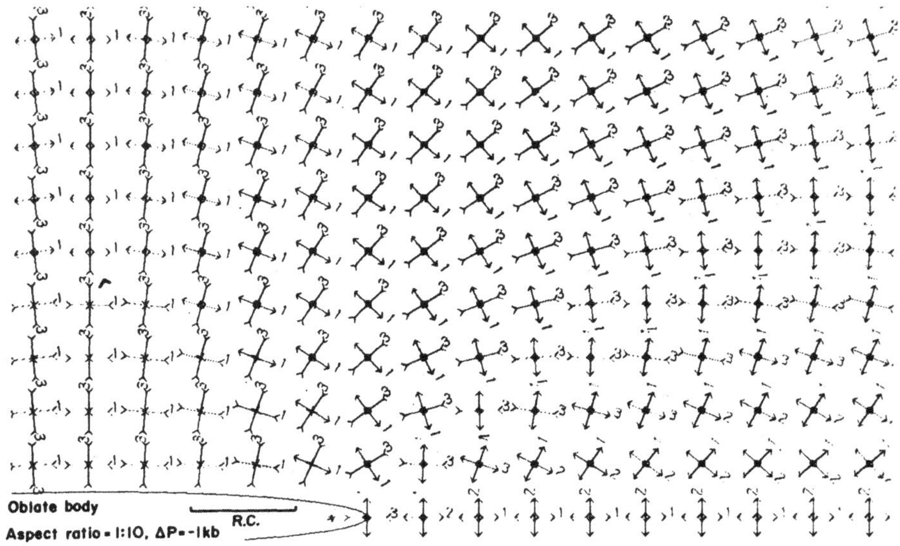


Fig. 2. Stress distribution around an oblate igneous mass of aspect ratio 1:10. 1, 2, and 3 are directions of maximum tensile (or minimum compressive), intermediate and minimum tensile (or maximum compressive) stress, respectively. Asterisk (*) marks the oblate end. The unit of scale is arbitrary.

Theory and experiments [9] show that the origin of a ridge-rift structure by asthenospheric upwell or intrusive wedging is initiated first by wide domal uplift, antiform or ridge in an elastic or visco-elastic crust prior to rift faulting [15]; the magnitude and nature of uplift is controlled primarily by flexural rigidity, presence and degree of anisotropy, and area and magnitude of stress concentration on the elastic plate. Stress contours (Fig. 1) and elastic displacement vectors indicate that an area of relative subsidence, basin or synform would occur in the centre of the domal uplift above the prolate end. This later evolves into an elongated rift valley (central trough) bounded by marginal, inward dipping (in the range 30° to 60°) funnel-shaped normal faults (fig. 4) as in the Red Sea area [16]. As crustal rocks have low tensile strength [17], high tensile stress concentration at the apex (Fig. 1) and high interstitial magmatic fluid pressure would produce extension and propagation of fractures immediately above the intrusion (Figs. 3, 4). However further up, these fractures tend to branch upward and outward into a set of en echelon extensional or normal faulted shear zones (Figs. 1, 3, 4). Further wedging produces rifting and the

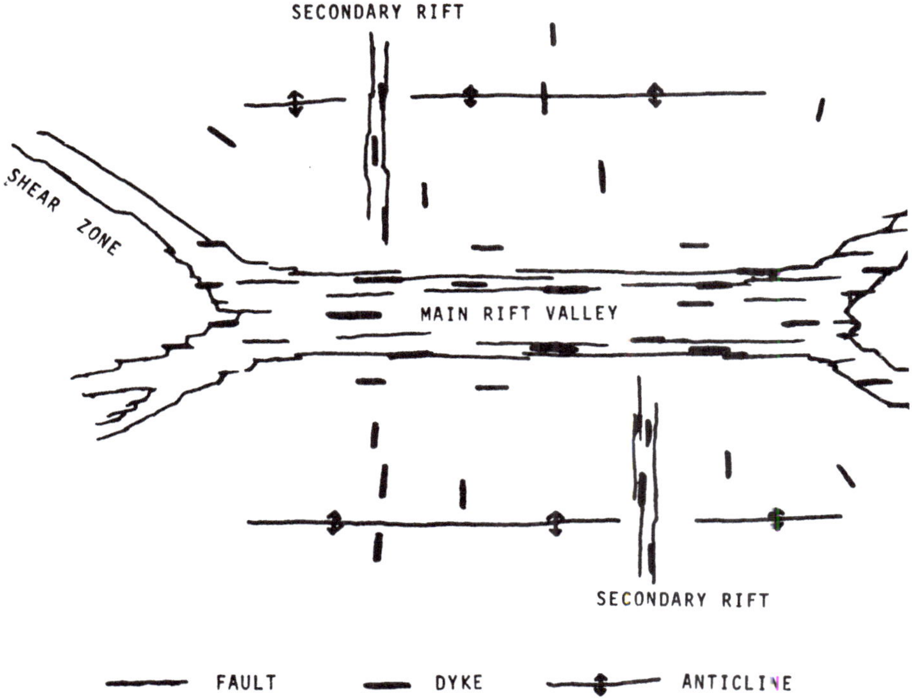


Fig. 3. Diagramatic plan view showing the development of main ridge-rift valley, branching of the rift valley into triple-junction or en echelon shear zones, normal faulting, parallel and perpendicular dykes, anticline between rift zones, and secondary rift due to stress imparted to the crustal plate by forceful magmatic wedging and/or by mantle upwelling.

following developments in the crustal plate: graben; graben and horst; block faulting; axial rift valley opening by magmatic intrusions as axial dykes (mafic or felsic) producing lateral stretching, further rifting and onset of spreading. This may be followed by marginal or flank uplift; central rift valley or marginal volcanism; possible block rotation; further development of complex graben and horst structures (like Rhine graben), possible sedimentation in the faulted block(s) or basin(s) with end-phase volcanism producing extensive plateau lavas by opening up normal faults. The development of an axial rift valley and its progressive subsidence and marginal uplift with lateral extension enhances block faulting and rotation. Fault geometry, in addition to stress configuration, is the key for block rotation, opening up

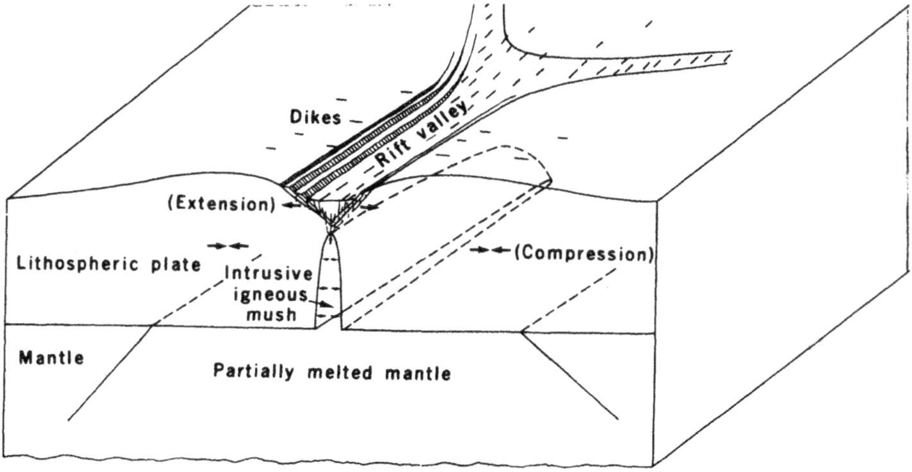


Fig. 4. Domal uplift and rift valley formation with normal faulting, and development of triple junction fracture geometry in the lithospheric plate due to wedging of igneous mass of aspect ratio 1:10 or higher. Wide domal uplift and relative subsidence in the central region are deduced from the distribution of elastic displacement vectors.

normal faults and to a great measure the timing and extent of volcanism [9]. Intrusion of igneous materials predominantly as sills and sheets in the layered elastic crustal plate during mantle upwelling or magmatic wedging favors greater uplift without rifting or prior to rifting by increasing ductility contrast and mean ductility and often leads to the formation of ridge-basins or ridge-rift valleys with no local signs of volcanism. Under such conditions, basin formation (inter-arc type) can also occur prior to rift valley formation [18]. This analysis indicates the probable development of marginal reverse faulting as associated with many rifts due to the effect of lateral compression (Fig. 2) with progressive wedging as postulated by Osmaston [19] and others [20]. This study also indicates that when a mantle wedge or upwell reaches an aspect ratio 1:10 or higher, and the excess intrusive pressure is in the order of 1 kb (i.e., the top of the wedge is at a depth of 10-30 km from the surface [21], the width of the rift valley system should be on the order of 20-60 km if the crust behaves as an homogeneous elastic plate. This is remarkably consistent with the general width of rift valleys [22].

An important inference from our theoretical studies [6,7 and 9] is that the initial development of ridge-basin and ridge-rift valley structures is directly related to extension resulting from excess magma or intrusive pressure. Later subsidence in the central basin or graben due to lowering of magma pressure immediately above the intrusion accentuates ridge-rift valley or graben structure. This is contrary to collapse hypothesis of rift valley formation by reduction of magma pressure as invoked by some authors [4].

For oblate magma bodies of aspect ratio 1:10 or higher, when magma pressure exceeds lithostatic stress, the maximum stress concentration at the oblate ends (Fig. 1) can initiate FFR type triple junction (F is fracture, shear fault or fault zone and R is ridge) fracture or rift geometry (Fig. 4). The initiation of FFR type triple junction (Figs. 3 and 4) begins with the tensional arm in front of the oblate end (where later intrusion of magmatic dykes may result in a ridge) branching into two shear zones at an angle of 45° to 60° with the magmatic stress axis in a homogeneous plate and/or en echelon arrays of tension fractures associated with or extending along the shear fracture zones. Bifurcation and branching of graben and rift valley(s) are observed in the continental crust as Rhine graben [13], the Red Sea into Gulf of Aqaba and Suez [9,16], the East African rift system [3], and also in areas where oceanic crusts are associated with magmatism such as East Pacific Rise, Mid-Atlantic Ridge, Carlsberge Ridge, etc. [23].

The Iceland rift zones [14], Hawaiian volcanic chains, ridges and rifts in the FAMOUS area of the mid-Atlantic ridge [4], and many active rift systems on the continental and ocanic crusts show en echelon arrangement of fractures, faults, fissures or rifts. Although Riedel shear [11] parallel to the axis may be an important mechanism for the origin of en echelon fracture or rift pattern, rock mechanics experiments [11] have shown that moderate fluid pressure and brittle to brittle-ductile transition favour the formation of en echelon pattern of fractures and faults in rocks. Thus, in areas of magmatic activities, fluid pressure (i.e., effective stress) and rock properties appear to exert an important control over the development of short en echelon ridge-rift structures. Variations in this pattern are largely controlled by heterogeneity and anisotropy of the crustal rocks and variation of stress conditions due to progressive wedging of magma bodies. Rifting by thinning and necking of the visco-elastic lithospheric plate can occur by viscous or plastic creep with increasing strain rate followed by rupture in areas of relative stationarity of the lithospheric plate over a hot spot (as in Africa).

The establishment of the clear relationship between alkaline basaltic magmatism and continental rifting [22] and the frequent association of continental rifts and almost all oceanic rifts with major high gravity anomalies and magmatic bodies at depth suggest that their origin is ultimately connected with deeper level and upper mantle rather than with surfacial crustal processes [24]. Despite the recognition of this connection, the question of whether rifting is the cause of magmatism or its effect has not yet been resolved. In the light of the mechanistic analyses summarized earlier, it can be concluded that stresses generated and imparted to the lithospheric plate by asthenospheric upwell, protrusion and magmatic wedge are capable of creating ridge-rift systems in the crustal plate. In this respect, asthenospheric upwelling and magmatic stress can be considered as an important and immediate cause of ridge-basin and ridge-rift structures. Rifting or graben formation and fault geometry on the other hand, in addition to spreading and its rate, control rift topography or bathymetry, the extent, initiation and time duration of volcanism, and the magnetic symmetry or asymmetry of the rift valley. The physical and rheological characteristics of rocks associated with rift zones or their change (from brittle-transition-ductile) from the surface vertically downward, and the stress differential largely determine the size and pattern of the rift system.

Excess magma pressure of an average of 1 kb or more in vertical dyke-like intrusions from the upper mantle pushes the elastic lithospheric plate almost horizontally apart forming rifts and rift valley structures; such a push at ridge centres and along the 70-100 km thick plate can be a strong plate driving force in addition to asthenospheric circulation.

## REFERENCES

1. R.W. Girdler, Nature, 217, 1102, 1968.
2. J.R. Cann, Nature, 226, 928, 1970.
3. B.H. Baker, P.A. Mohr and L.A. Williams, Geol. Soc. Am. Spec. Pap., 113, 67, 1972.
4. R.B. Whitmarsh and A.S. Laughton, Deep Sea Res., 23, 1005, 1976.
5. S.A. Fedotov, Bull. Volc. 39, 1, 1975.
6. H. Koide and S. Bhattacharji, Econ. Geol., 70, 781, 1975.
7. H. Koide and S. Bhattacharji, Science, 189, 791, 1975.
8. H. Ramberg, Phys. Earth Planet. Inter., 5, 45, 1972.
9. S. Bhattacharji and H. Koide, Nature, 255, 21, 1975.
10. W.J. Morgan, Am. Assoc. Pet. Geol., 56, 203, 1972.
11. H. Koide and S. Bhattacharji, Proc. 2nd Int. Conf. Basement Tect., (in press).
12. N. Hast, Phil. Trans. Roy. Soc. Lond., Sr. A.274, 409, 1973.
13. J.H. Illies and G. Greiner, Geodynamics: Progress and Prospects, Am. Geophy. Union, Washington, DC, 1976.
14. G.P.L. Walker, Nature, 255, 468, 1975.
15. R.W. Girdler, Geodynamics: Progress and Prospects, Am. Geophys. Union, Washington, DC, 1976.

16. R.W. Girdler, Hot Brines and Recent Heavy Metal Deposits in the Red Sea, Springer, 1969.
17. W.F. Brace and E.G. Bombolakis, J. Geophys., 68, 3709, 1963.
18. S.S. Sun and G.N. Hanson, Geology, 3, 297, 1975.
19. M.F. Osmaston, Tectonophysics, 11, 387, 1971.
20. K. Deffeyes, The Megatectonics of Continents and Oceans, Rutgers University, New Brunswick, N.J., 1970.
21. J.L. Roberts, Mechanics of Igneous Intrusions, Gallery Press, Lond., 1970.
22. V.V. Beloussov, The Earth Crust and Upper Mantle, Am. Geophys. Union, Monogr., 13, 1969.
23. B. Isacks, J. Oliver and L.R. Sykes, J. Geophys. Res., 73, 5855, 1968.
24. I.B. Ramberg, Norg. geol. Unders., 325, 1, 1976.

# EXPERIMENTAL MODEL STUDIES OF RIFT-VALLEY SYSTEMS

Hans Ramberg

Geologiska Institutionen
Box 555, S-751 22 Uppsala, Sweden.

ABSTRACT. As emphasized by H. Cloos in his classic publication [1] 1939 on rift systems and volcanism, the major rift systems occur within gentle dome-shaped regions of the Earth's crust.

Later geophysical studies have indicated the presence of abnormal mantle materials below the rift systems, see e.g. Illies and Mueller [2]. It lies close at hand to conclude that the evolution of the rift valley is somehow connected with the body of abnormal mantle materials.

Cloos considered the tension and the rifting to be the result of the gentel bending of the crust, much as tension develops in the convex region of a bent elastic plate.

In a series of dynamic model experiments relating to a number of tectonic phenomena in which gravity plays a crucial role, the writer found that tension in the model crust was not so much caused by the bending as it was created by the lateral spreading of the subcrustal body of abnormal mantle [3,4,5].

When a plastic buoyant mass rises, either from the low velocity layer or from deep in the mantle (mantle diapir), and the body meets strata that are less dense than the body itself, then the body tends to spread laterally within or underneath the low-density strata. The spreading body creates tensile stress in horizontal direction in the overlying crust due to friction coupling. Since solid rocks are weak in tension, tensile fractures and separation of the walls are apt to happen. Crustal fragments which form between the more or less parallel fractures will sink down in the abnormal mantle, the amount of subsidence being determined

*I. B. Ramberg and E.-R. Neumann (eds.), Tectonics and Geophysics of Continental Rifts, 39-40.*

by the difference in density and the shape of the fragments. This picture of the evolution of rift system above an ascending and spreading abnormal (low density) portion of the mantle more or less accurately was imitated in many experimental models run in a centrifuge in which the strong centrifugal force played the same role in the models as does gravity in the natural structures. Description of some of the models are found in the cited publications by the author.

It is interesting that the experimental models demonstrate that the rising and subsequently spreading mass of abnormal mantle must be considerably more viscous than a silicate melt - it is either completely crystalline but yet plastic, or it is a mixture containing only a small portion of melt. The experiments indicate that a liquid body of similar horizontal dimensions as a rift-valley system will cause a complete collapse of the overlying crust (unless the liquid is more dense than its overburden), the crust sinking to the bottom of the huge chambre and essentially all the liquid being forced up to the surface (see Newall and Rast [6], p. 261-286). This is of course not what one sees in the rift systems now existing in the continent. Such a situation may have existed in times when the Earth was at higher temperature, but rift systems now accessible for field study are more in accord with models having an essentially solid but plastic subcrustal "cushion" from which minor amounts of melt can be squeezed to feed the volcanoes in the rift valley.

## REFERENCES

1. H. Cloos, Geol. Rundsch. 30, 401 and 637, 1939.
2. J.H. Illies and St. Mueller, Graben Problems, Upper Mantle Project, Report No. 27, E. Schweizerbart'sche Verlagsbuchhandlung, Stuttgart, 1970.
3. H. Ramberg, Bull. Geol. Inst. Univ. Uppsala, 42, 97, 1963.
4. H. Ramberg, Gravity, Deformation and the Earth's Crust, Academic Press, London, 1967.
5. H. Ramberg, Lithos, 4, 259, 1971.
6. G. Newall and N. Rast, Geol. Jour. Spec. Issue, 2, 1970.

# RATES OF SEDIMENTATION AND VERTICAL SUBSIDENCE IN NEORIFTS AND PALEORIFTS

J.J. Tiercelin and H. Faure

Laboratoire de Géologie du Quaternaire,
Centre National de la Recherche Scientifique, Luminy,
13288 Marseille, Cédex 2, France.

ABSTRACT. The variation in subsidence rate during a rift's active life is probably a good indication of the rifts geodynamical history and can contribute towards an understanding of the mechanism responsible for its evolution.

The rate of sedimentary accumulation in the bottom of an active neorift is dependant on many factors, not only ones such as petrology and climate, but also the morphology and topography of the rift margins. The distribution of surface relief is directly related to the relative vertical movement between the floor and shoulder of the rift. In many places the infilling of neorifts and palaeorifts has been studied in detail during, for example palaeontological and anthropological investigations of the neorift of East Africa and in oil prospecting in palaeorifts of the Atlantic and North Sea margins. Once a sufficiantly good chronology is available, mean sedimentation rates may be calculated and in some cases the variation of these rates through time may be expressed graphically.

The nature of the sedimented material and its depositional environment must be carefully evaluated in an interpretation of geodynamical conditions. It seems useful here to present some preliminary results of rates measured during East African rift studies and to compare them with calculated rates.

## 1. RATES OF SEDIMENTATION IN THE EAST AFRICAN RIFT

The Afar depression (Ethiopia and Republic of Djibouti) is the site of the junction of three major rift zones, two of them oceanic

*I. B. Ramberg and E.-R. Neumann (eds.), Tectonics and Geophysics of Continental Rifts, 41-47.*

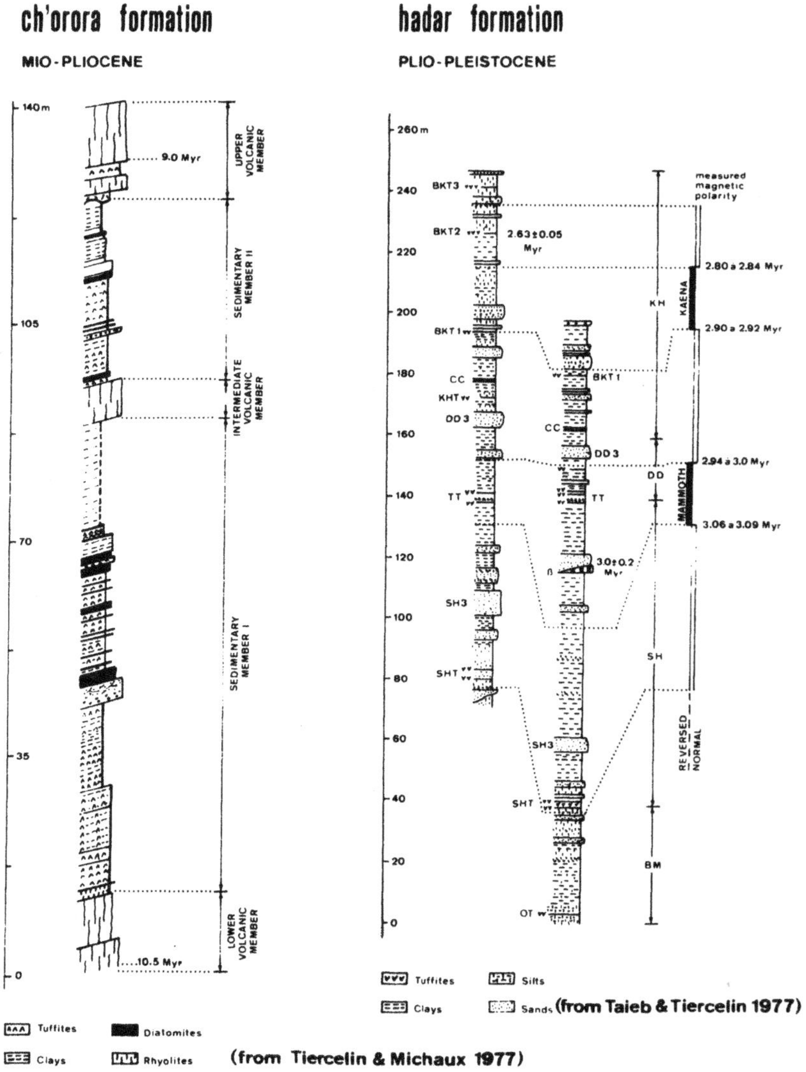


Fig. 1: Geological columns of the Ch'orora Formation and the Hadar Formation, Afar Depression, Ethiopia. Geochronologic and palaeomagnetic datas.

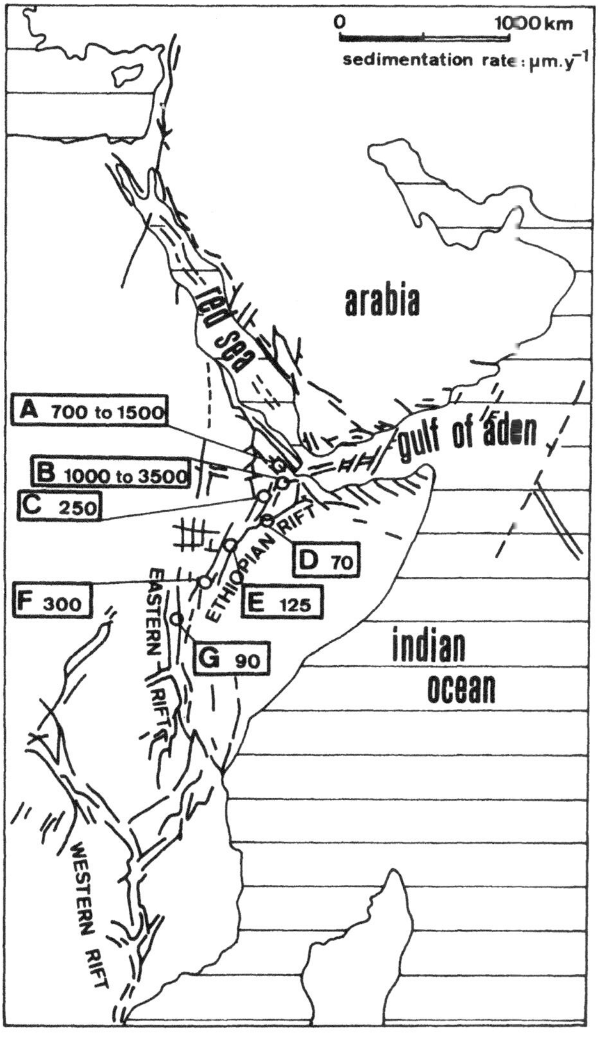


Fig. 2: Major Faults of East African Rift [10]. Location of sedimentary formations (A: Northern Danakil depression; B: Northern Danakil depression; C: Hadar Formation; D: Ch'orora Formation; E: Gademotta Formation; Omo basin; G: Northern Kenya, Lukéino Formation) and sedimentation rates (in µm $y^{-1}$) corresponding.

(Red Sea and Gulf of Aden) and one continental (Ethiopian rift). The latter represents the northern part of the East African rift system. Between 10° and 12°N, on the western margin of the Afar in the Hadar area there is a large sedimentary basin of Plio-Pleistocene age. Here the International Afar Research Expedition has found a rich vertebrate fauna, including Hominid remains.

This marginal basin is closely comparable, from a structural viewpoint, with the Mio-Pliocene basin between 9° and 9°15'N in the southeastern Afar which is delimited by the occurrence of the Ch'orora Formation. The continental sediments in these two basins are the oldest so far discovered in the Afar depression.

The Hadar Formation with an exposed thickness of 280 m in the central part of the basin, has been conveniently divided into four members, viz. the Basal Member (BM) Sidi Hakoma Member (SH), Denen Dora Member (DD) and Kada Hadar Member (KH), separated respectively by major tuff horizons (SHT, TT and KHT) as shown in the geological column (Fig. 1) [1,2]. The Hadar Formation comprises of lacustrine, lake margin and associated fluvial deposits (clays, silts, sands and conglomerates). Geochronological data pertinent to establishing the age of the Formation are provided by a primary unreworked tuff (BKT 2; K-Ar 2.63±0.05 m.y., zircon fission track 2.58± 0.23 m.y.) in the Dada Hadar Member, and one basalt flow (Kada Damoum basalt; K-Ar 3.0 ± 0.2 m.y.) in the Sidi Hakoma Member [3].

According to this geochronology and relevant palaeo-magnetic data, the mean sedimentation rate for the Hadar Formation, between 3.3 and 2.6 m.y., was 250 $\mu m\ y^{-1}$.

The Ch'orora Formation has an exposed thickness of 140m along the southeastern Ethiopian Escarpment. The lowest volcanic unit here is the Bacca Basalt, overlain by rhyolitic ignimbrites followed by the Ch'orora Formation, which in turn is succeeded by further layers of ignimbrites. Southeast of the Bumbi Mountains the Ch'orora Formation sediments are diatomites, clays, conglomerates and well-sorted and graded pumice deposits [4,5].

K-Ar age determinations of the Bacca Basalt below the Ch'orora Formation and of an ignimbrite overlying these sediments, yielded values of 10.5 and 9.0 m.y. respecitvely [6], see Fig. 1. The mean sedimentation rate of the Ch'orora Formation was thus found to be 70 $\mu m\ y^{-1}$.

Other Examples of sedimentation rates have been calculated. In the Omo basin (south Ethiopia), the mean rate was 300 $\mu m\ y^{-1}$ during the Plio-Pleistocene [7]. In the Baringo area of the northern Kenya rift valley, the Lukeino Formation reveals a sedimentation rate of 90 $\mu m\ y^{-1}$ between 6.7 and 5.4 m.y. [8]. In the

Lake Ziway district of the Ethiopian rift the sedimentation rate of the Gademotta Formation lacustrine deposits was 125 µm $y^{-1}$ between 280,000 and 117,000 years BP [9].

Rates calculated from radiocarbon datings of partially compacted lacustrine deposits from the upper Pleistocene and Holocene (70,000 to 5,000 years BP) are of the order of 700 to 1500 µm $y^{-1}$.

| LOCALITIES | FACIES | THICKNESS (m) | SEDIM. RATE (µm.y⁻¹) | |
|---|---|---|---|---|
| NORTH DANAKIL DEPRESSION | CLAYS | 50 to 100 | 700 to 1500 | A |
| NORTH DANAKIL DEPRESSION | EVAPORITES | 2000 to 7000 | 1000 to 3500 | B |
| DANAKIL DEPR. HADAR FORMATION | CLAYS, SANDS | 280 | 250 | C |
| DANAKIL DEPR. CH'ORORA FORM. | DIATOMITES, CLAYS, SANDS | 100 | 70 | D |
| ETHIOPIAN RIFT GADEMOTTA FORM. | SILTS, SANDS | 20 | 125 | E |
| ETHIOPIAN RIFT OMO BASIN | CLAYS, SANDS | 1100 | 300 | F |
| NORTHERN KENYA LUKEINO FORM. | DIATOMITES, CLAYS, SILTS | 110 | 90 | G |

1 2 3 4 5 6 7 8 9 10 Myr

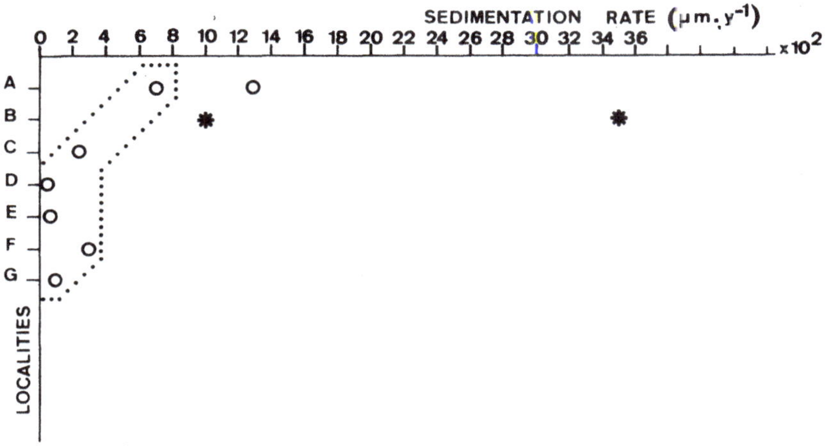


Fig. 3: Rates of sedimentation in the East African Rifts.
o - non-evaporitic rocks
* - evaporitic rocks

## 2. COMPARISONS BETWEEN RATES OF SEDIMENTATION AND SUBSIDENCE

These values for sedimention rates, calculated for the East African rift, may be compared with rates of vertical displacement during the Pleistocene along major faults in the northern or Ethiopian rift, i.e. 200 to 500-1000 $\mu m\ y^{-1}$.

They are also of the same order of magnitude as the continental sedimentation of certain fluvio-lacustrine facies in the Mesozoic proto-Atlantic rift, i.e. 500 to 1000 $\mu m\ y^{-1}$, during the Neocomian and Aptian of the Gulf of Guinea [11,12]. In this respect mean sedimentation rates are probably chiefly related to relative vertical movement between the floor and shoulders of the rift.

Rates of precipitation of halite in the Northern Danakil rift were one or two orders of magnitude greater during the Pleistocene i.e. $10^4$ to $35 \times 10^4$ $\mu m\ y^{-1}$. Halite deposition seems to be determined more by seawater supply and evaporation rates than by subsidence. For this reason the Aptian evaporites of the proto-Atlantic rift probably represent a short time interval - less than $10^5$ or $10^6$ years - in the stratigraphical succession.

Marine sedimentation rates in both Mesozoic and Cenozoic rifts varied greatly, from off-shelf to prograding deltaic, but rates of vertical accumulation are good indicators of subsidence rates.

Comparison of sedimentation and subsidence rates in palaeorifts and neorifts taking into account the environmental significance of the deposits, seems to indicate great similarity of environment during Mesozoic and Cenozoic. Interpretation of this is probably a comparable geodynamic evolution in the break up of continental crust under similar latitudinal position.

## REFERENCES

1. M. Taieb, D.C. Johanson, Y. Coppens and J.L. Aronson, Nature, **260**, 289, 1976.
2. M. Taieb et J.J. Tiercelin, La formation d'Hadar (Afar, Ethiopie): Stratigraphie et paléoenvironnements sédimentaires. VIII$^e$ Congr. Panafr. Prehist. Et Quat., Nairobi (in press).
3. J.L. Aronson, T.J. Schmitt, R.C. Walter, M. Taieb, J.J. Tiercelin, D.C. Johanson, C.W. Naeser et A.E.M. Nairn, Nature, **267**, 327, 1977.
4. O. Sickenberg et M. Schönfeld, The Chorora formation - Lower Pliocene limnical sediments in the southern Afar (Ethiopia) in: Afar Depression of Ethiopia (1), ed. by A. Pilger and A Rösler, Schweiserbart'sche, Stuttgart, 410, 1976.

5. J.J. Tiercelin and J. Michaux, Un exemple de remplissage de rift dans l'Est africain: la formation de Ch'orora, Ethiopie: Stratigraphie et mammiféres fossiles. 5[e] Réunion annuelle des Sciences de la Terre, Rennes, 1977.
6. K.Kunz, H. Kreuzer and P. Müller, Potassium-Argon age determinations of the Trap basalt of the south eastern part of the Afar Rift, in: Afar Depression of Ethiopia (1), ed. by A. Pilger and A. Rösler, Schweizervart'sche, Stuttgart, 416, 1976.
7. Y. Coppens, F.C. Howell, G.L. Isaac and R.E.F. Leakey (Eds.), Earliest Man and Environments in the Lake Rudolf Basin, in: Prehistoric Archology and Ecology Series, ed. by K.W. Butzer and L.G. Greeman, 1976.
8. M. Pickford, Nature, **256**, 279, 1975.
9. F. Wendorf, R.L. Laury, C.C. Albritton, R. Schild, C.V. Haynes, P.E. Damon, M. Shafiqullah and R. Scarborough, Science, **187**, 740, 1975.
10. Carte tectonique internationale de l'Afrique, 1/5000.000, ASGA-UNESCO, 1968.
11. J.R. Delteil, J. Le Fournier and J. Micholet, Schéma d'évolution sédimentaire d'une marge continentale stable: Exemple - type du golfe de Guinée, de l'Angola au Cameroun. IX[e] Congrés internationale de Sédimentologie, Nice, 1975.
12. P.A. Rona, Geol. Soc. Am. Bull., **84**, 285_, 1973.

# DEEP STRUCTURE AND EVOLUTION OF THE BAIKAL RIFT ZONE

N.A. Logatchev, V.A. Rogozhina, V.P. Solonenko
and Yu.A. Zorin
Institute of the Earth's Crust
Siberian Branch of the USSR Acad. Sci.
Irkutsk

ABSTRACT. The varying structural pattern of the Baikal system of rift valleys and faults is defined by interaction of the rift-forming stress field and the structural grain of the basement. The most prominent feature is the central segment of the rift zone which embraces the deepest and largest Baikal depression and is confined to the suture zone between two major lithospheric plates, the Precambrian Siberian platform and its folded border, the Sayan-Baikal highlands.

The mechanism of the Baikal rift zone formation was a complex process with alternating phases of crustal fracturing and plastic extension. In addition to the anomalous upper mantle layer beneath the Baikal rift zone, a large-scale anomalous mantle region has recently been discovered, whose lower boundary descends to 300 - 700 km below the rift zone and adjacent mountain areas.

The high seismicity of the Baikal rift zone is mainly related to the recent tectonic activity of the rift valleys. The latter form a distinct evolutionary succession: incipient - embryonic - mature - aborted rift depressions.

## 1. INTRODUCTION

Among the known continental rift systems the Baikal rift zone is the one with the most specific structural pattern. Its general strike changes at three sites along its length (Fig. 1). Until recently this phenomenon was explained by the structural control exercised by the Siberian platform. In reality, such control is evident only for the South Baikal depression. Further to the west

*I. B. Ramberg and E.-R. Neumann (eds.), Tectonics and Geophysics of Continental Rifts, 49-61.*

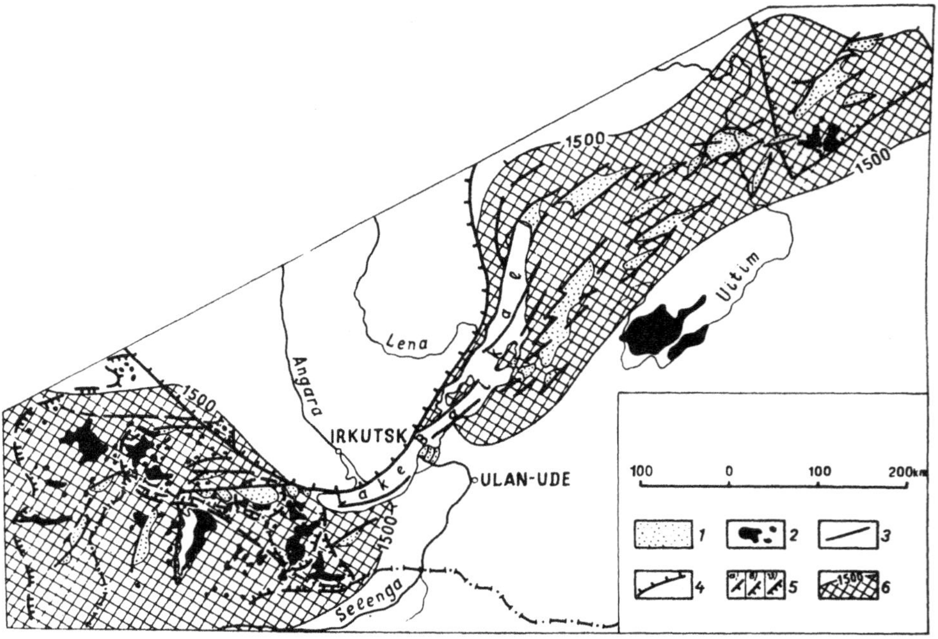


Figure 1. Schematic diagram of the neotectonic structure of the Baikal rift zone. 1 - rift valleys; 2 - Cenozoic volcanics; 3 - major faults (newly formed and rejuvenated); 4 - margin of the pre-Cambrian Siberian platform; 5 - areas of tectonic activity during: a) Upper Cretaceous; b) Miocene-Lower Pliocene; c) Upper Pliocene-Pleistocene; 6 - 1500 meter isoline of the domal uplifts.

the rift zone deviates from the platform margin and then crosses the structural grain of the Late Precambrian Baikal fold belt at right angles and penetrates into the Caledonian fold system of North Mongolia.

At the northern termination of the South Baikal depression the rift valleys begin branching. In the Stanovoy highland the rift zone follows first the Baikal fold belt, and then the ancient Muya massif, before penetrating the Archean Aldan shield and, farther east, the pre-Riphean fold belt of the Stanovoy range.

Although these structural features of the Baikal rift zone were recognized previously [1], they were not convincingly explained by either the data from geophysical investigations or studies of the geological structure.

## 2. DEEP STRUCTURE

According to the new geophysical data the depth to the base of the crust in the Baikal rift zone varies from 34 to 46 km, the shallowest depth being registered in the deepest part of the Baikal depression and the greatest beneath the surrounding mountains [2,3]. As evidenced by gravitational data, thinning and attenuation of the crust is a general feature of all the major rift depressions of the world [4].

Recently obtained results on the structure of the upper mantle cast new light upon the problem of the deep structure of the Baikal rift zone. The data from seismic stations in East Siberia, compared with Jeffereys' godograph prove that P-waves of remote earthquakes reach the station with a delay which depends on the position of the epicentre relative to the station. An analysis of these delays suggests that they are caused by the existence, in the upper mantle beneath the vast South Siberia and Mongolia mountain region, of an extra-large body of anomalous low-velocity mantle. At the stations situated in the Baikal rift zone the travel time of P-waves is increased for all the earthquakes whose rays penetrate into the mantle. This indicates that the low velocity area within the limits of the Baikal rift zone is situated just beneath the Earth's crust. This conclusion helps us to estimate its size and to define its form. By means of Jeffreys' velocity model, trajectories of seismic rays in the mantle for the waves registered at the stations located in the rift zone are plotted on a few vertical sections.

Assuming that the velocity of P-waves in the area of the anomalous mantle is diminished (compared to the normal one) by a constant value (0.3 km/sec), the length of the intercepts of the seismic rays where the delay takes place have been estimated. The terminations of such intercepts plotted against the base of the crust roughly correspond to the points on the outer boundary of the low velocity inhomogeneity (the anomalous mantle area). On the basis of these data a schematic diagram of the morphology of the anomalous mantle zone has been compiled (Fig. 2). The projection of this area onto the Earth's surface embraces not only the Baikal rift zone, but also the mountain areas of the East and West Sayan, the Soviet and Mongolian Altai, the highest part of the Gobi Altai, the Khangay and the Khentey. The upper boundary of the anomalous mantle zone rises to the base of the crust only beneath the rift zone and (probably) the East Sayan. Beneath the other mountain ranges the upper boundary of the low velocity inhomogeneity deepens to 100 km and more (Fig. 2). Beneath the greater part of the rift zone the lower boundary of the anomalous mantle is located at 400 km depth. To the south it goes down to 500 km and to the southwest it plunges to 800 km.

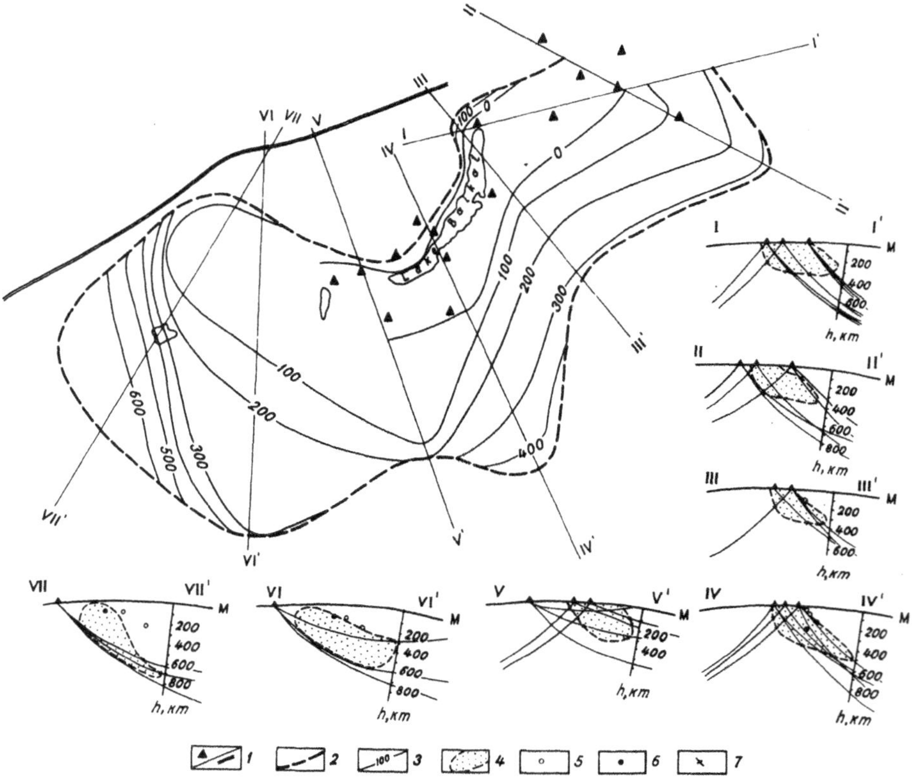

Figure 2. Structure of the upper boundary of the low velocity layer under the Baikal rift zone and adjacent areas. 1 - seismic stations and profiles; 2 - boundary of low-velocity layer projected onto the Earth's surface; 3 - isolines of the upper boundary with the crust (estimated in km); 4 - cross-section of the low velocity layer; 5 - point obtained from an auxiliary ray crossing the cross-section plane outside the low velocity limits; 6 - the same, but passing through the low velocity layer; 7 - boundary of the ray intercept where delay occurs.

If the gravitational influence of the Earth's crust is ignored, it appears that the upper part of the anomalous mantle region in the rift zone is inhomogeneous with respect to density. The maximum density deficiencies are registered in the vicinity of Cenozoic basalt fields (Figs. 1 and 3). Evidently, the density decrease is connected with a relatively small enrichment of the mantle material with basalt. The density of the upper part of the low velocity inhomogeneity down to 100 km is 0.03-0.04 $g/cm^3$ less than that of the normal mantle; at deeper levels the density contrast probably decreases to 0.005 $g/cm^3$. We believe that the anomalous mantle zone was formed by the rise of mantle material from considerable depths. In its topmost part, light differentiates accumulated and influenced the base of the Earth's crust. In particular, the rise of mantle material caused the isostatic uplift of the Sayan-Baikal dome. Spreading in different directions (to attain a minimisation of gravitational energy), the anomalous material caused extensional stresses in the crust which led to rift valley formation (Fig. 4).

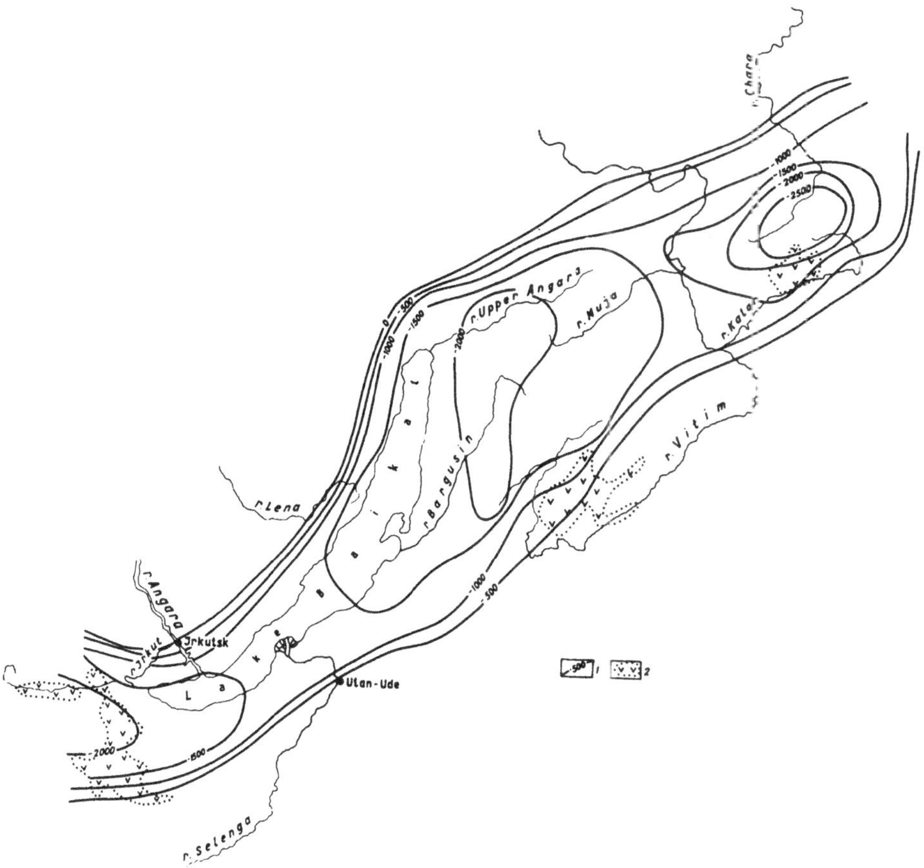


Figure 3. Distribution of the anomalous masses within the upper mantle under the Baikal Rift Zone.
1 - isolines of anomalous mass quantity on a square unit ($t/m^2$); 2 - fields of Cenozoic basalts.

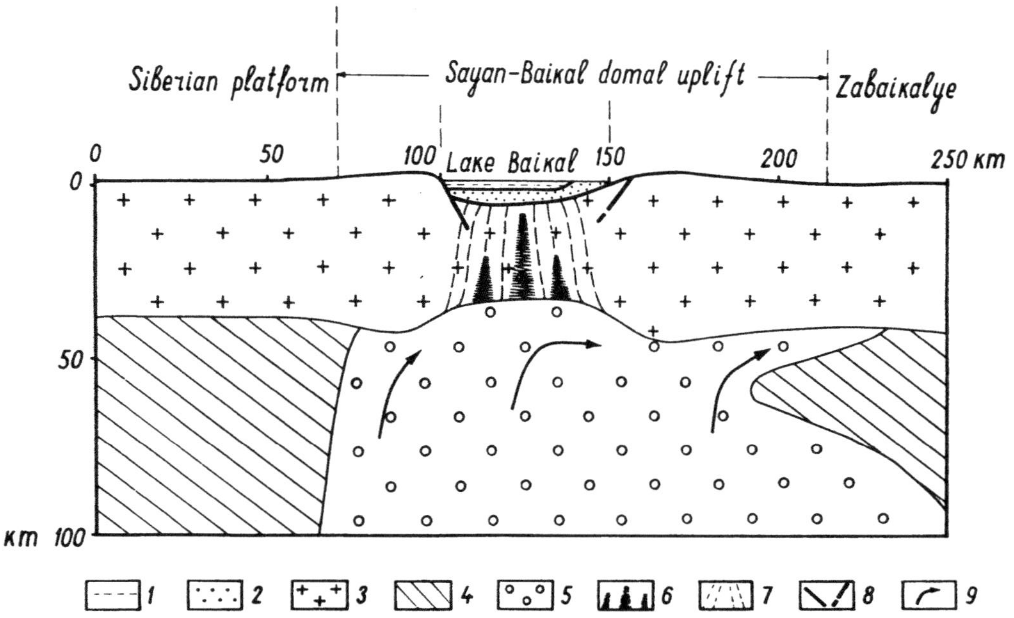


Figure 4. Deep structure of the central segment of the Baikal rift zone. 1 - water in Baikal; 2 - sediments; 3 - crust; 4 - normal mantle; 5 - anomalous mantle; 6 - mafic and ultramafic intrusions; 7 - heated region of the crust; 8 - faults; 9 - convective flow of the anomalous mantle substance.

If this model holds true, the Baikal rift zone should be considered as an isolated area, not connected with the world rift system, because the anomalous mantle region is spatially limited.

The Sayan-Baikal domal uplift is a structural unit of the first order, with variable strike. Except for the western flank depressions, all the other rift depressions extend in a northeasterly direction. In the Sayan part of the uplift, typical rift valleys are absent. However, because seismic data show the anomalous mantle zone extending upwards to the base of the crust the rifting had to embrace this region as well. The changes in strike of the domal uplift are accompanied by essential changes in the earthquake foci mechanism. In its Baikal-Stanovoy segment, the extensional stresses are oriented across the structural forms, whereas in the Sayan-Khamar-Daban part the stresses are compressive. Vast areas of Mongolia, China and the southeastern USSR are characterized by such compressive stresses.

If the extensional stresses at the base of the crust were created by the spreading of anomalous mantle, combined with the greater regional compressive stresses, the vector of which is oriented NE - SW then one can explain specific structural features of the Baikal rift zone as well as the changes in the earthquake foci mechanism. For example, whereas compressive stresses prevent the development of rift valleys in the Sayan-Northern Mongolia region, rifting has readily proceeded in the Baikal-Stanovoy upland where the local extension is oriented normally to the vector of regional compresssion.

## 3. NEOTECTONIC MOVEMENTS AND THEIR HISTORY

Besides relating to the general stress field, the structure of the Baikal rift zone depends essentially on the spatial arrangement of the structural units of the basement. The neotectonic structure of the Sayan-Baikal upland, which includes rift faults and valleys, is defined by two large semiseparate uplifts (the Sayan-Khamar-Daban uplift and the Baikal-Stanovoy uplift). In the saddle between them (Fig. 1) the South Baikal depression, the largest and the deepest among the Baikal rift zone valleys, is situated in close proximity to the Siberian platform margin. Although more common in the uplifted areas, the depressions are smaller and more widely scattered. Depressions are more common in the northeast and less so in the southwest. This pattern reflects the configuration of the uplifts themselves [5]. Such a distribution of the rift faults and valleys results from an interaction of the basement structure with regional extension and the extension caused by dome-like warping of the lithosphere. It is no accident that rift structures are more prevalent and their total width greater on the most uplifted flanks of the rift zone.

The regional extension of the crust also leads to dispersion of the rift valleys in the areas where marginal faults are oblique to the general strike of the rift zone and to the regional extension vector. An en echelon arrangement of depressions occurs in such regions and displacements along the faults are of strike-slip type.

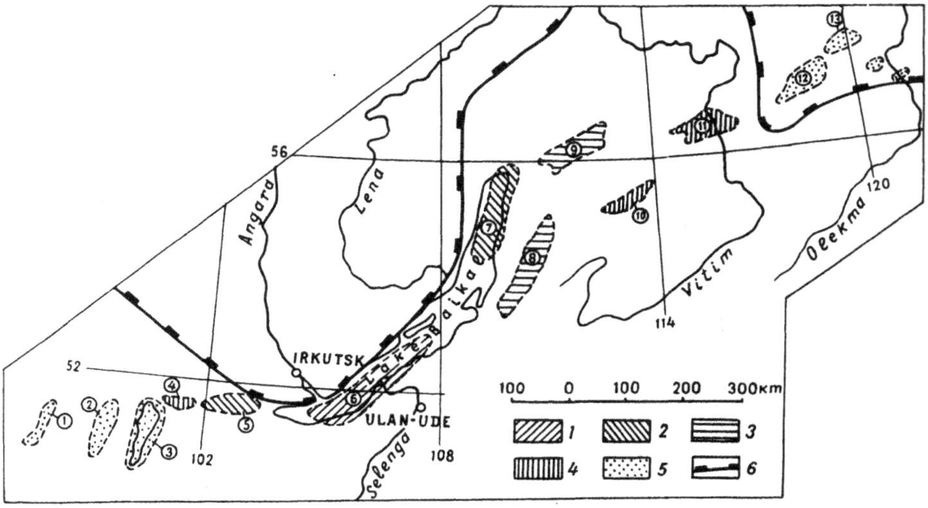


Figure 5. Probable sequence of initial formation of the largest rift valleys of the Baikal system.
1 - Eocene; 2 - Oligocene - Early Miocene; 3 - Miocene; 4 - Late Miocene - Early Pliocene; 5 - Pliocene - Quaternary; 6 - southern boundary of the pre-Cambrian Siberian craton. Numbers in circles designate major proto-rift depressions: 1) Busingol-Belin, 2) Darkhat, 3) Khubsugul, 4) Khoitogol-Mondy, 5) Tunka-Tora, 6) South Baikal, 7) North Baikal, 8) Barguzin, 9) Upper Angara, 10) Tsipa-Baunt, 11) Muya; 12) Chara; 13) Tokka.

Among the depressions of the rift zone, the South Baikal depression is the most prominent in structural and geomorphological features, a result of its location at the contact of the two large lithospheric plates of East Siberia (the Siberian platform and the neighbouring heterogenous fold system). The physical properties of this ancient, deep-seated suture zone, caused the lithosphere to fail when subjected to extension. This geological situation provided maximum concentration of rift-forming stresses in a comparatively narrow strip, which caused an earlier proto-rift valley formation at this place (the future Baikal depression). With time this process extended to the west and northeast (Fig.5) moving away from the Siberian platform margins, in accordance with the tectonic stress field. The longitudinal spread of the rift zone has been established beyond doubt, although we have no grounds to consider the propagation in both directions from the central segment to have been sunchronous processes. On the contrary the higher seismic activity on the eastern flank, as compared with the western one, testifies to possible differences [1].

Two main stages in the development of the Baikal rift zone are distinguished [6,7,8]. The early stage (Eocene - Early Pliocene) was characterized by weak and moderate tectonic movements with the prevalence of slow plastic deformations of the basement. The subsidence (totalling 1-3 km) of the proto-rift valleys of this stage was independent of the rate of uplift of the territory [1]. Movements along the faults, if any, were slow, without creating sharp topographical contrasts. The general morphology of the zone was defined by a combination of wide shallow basins with low erosional and volcanic plateaux.

At the later stage (Middle Pliocene-Recent) a six-fold increase of the tectonic deformation rate, as compared with that of the previous stage, led to crustal destruction, accompanied by displacements along faults. Simultaneously, sudden depending of the proto-rift depressions took place against the background of the general uplift of the rift zone. Numerous new rift depressions were formed as a result of further breaking up of the crust during extension [1,6]. There are no other continental rift zones with so numerous recent rift depressions of different evolutionary stages.

## 4. SEISMOTECTONICS

The emergence of the proto-rift valleys started with gentle warping and then doming of the Sayan-Baikal region. At a later stage, rupturing of the crust began and intensified. Both asymmetric and true grabens are found among present rift valleys. They form the following evolutionary succession (Fig. 6): incipient - embryonic-nature (mono- and polystructural) - aborted rift valleys [1].

Incipient (in statu nascendi) rift depressions are upland grabens formed by recent seismotectonic movements. Some of them began to develop in the Late Pleistocene, but their most active formation period was the Holocene. The boundary faults are newly formed or have been inherited from pre-rift faults. These rift

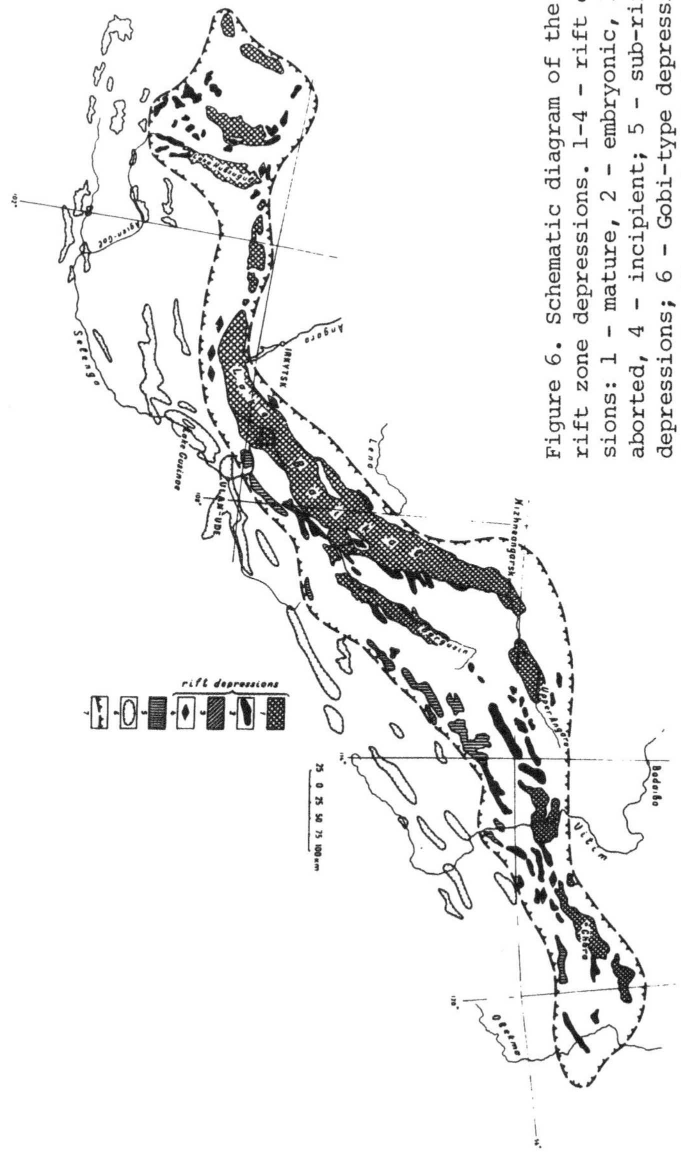


Figure 6. Schematic diagram of the Baikal rift zone depressions. 1-4 - rift depressions: 1 - mature, 2 - embryonic, 3 - aborted, 4 - incipient; 5 - sub-rift depressions; 6 - Gobi-type depressions; 7 - Baikal rift zone boundaries.

valleys (0.6 to 2 km long, from 10 to 800 m wide, and from a few meters to 100 m deep) are miniature morphological expressions of vigorous seismogenerating movements of major structures. They are encountered in the regions with traces of the strongest earthquakes (magnitudes up to 7.75). Numerous earthquakes with focal depths of 25 to 30 km have recently been registered there.

Embryonic rift valleys have been forming since the Pleistocene. They are 25-30 km long and up to 6 km wide. The thickness of Pleistocene-Holocene sediments, at places including basalt flows, amounts to 300-400 m. Chains of embryonic trenches are found in the eastern part of the rift zone. In spite of their small size they sometimes have intra-depressional cross uplifts, well expressed block differentiation on the base, one- and two-sided fault boundaries, i.e. all the morphotectonic features typical of the mature rift valleys. The present seismic stress field extends outwards beyond the limits of the embryonic rift valleys and downward to considerable depth. Earthquake foci have been registered at depths of up to 40 km, instead of the 5-15 km focal depths typifying earthquakes in large mature depressions. Embryonic rift valleys continuously widen at the expense of the adjacent mountains, and deepen simultaneously. This phenomenon was observed during the last strong earthquake east of the Vitim River (27.VI.1957, M=7,9; $I_o$=X-XI) when the Namarakit embryonic depression subsided by 5-6 m and widened westwards [9].

Mature depressions of the Baikal rift zone are some tens of kilometers to 400 km long and 10 to 70 km wide. The amplitude of vertical displacement varies from 2 to 7 km aggregate displacement may locally amount to 8.5 km over a distance of a few kilometers. These depressions are mainly filled with sediments; the Baikal and Khubsugul depresssions contain large lakes. The sediments attain a thickness of 6 km in the South Baikal depression [4]. It is actively widening: during the earthquake of 12.I.1862 a crustal block of about 260 $km^2$ subsided in the epicentral zone and the lake area increased by 203 $km^2$. The expansion of the depression at the expense of the surrounding mountains is confirmed by parallel faults with oblique piedmont valleys between them and blocks that have subsided during strong earthquakes [9,10].

Dry rift valleys, as well as the Baikal Lake depression, are mostly polystructural. They have been formed by the transverse or longitudinal margin and coalescing of smaller depressions through the subsidence of the cross uplifts which previously had separated them. Along the rift axes, the Cenozoic sediment thickness reaches 3 km: The thickness decreases to about 0.5 km in the flank depressions.

Aborted depressions are morphologically similar to embryonic ones. They occur mainly in the southwestern part of the rift zone why their development ceased remains unclear.

Seismotectonic investigations in the Baikal rift zone by "paleoseismogeological" methods, which enable one to extract information concerning strong earthquakes that occurred a few thousand years before the present, have demonstrated the instability of this region. Three periods of seismic revival have occurred, each followed by changes in volcanic activity. They took place in the Early Pleistocene, the last interglacial and Middle Holocene, respectively. At the western termination of the rift zone recent seismic acitvity is weaker than in its northeastern part, where the rifting process is very active and spreading both distally and laterally.

The rift zone axis deviates from the Chara rift valley to the east and the system ends in the Tokko depression - as if it cannot overcome the resistance of the Aldan shield. East of the Chara graben two branches of embryonic depressions penetrate into the Stanovoy range. They are only vaguely expressed morphologically but are actively developing: during the last 10 years 3 earthquakes with M=6,5-7 took place in this zone and some traces of stronger earthquakes are found there.

Near the southern edge of the Aldan shield, as well as in the East Sayan and Mongolia south of the Khubsugul rift depression, rifting extensional stresses change into compressive ones [11]. Thus, the Baikal rift zone is a developing geodynamic structural system isolated from the world rift system by vast fields of tectonic compression.

## 5. CONCLUSION

Though the Baikal rift zone possesses many distinctive features, its general evolution in many respects resembles that of other continental rifts. Local energy sources played a decisive role in its formation, whereas its scattered arrangement (faults, rift depressions and volcanic fields) is determined by close interaction between the rifting mechanism and the irregular structural grain of the pre-rift basement. The Baikal rift valleys began to form not later than in the Eocene,with deep opening of the upper crust at the south-eastern edge of the Siberian platform in the region of the South Baikal depression; subsequent propagation of rifting took place to the west and northeast of this earliest and largest depression.

Rifting in the Baikal region is caused by local upvelling of mantle material induced by gravitational convection. The rift-forming extensional stress field is due to slow outward displacement of the anomalous upper mantle substance in a region from the margin of the Siberian platform to the southeast. Seismotectonics and focal mechanism studies of earthquakes support the idea that the Baikal rift zone is a mobile geodynamic belt isolated at both of its terminations from the world rift system by vast fields of compression.

REFERENCES

1. V.P. Solonenko, Seismotectonics and the recent development of the Baikal rift zone structure, in: Baikal Rift, Nauka, Moscow, 1968. In Russian.
2. N.I. Puzurev, M.M. Mandelbaum, S.B. Krylov, B.P. Mishenkin, G.V. Krupskaya and G.V. Petrik, Geol. Geofiz., 5, 155, 1974. In Russian with English summary.
3. V.V. Nikolayev, V.P. Solonenko and S.D. Khilko, Evolution of fifting in the north-east of the Baikal rift zone, in: Baikal Rift, Nauka, Sibirian Branch, Novosibirsk, 1975. In Russian.
4. Yu.A. Zorin, Recent structure and isostasy of the Baikal rift zone and the adjacent territories, Nauka, Moscow, 1971.
5. N.A. Logatchev, V.I. Galkin and G.S. Goldyrev, Sayan-Baikal-Stanovoye upland, in: The uplands of the Pribaikalye and Zabaikalye areas, Nauka, Moscow, 1974. In Russian.
6. N.A. Florensov, Mesozoic and Cenozoic depressions in Pribaikalye, Publ. House of the USSR Ac. of Sci., Moscow-Leningrad, 1960. In Russian.
7. N.A. Logatchev, Sedimentary and volcanogenous formations of the Baikal rift zone. In: Baikal Rift, Nauka, Moscow, 1968. In Russian.
8. N.A. Florensov, N.A. Logatchev and V.P. Solonenko, Cenozoic volcanism of rift zones, in: Int. Geol. Congr. Rep. XXIII Sess., Abstr., Academia, Prague, 1968.
9. V.P. Solonenko, A.A. Treskov, R.A. Kurushin et al., Living tectonics, volcanoes and the seismicity of the Stanovoye Upland, Nauka, Moscow, 1966. In Russian.
10. V.P. Solnenko, A.A. Treskov, V.M. Zhilkin et al., Seismotectonics and seismicity of the Pribaikalye rift system, Nauka, Moscow, 1968. In Russian.
11. L.A. Misharina, Stress fields in the Earth's Crust in the rift zones, Nauka, Moscow, 1967. In Russian.

# TWO STAGES RHINEGRABEN RIFTING

J.H. Illies

Geologisches Institut, Universität Karlsruhe,
D-75 Karlsruhe, Germany F.R.

The formation of the Rhinegraben was chiefly governed by the coincidence of 3 essential circumstances.

(1) The regional setting of the graben appears to be mainly basement controlled. As stated by Stille [1], a coherent weakness belt traverses the basement of Western Europe from southern Norway down to southern France ("Mediterranean-Mjøsa zone"). In the Rhinegraben area this belt is indicated by a swarm of lineaments, shear zones, and dykes, parallel to the NNE trend of the graben [2]. These pre-existent features, most of them formed during the Asturian phase of the Hercynian orogeny, had paved the way for rifting and locally attracted fault progression during the break-up of the graben in Eocene times.

(2) Rhinegraben formation was mantle controlled. Underneath the southern graben segment, including its shoulders, a mantle diapir or plume ascended, coming closest to the surface below the Miocene composite volcano of the Kaiserstuhl, where the Moho lies at a depth of about 24 km [3]. The rift valley traces the crest line of this mantle body; mantle upwelling may have been additionally energized by crustal rifting. The isobases of Cenozoic shoulder upwarping to a large extent follow the contours of the asthenolitic body [4]. The first graben subsidence during the Eocene started on top of the diapir, and the depocentre shifted progressively northward during further rift valley accretion. It is assumed that mantle updoming supported crustal rifting [5] just as gravity sliding from the mantle bulge supported wedge block subsidence.

(3) Rifting was stress field controlled. The mid-Eocene to to Lower Miocene period of tensional rifting and fault trough sub-

*I. B. Ramberg and E.-R. Neumann (eds.), Tectonics and Geophysics of Continental Rifts,* 63-71.

sidence (Fig. 1) had evolved under regional stress conditions with the component of maximum horizontal compression oriented about parallel to the trend of the rift valley [6]. The lateral yielding of the framing blocks necessary for wedge block subsidence, tilt block rotations, and crustal spreading, demanded a stress regime with its respective horizontal components fitting both the given crustal anisotropy and the crest line of the mantle bulge underneath. On the other hand, from mid-Pliocene to Recent times the rifting was governed by a regional stress field whose component of sinistral shear was about parallel to the axis of the rift valley. This reactivated the vectorial anisotropy of the Hercynian shear zone. Consequently, the original tensional graben structure was remodelled into a sinistral strike-slip belt [6].

The geological history of the Rhinegraben is first and foremost seen as the result of mutual mechanical interactions of pre-existing basement features, mantle diapirism and varying stress conditions.

During mid-Cretaceous times widespread volcanic activity mainly with olivine-nephelinitic products preceded the break-up of the crust [8]. About 45 m.y. later, in Middle Eocene, surficial faulting, fault trough subsidence and sediment infilling of a freshwater to brackish lagoon commenced [8]. During the Upper Eocene to Lower Oligocene interval the first climax of taphrogeosynclinal subsidence and subsequent shoulder upwarping occurred. After that, during Middle Oligocene times, a marine transgression flooded the Rhinegraben and the adjacent Hessen depression. A narrow graben sea was formed that split Europe from the molasse foredeep in the south up to the eastern part of the North Sea basin in Lower Saxony. Farther south towards the Mediterranean the Middle Oligocene transgression had invaded the Bresse and temporarily the Limagne graben. During the Upper Oligocene the subsidence had slowed down, and in the generally regressive period of the Lower Miocene the depocentre of the Rhinegraben shifted northward, exhibiting high sedimentation rates and intense growth faulting in the Heidelberg segment. In Middle Miocene to Lower Pliocene times grapen subsidence died out. Fluvial erosion was subsequently active over wide parts of the graben floor [8].

During mid-Pliocene, graben tectonics reawoke, henceforth controlled by a sinistral shear motion [6]. Conditioned by the crooked course of the Paleogene-formed rift valley, the reactivation of a strike-slip motion produced extension shear or compression shear in the individual graben segments (Fig. 2). In the northern or Heidelberg segment the local graben trend relative to the axis of shear supported lateral extension and subsidence. The depocentre during the Upper Pliocene to Quaternary is therefore found in this segment, reaching a maximum thickness of more than 1000 m of sediments [9]. The Recent rates of subsidence attain about 1 mm/year [10]. In the central or Baden-Baden segment the shear motion produced compression. The eastern master fault, primarily a dip-slip element, has been overprinted by thrusting, and wide parts of the graben floor were subjected to uplift and fluvial erosion [6, 11]. This is the only segment where excessive heat flow values are observed [12]. Presumably, the compression shear excites convection of sealed-in thermal waters in deeper aquifers. In the southern or Freiburg segment extension shear prevails. Locally, the rigid block of the Miocene composite volcano, the Kaiserstuhl, situated in the centre of the graben, hampers shear motion. A belt of Pleistocene to Recent subsidence [13, 14] swings around it to the west and enhanced macroseismic activity is concentrated along its western and northern front [15]. Opposite, along its eastern and southern side, extremely low excess stresses were determined [6]. This region, the tilt block area of the Freiburg embayment, is found to be situated on the lee-side of active shear stresses.

All over the graben fault zones of Pleistocene to Recent activity shows a preferred strike direction of about 170°, i.e. oblique to the about 20° average trend of the whole rift valley. These fault zones, some of them seismically active, sometimes show indications of second order shearing in the sense of Riedel [16]. In his 1929 publication this author first mentioned the Rhinegraben as an example of that mechanism today termed Riedel shear.

Some of the main events of the Rhinegraben tectonic history coincide with phases of the Alpine orogeny. E.g. the climax of tensional Rhinegraben rifting in Upper Eocene to Lower Oligocene came to pass simultaneously with the culmination of Alpine compressional deformations [4]. Moreover, the general "Pyrenean" trend of contemporaneous Alpine plate collision was about normally oriented to the trace of Rhinegraben rifting in the adjacent foreland. Compressional plate collision and tensional intraplate rifting normal to the collision front are seen as controlled by one and the same regional stress field [6].

Alpine folding and thrusting came to an end in late Miocene to early Pliocene times. A general uplift, first indicated during Oligocene, had increasingly substituted the previous stage of compressive deformations. The Recent tectonic behavior is that of epeirogenic updoming, rather than plate tectonic compression. Geodetically, the uparching in the Central Alps of Switzerland attains 1.7 mm/year [17]. The mountain range, which during Upper Cretaceous to Miocene was the main front of plate collision, has been overshadowed by the Recent plate tectonic advance which became more and more localized to active island arcs farther south, such as the Cretic, the Liparic and the Balearic. Notwithstanding this plate tectonic reorientation, excessive horizontal stresses up to 350 bars had been determined in the Western Alps at several places

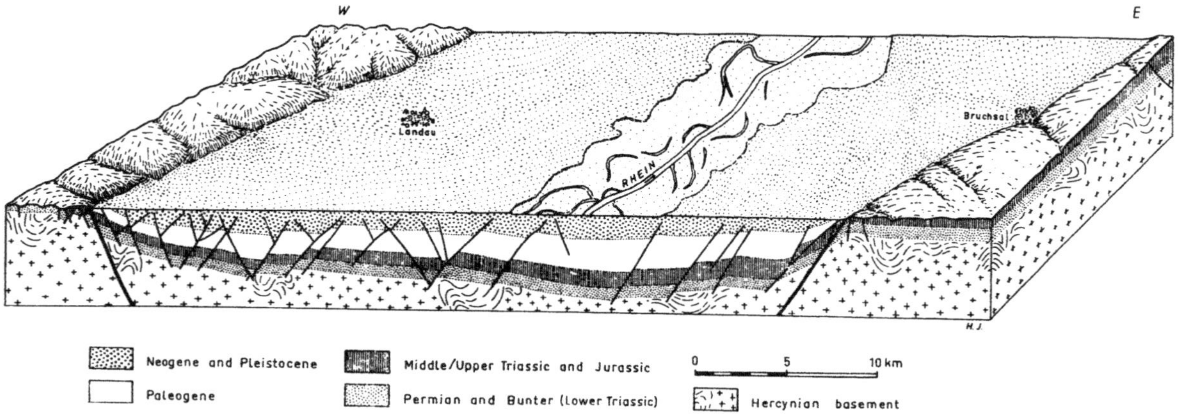


Fig. 1. Block diagram of a Rhinegraben segment north of the city of Karlsruhe. Converging master faults and sets of internal dip-slip faults of a mostly antithetic character reveal dilative motion of the framing blocks away from the graben axis. The observed fault pattern is mainly a result of the mid-Eocene to Lower Miocene stage of tensional rifting and crustal spreading.

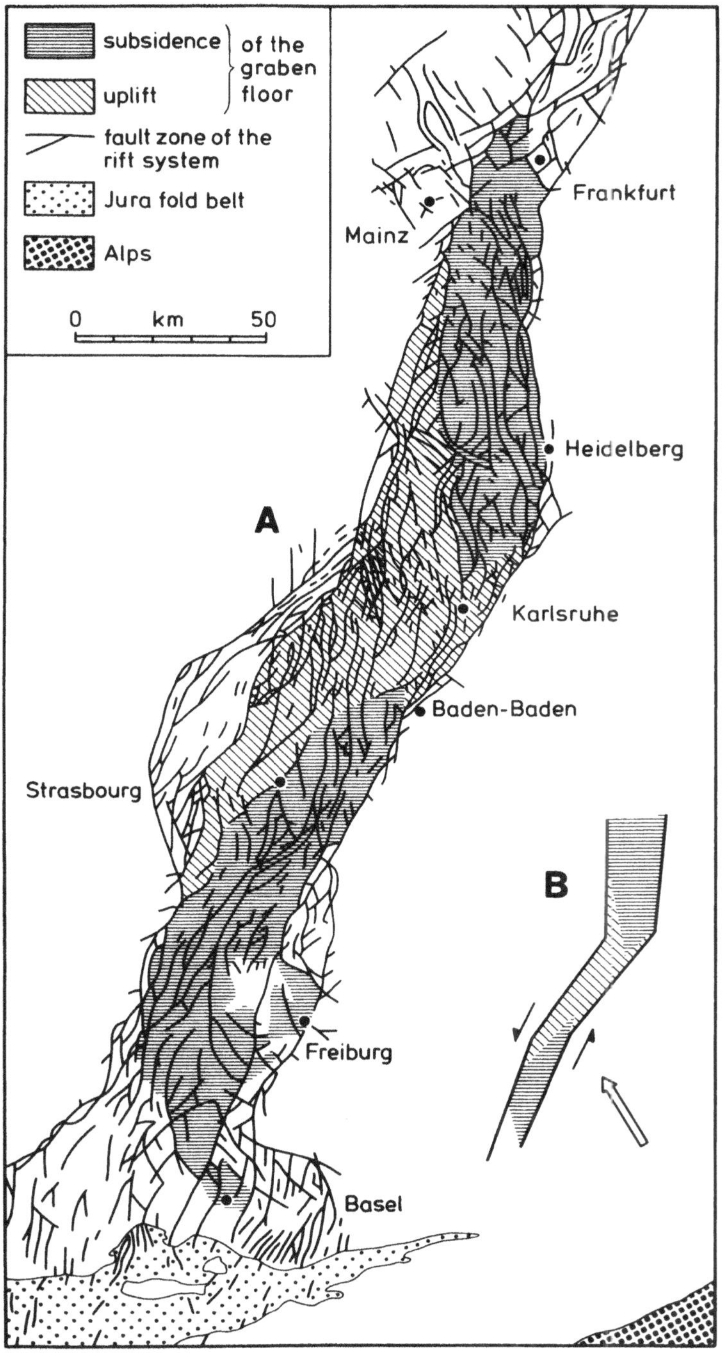


Fig. 2. A. Recent tectonic activity of the Rhinegraben is controlled by an axial sinistral shear motion. Due to the uneven course of the graben margins the central segment is experiencing compressional shear and uplift. In the northern as well as in the southern part of the graben extension shear and consequent subsidence are observed.
B. Schematic model to illustrate horizontal and vertical block motions along Rhinegraben.

by means of overcoring tests [18]. These extraordinarily high stresses are in part caused by topographic effects of the mountain range. An additional share of stresses may be ascribed to high denudation rates [19] which caused unloading and subsequent rock destressing and expansion, as well as expansion due to phase transformations of minerals in the underlying crust. The stress trajectories in the Western Alps appear dependent on the course of the isobases of recent uplift that favours the concept of a mainly self-induced character of the stresses determined.

The rising and lateral expanding mountain body meets a northern frame of foreland blocks, which is underlain by a strongly consolidated Hercynian basement. Within this stable abutment the Rhinegraben is a weak zone predestined to force the frame open and to be a centre for the strain relief of Alpine stresses. Since the general trend of the graben approximately fitted the direction of the sinistral shear component of the changed regional stress field, graben tectonics have been revived by shear since mid-Pliocene times. East of the Rhinegraben, in the area of the Black Forest and the adjacent platform of the Swabian Jura, some NW-SE trending basement fractures have been remodelled into dextral strike-slip zones [6]. The crustal mobility north of the Alps was considerable increased by these shear displacements, conjugate to the sinistral shear belt of the graben. In particular, the Alpine front segment appropriate for a northward yielding became widened to form a fan-shaped pattern of seismotectonic activity linking together the Western Alps and the Rhinegraben (Fig. 3).

At its northern end the Rhinegraben proper is intersected transversally by the Hercynian block of the Rhenish massif. This intersection had forced an about 20 km eastward offset of the Paleogene fault trough axis from the Rhinegraben to its primary northward appendage, the Hessen depression. The offset, which exhibits the typical dog-legged features, cuts off the slide-bar mechanism of Recent Rhinegraben tectonics. From here on the continued mobile belt of seismotectonic fault activity is diverted towards the NNW, straight into the Rhenish massif [8, 20]. This trend is parallel to today's direction of maximum horizontal compression (Fig. 3). Regardless of numerous Quaternary active normal faults and a widespread Pleistocene to Holocene volcanic activity, graben features proper remain absent in this segment. Thick series of strongly folded Devonian shales and slates, with a dense pattern of jointing, cleavage and bedding planes, left the crust vulnerable to plastic rather than rigid deformation.

The mobile belt, its whole course trending nearly parallel to The Rhine river, encountered another pre-existent breaking point near Bonn, that once again fitted the direction of maximum horizontal compression. It is the fault bundle of the Lower Rhine Embayment, first active as cross fractures during Hercynian folding and

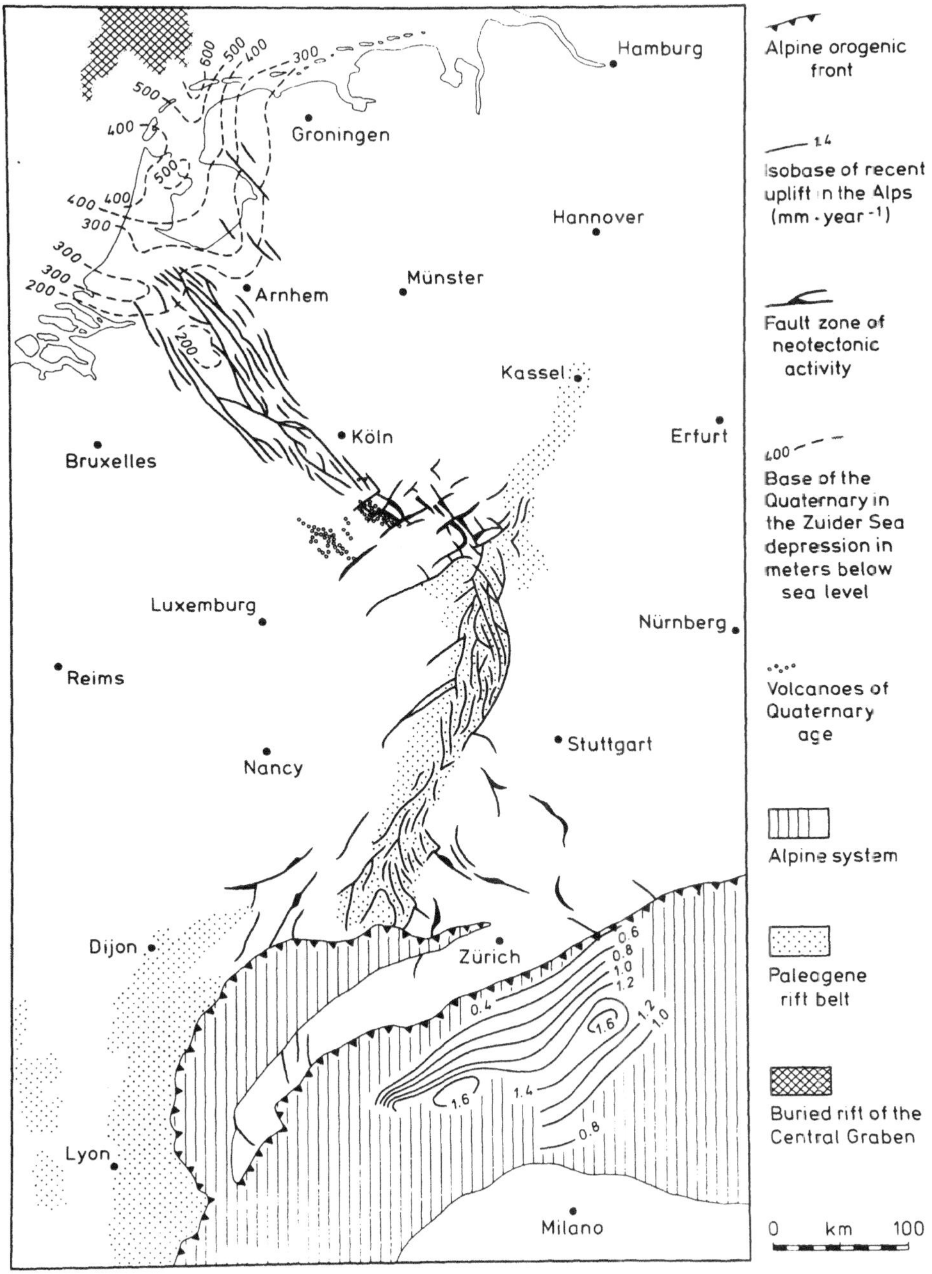


Fig. 3. An active subplate boundary links Western Alps to North Sea basin.

later, in Upper Cretaceous and Miocene times, reactivated by tilt block tectonics. This fault zone underwent tensional rifting, high rates of vertical movements and a funnel-like widening during the Pliocene to Recent period [21]. Towards its southern end this segment is seismically active, and the late Pleistocene Rodderberg volcano marks the hinge of the fan-shaped rift valley. Farther north in the Netherlands the neotectonic activity of the rift valley terminates near Utrecht.

At the same latitude, but with about a 60 km eastward offset, the trough of the Zuider Sea depression begins near Arnhem [22] where the Quaternary base is down-dropped to a depth of about 600 m (Fig. 3). The depression leads to the southern end of the Central Graben of the North Sea basin. The latter has remained and extinct rift valley since end of the Meozoic [23]. When active rift valley propagation had reached the area of the North Sea coast and a fit had occurred of the stress pattern and the given ruptural anisotropy, the buried rift belt became rejuvenated at its southern end in a similar way as other weakness zones did farther south. Violent storm tides during the 13th century, mainly that of 1287, has submerged the marshes by creating the Zuider Sea lagoon [24]. The actual coast line approximately traces the contours of the tectonic depression.

Since Pliocene times the Alpine mountain range has been rapidly uplifted, through isostatic readjustment of the compression-produced crustal bulge and the subsequent processes of denudation and unloading. These processes which were mainly epeirogenetically controlled, created regional stress conditions, completely different from those previously generated by plate collision and subduction. A consequent rotation and alteration of the main stress vectors forced the block mosaic of the foreland to react with changes of the kinematics along individual block boundaries [6]. The Paleogene Rhinegraben rift system (Fig. 1) was a tensional structure, its trend, fault pattern and mantle rise were an intraplate reaction to the contemporanous main stage of Alpine plate collision. This structure was remodelled into sinistral shear zone. Other Paleogene rifts (e.g. Limagne graben, Bresse graben and Hessen depression), became extinct rift valleys in early Miocene times as a result of the changed stress conditions. Instead of the Paleogene formed Rhinegraben rift system, a new subplate boundary has been created by rift valley accretion. Its course was determined by the coincidence of the pre-existent crustal anisotropy and the regional stress pattern. The Rhinegraben in a narrower sense (between Basel and Frankfurt) remained the only segment where a Paleogene formed graben coincides with the course of the Pliocene-Recent time active subplate boundary.

REFERENCES


1. H. Stille, Abh. Preuss. Geol. Landesanst., N.F., 95, 110, 1923-25.
2. H. Illies, Geol. Rundsch., 52, 317, 1962.
3. J.B. Edel, K. Fuchs, C. Gelbke and C. Prodehl, J. Geophys., 41, 333, 1975.
4. H. Illies, Geol. Rundsch., 64, 677, 1975.
5. H. Cloos, Geol. Rundsch., 30, 401, 1939.
6. H. Illies and G. Greiner, Oberrhein. geol. Abh., 25, 1, 1976.
7. I. Baranyi, H.J. Lippolt and W. Todt, Oberrhein. geol. Abh., 25, 21, 1976.
8. H. Illies, Geologie en Mijnbouw, 56, 1977 (in press).
9. F. Doebl, The Tertiary and Pleistocene Sediments of the Northern and Central Part of the Upper Rhinegraben, in: The Rhinegraben Progress Report 1967, ed. by J.P. Rothé and K. Sauer, Freiburg and Strasbourg 1967.
10. F. Schwarz, Z. f. Vermessungswesen, 101, 14, 1976.
11. Nivellementnetze in Rheinland-Pfalz, Maβstab 1:250 000, Landesvermessungsamt, Koblenz 1977.
12. R. Haenel, Geol. Rundsch., 65, 797, 1976.
13. J. Bartz, Die Mächtigkeiten des Quartärs im Oberrheingraben, in: Approaches to Taphrogenesis, ed. by H. Illies and K. Fuchs, Scweizerbart, Stuttgart 1974.
14. H. Mälzer and H. Schlemmer, Tectonophysics, 29, 275, 1975.
15. R. Lais, Gerlands Betr. z. Geophys., 12, 45, 1912.
16. W. Reidel, Centralbl. f. Min., Geol. u. Paläont., (B), 1929, 354, 1929.
17. Rezente Krustenbewegungen, Schweizerisches Landesnivellement 1976.
18. H. Illies and G. Greiner, Geol. Soc. Am. Bull., 89, 1978 (in press).
19. H. Jäckli, Eclog. Geol. Helv., 51, 354, 1958.
20. L. Ahorner, Tectonophysics, 29, 233, 1975.
21. L. Ahorner, Eiszeitalter u. Gegenwart, 13, 24, 1962.
22. Atlas van Nederland, Blad II-4, Tektoniek, Topographische Dienst, 1975.
23. P.A. Ziegler, Am. Assoc. Petrol. Geol. Bull., 59, 1073, 1975.
24. Ir. T. Edelman, De historische veranderingen in de natuurlijke gesteldheid van het Nederlandse waddengebied, in: Het Waddenboek, ed. by Nederlandse Geologische Vereniging, Zutphen, 1964.

# THE RIO GRANDE RIFT AND A DIAPIRIC MECHANISM FOR CONTINENTAL RIFTING*

R.J. Bridwell

University of California, Los Alamos Scientific Laboratory, Los Alamos, New Mexico, 87545, USA

ABSTRACT. Continental rifts have been characterized by en echelon valleys which average 50 km in width, are bounded by extensional normal faults and act as crustal lineaments of great length transecting broad uplifted arches. Gravity, heat flow and volcanism, relatively well known for the Rio Grande Rift, provide models of the lithosphere which include variations in thickness, temperature, stress and viscosity. Inclusion of the third dimension in the process of continental rifting implies a thinning of the lithosphere through intrusion of a high-temperature, less-dense, low-viscosity mantle diapir. This mass moves buoyantly upward, creating uplift, shear-stress concentrations in the crust beneath the rift, normal faulting and attendant surface volcanism.

## 1. INTRODUCTION

This paper presents new interpretations of the concept of rifting along the Rio Grande Rift through an integrated series of geophysical models of the third dimension - the lithosphere and asthenosphere. Gravity models provide a preliminary estimate of an attenuated lithosphere, two-dimensional heat flow models provide a temperature distribution and steady-state creep models provide estimates of the shear-stress and effective viscosity in a tectonically thinned lithosphere. A mechanism of rifting consistent with geology and geophysics along the rift is proposed.

* Work supported by U.S. Energy & Research Development Administration under contract W-7405-ENG-36. I thank J. Callender and T. Shankland for review and N. Carter for a preprint of [15].

*I. B. Ramberg and E.-R. Neumann (eds.), Tectonics and Geophysics of Continental Rifts, 73-80.*

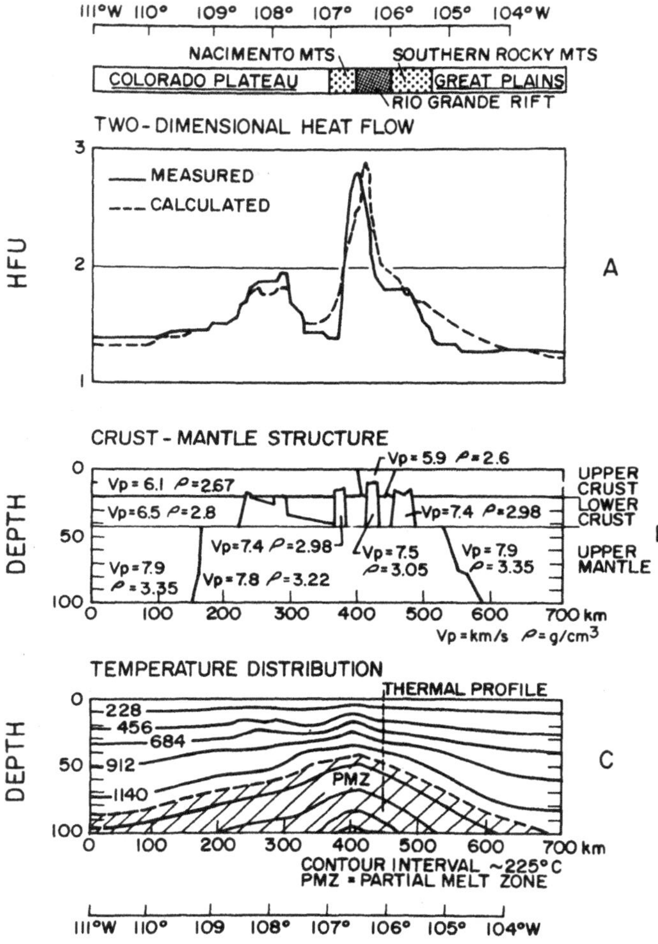


Fig. 1. Two-Dimensional Heat Flow Fit (A), Structure of the Lithosphere (B), and Steady-State Geothermal Model (C).

## 2. GRAVITY, THERMAL, AND STRESS MODELS OF THE NORTHERN RIO GRANDE RIFT

Long wavelength, negative Bouguer gravity anomalies are associated with the Rio Grande Rift and adjacent provinces. Gravity models, constructed along 36°N latitude, define an intrusion of low density material into a thinned lithosphere in Fig. 1 B [1]. The significant feature of the Bouguer low is a large mass deficiency required to explain the negative anomaly. Deep seismic soundings at latitude 34.5°N indicate the Moho shallowing to 36 km beneath the rift and intrusive bodies in the lower crust [2]. Microearthquake surveys indicate a magma chamber at the shallow depth of 18 km [3]. Such bodies are considered typical of petrogenetic equivalents intruding the lower crust in Fig. 1 B.

General trends of heat flow for the Rio Grande Rift, Colorado Plateau, and Southern Rocky Mountains are based on reduced heat flow [4] and numerous surface heat flow values [5,6]. Surface heat flow provides thermal boundary conditions for finite element models of the steady-state temperature distribution shown in Fig. 1 C. Modelling parameters are specified in [1].

The significant feature of the thermal model is a variation in geothermal gradients. If the mantle contains 0.1% $H_2O$ [7], $1200^{o}C$ can be interpreted as the melting isotherm. Hence this isotherm indicates the lithosphere-asthenosphere boundary. Clearly, there is an asymmetry to the thermal base of the lithosphere with considerable thinning beneath the rift. The thermal profile in Fig. 1 C is chosen at a surface heat flow of 2 HFU for comparison with other rift geotherms.

Two categories of thermal models provide geotherms for the rift. Theoretical geotherms are fits to measured surface heat flow assuming spatial distribution of thermal conductivities from gravity or sparse seismic refraction models. Experimental geotherms utilize pyroxene geothermometry to define temperature and pressure of mantle xenoliths [8,9]. Regional geotherms are summarized in Fig. 2. The experimental geotherm from the southern rift is a minimum geotherm [9]. It is similar to the high-temperature oceanic geotherm [10].

Theoretical geotherms are somewhat higher than geotherms from petrologic constraints. These differences can result from, 1) two-versus-one dimensional fits and, 2) obvious differences in crust-mantle structure at $36^{o}N$ and $32^{o}N$ [6] from gravity models. This divergence may be resolved with new pyroxene geothermometry from numerous xenolith localities recently discovered along the central and northern rift [11].

Steady-state flow of upper mantle rocks at temperatures greater than 0.4-0.5 of the melt temperature $T_m$ is a thermally-activated rate process related to temperature and stress by a Weertman equation [12]

$$\dot{\varepsilon} = A \exp(-Q/RT)\tau^{n}. \qquad (1)$$

In this equation three material constants A, Q, and n are determined experimentally. A is a slightly temperature dependent constant, Q is creep activation energy, and n is an exponent ranging from 2 to 4 for upper mantle rocks [12]. $\dot{\varepsilon}$ and $\tau$ are strain rate and shear stress. $\dot{\varepsilon}$ is determined from geologic deformation rates and the

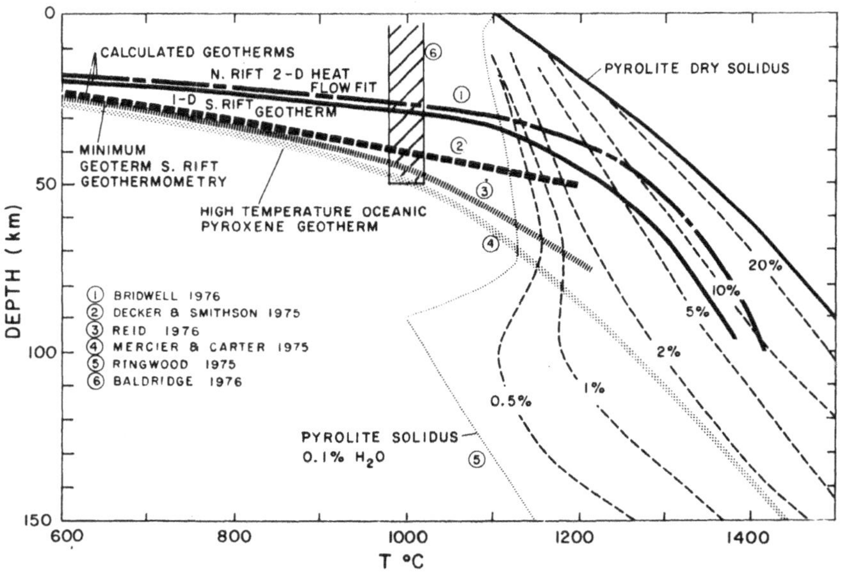


Fig. 2. Theoretical and Experimental Geotherms for Rio Grande Rift.

equation can be solved for $\tau$ if T is known. The effective viscosity is one third the shear stress divided by strain rate.

Recent experimental studies provided activation energy Q and stress exponent n under dry and wet conditions for Mt. Burnet dunite, a rock considered typical of the upper mantle [12]. Experimental constants for dry dunite are $A=1.7 \times 10^9$ kbars $^{n-1}$, Q= -130 kcal mole$^{-1}$, and n=3.2. Recent studies of natural strain rates from geodetic measurements of surface displacements along the San Andreas fault [13] and glacial rebound [14] have the range $10^{-13} > \dot{\varepsilon} > 10^{-15} s^{-1}$, implying an average geologic strain rate of $10^{-14} s^{-1}$. The temperature distribution will be defined by the two-dimensional heat flow fit [1].

Nonlinear steady-state shear stresses in the thermal regime $0.4<T<T_m$ are shown in Fig. 3. Stress variations range from 1700 to 3 bars through the upper mantle to the base of the lithosphere. Elastic stresses of the crust are implied to exceed 1700 bars whereas zero shear stresses in the upper mantle imply hydrostatic pressures at the base of the lithosphere.

Effective viscosities range from $>10^{22}$ to $<10^{19}$ poises. The lower value of $<10^{19}$ is consistent with calculated values from

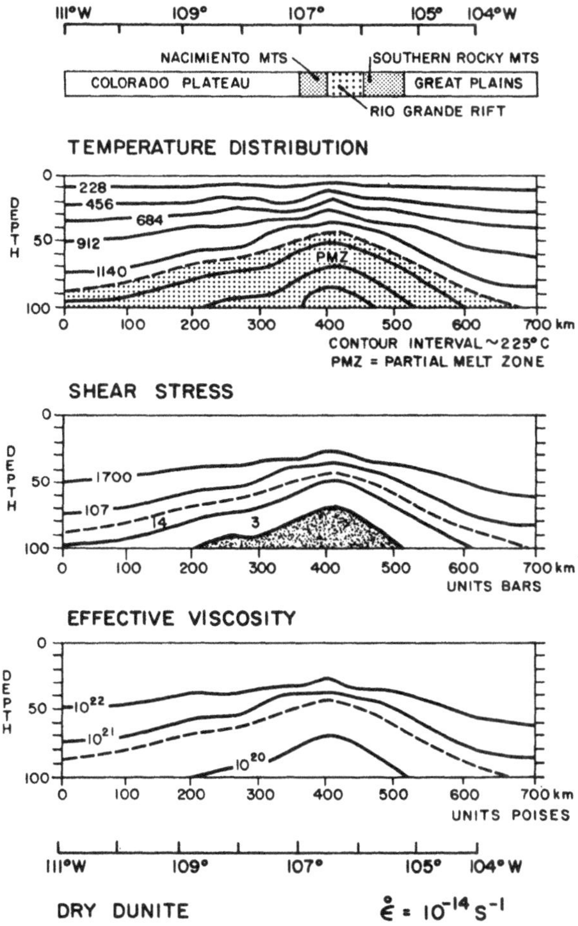


Fig. 3. Shear Stress and Effective Viscosity for Rio Grande Rift.

glacial rebound [14].

Steady-state flow stresses beneath Kilbourne Hole and Mt. Taylor, New Mexico, have been estimated using mechanical creep textures derived from natural peridotite kimberlites equilibrated at depths given by pyroxene geothermometry [15]. A comparison of theoretically predicted nonlinear steady-state flow stresses and estimated mechanical stresses is shown in Fig. 4. Mechanical stresses result from analysis of dry and wet rocks [12] and the northern rift prediction uses dry constants and theoretical geotherms fitting the two-dimensional surface heat flow. Theoretical

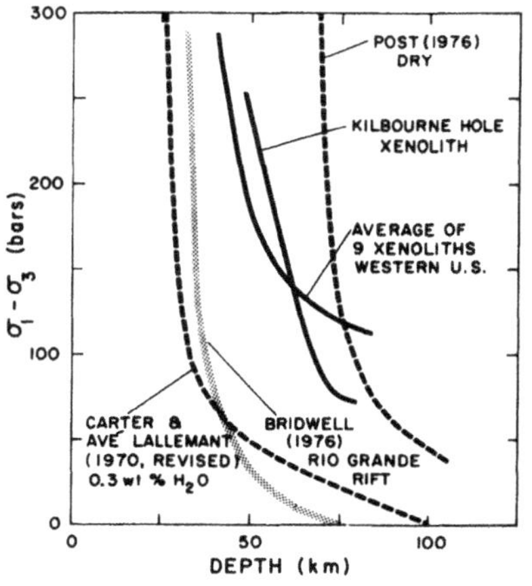


Fig. 4. Comparison of steady-state flow stress from theoretical analysis (dotted) and natural kimberlite zenoliths (dashed and solid). Modified from [15].

geotherms are shallower than the high temperature oceanic geotherm of Fig. 3 from [15]. Thermal scaling in Eq (1) decreases shear stress beneath the rift relative to estimated mechanical stresses. The effect is to move the 100 bar stress contour upward from depths of 80 to 50 km. The comparison is considered reasonable since the geotherms result from a fit to two-dimensional heat flow, surface volcanism clearly occurred in the rift during the last 3-10 m.y., and the heat source required to fit the thermal anomaly of the rift provides calculated geothermal anomalies in 5-10 m.y. [1].

## 3. SUMMARY OF GRAVITY, THERMAL, AND STRESS MODELS

Exciting implications concerning genesis of the Rio Grande Rift result from the geophysical models. The gravity model is based on the fundamental tenant that the negative Bouguer anomaly is caused by a low-density mass in the upper mantle beneath the rift. As with all gravity models, geometry is only crudely approximated. The thermal model satisfies three constraints. Its geometry and material properties are consistent with gravity, it provides a fit to the two-dimensional surface heat flow, and geothermal gradients beneath the rift are within the range of sparse petrologic controls. The stress and viscosity model is consistent with mechanical stress estimates from upper mantle xenoliths.

Both gravity and thermal models imply a fundamental change in the nature of the upper mantle beneath the rift. The lithosphere is thinned through intrusion of a low-density mass. Thermal contours exhibit quite shallow gradients beneath the rift relative to more stable adjacent regions. Hence the low density mass contains unusually large quantities of heat in the upper mantle. The stress model exhibits a steep gradient indicating thermal shear-stress concentrations in the upper mantle. Attendant viscosity variations imply a low viscosity to the low-density mass. Each of these characteristics aids in creating a gravitational instability in the upper mantle. The hot, low-density low-viscosity mass would reduce the instability by moving upward.

Genesis of a large pervasive feature such as a continental rift requires definition of the vertical dimension. This paper suggests that continental rifts are essentially a surface manifestation of deep-seated convective processes involving the upper mantle and perturbations in underlying zones of low viscosity. This dimension would be characterized by significant thinning of the crust and thickening of the upper mantle through diapiric intrusion of hot, less-dense low-viscosity material. Surface uplift, thinning of the crust, and extensional normal faulting due to steep shear-stress gradients are a response to physical processes of convective mass and heat transfer in the upper mantle. Hence a new definition for an active continental rift might contain, 1) the broad topographic arch, 2) elongate fault-bounded rift valleys created by extension, 3) associated surface volcanism, 4) high surface heat flow in and near the rift valleys, and 5) hot convective intrusion of material from the asthenosphere into the overlying lithosphere.

A preferred genetic sequence outlining physical processes of rifting consistes of, 1) initiation of a thermal pulse in the asthenosphere which convects mass and heat upward, creating magma chambers in the plastic upper mantle, 2) buoyant uplift of the overlying lithosphere to form broad arches, 3) thinning of the elastic crust due to enhanced shear stress gradients, 4) initiation of extensional normal faulting along a 'keystone' crest, 5) magmatic injection and extrusion of volcanics along fault zones, 6) accelerated uplift as a response to reduction of strength during both faulting and increased temperature, and 7) development of deep, elongate, clastic-filled rifts whose graben bottoms are moving downward relative to adjacent arches but whose total motion is probably upward relative to convecting masses below.

REFERENCES

1. R.J. Bridwell, New Mex. Geol. Soc. Guidebook, 26, 283, 1976.

2. J. Oliver and S. Kaufman, Geotimes, 20, 1976.
3. A.R. Sanford, O.S. Alptekin and T.R. Toppozada, Seis. Soc. Am. Bull., 63, 2021, 1973.
4. R.F. Roy, D.D. Blackwell and E.R. Decker, in: Nature of the Solid Earth, ed. by E. Robertson, McGraw-Hill, New York, 1972.
5. M.C. Reiter, C.L. Edwards, H. Hartman and C. Weidman, Geol. Soc. Am. Bull., 86, 811, 1975.
6. E.R. Decker and S.B. Smithson, J. Geophys. Res., 80, 2542, 1975.
7. A.E. Ringwood, Composition and Petrology of the Earth's Mantle, N.Y. McGraw-Hill, 618 pp., 1975.
8. W.S. Baldridge, Abs. with Prog., Rocky Mtn. Sec., Geol. Soc. Am. 29th Ann. Mtg., 567, 1976.
9. J.B. Reid, Jr., Abs. with Prog., Rocky Mtn. Sec., Geol. Soc. Am. 29th Ann. Mtg., 621, 1976.
10. J.C. Mercier and N. Carter, J. Geophys. Res., 80, 3349, 1975.
11. W.S. Baldridge, personal communication, 1977.
12. N. Carter, Rev. Geophys. Space Phys., 14, 301, 1976.
13. C.A. Whitten, EOS Trans. AGU, 37, 393, 1956.
14. R.I. Walcott, Quat. Res., 2, 1, 1972.
15. J.C. Mercier, D.A. Anderson and N. Carter, Stress in the Lithosphere: Inferences from Steady State Flow of Rocks, Pageoph., in press, 1977.
16. R.L. Post, Jr., High Temperature Creep of Mt. Burnet Dunite, Tectonophysics, in press, 1976.

# OWENS VALLEY - A MAJOR RIFT BETWEEN THE SIERRA NEVADA BATHOLITH AND BASIN AND RANGE PROVINCE, U.S.A.

Michael F. Sheridan

Department of Geology, Arizona State University
Tempe, Arizona 85281

ABSTRACT. Owens Valley is a graben in eastern California that extends approximately 250 km northward from the Garlock fault to Mono Lake. The rift zone is 10 km wide with a vertical displacement of more than 5 km. The graben contains from 1.5 to 2.5 km of sediments, but the mountain fronts rise 2.75 km above the valley floor to an elevation above 4,000 m. This rift valley is thought to have formed within the last 3 million years. Seismicity and faulting are still active: the last major fault occurred in 1872 and showed a maximum displacement of 5.5 m.

Quaternary volcanic features associated with the rift include: 1) Long Valley, a 17 by 32 km rhyolitic caldera at the northern end of the rift, that formed 0.7 m.y. ago with the eruption of 600 $km^3$ of Bishop Tuff, 2) Mono Craters rhyolitic ring structure north of Long Valley, 3) Big Pine basaltic field at the midsection of the rift, and 4) Còso volcanic field that may represent incipient collapse at the southern end of the rift.

This region is characterized by bimodal basalt-rhyolite volcanism. The magma for the Long Valley and Mono Craters silicic centers was probably derived from partial melting of a detached subducted plate. The basaltic fields have a mantle origin related to continental fragmentation and development of a thin crust above an anomalous asthenosphere.

## 1. INTRODUCTION

The western United States has a history of tectonism and volcanism that extends from the mid-Tertiary through the present.

*I. B. Ramberg and E.-R. Neumann (eds.), Tectonics and Geophysics of Continental Rifts,* 81-88.

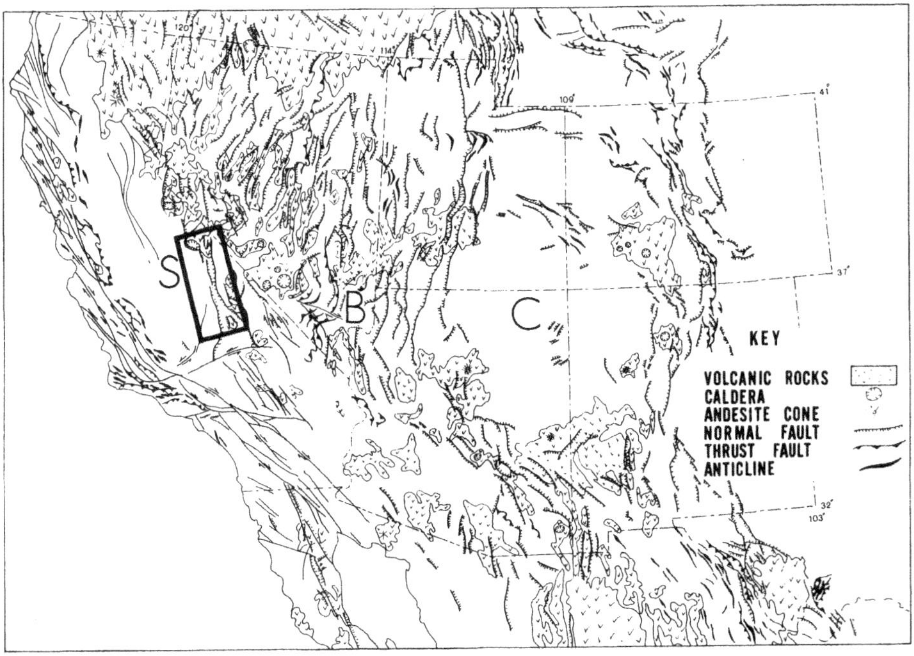


Fig. 1. Tectonic setting of Owens Valley (outlined by dark line). Sierra Nevada (S), Basin and Range (B), and Colorado Plateau (C) indicated. Modified from [2].

Owens Valley rift is a singular feature noted for Quaternary tectonism and volcanism.

### 1.1 Purpose

There are a number of papers dealing with tectonism and/or volcanism in Owens Valley, but most predate or ignore the plate tectonic framework of western United States. Here the features of Owens Valley are summarized in light of a current plate model [1].

## 2. TECTONIC SETTING

Owens Valley is a 250 km long graben that lies on the boundary between the Sierra Nevada and Basin and Range provinces in eastern California (Fig. 1). The prominent structural element to the west is the San Andreas fault that has undergone up to 350 km of right-lateral displacement in the past 24 m.y. [1]. Because calculated plate motions indicate that 1400 km of offset should have occurred during this period, Atwater [1] favors a broad distribution of NNW faults with right-lateral displacements to absorb the transform system movement. Extending for 150 km to the east of Owens Valley are several fault zones with compatible orientations and displacements: Panamint fault zone, Death Valley fault zone, and Las Vegas shear zone.

The Sierra Nevada to the west of the rift is characterized by a single block of batholithic terrain comprising a thick crust (60 km) with a low heat flow [3]. The Basin and Range basement to the east is composed of complexly faulted Paleozoic and Mesozoic geosynclinal to shelf sediments that form a thin crust (20 km) above an anomalous asthenosphere with a high heat flow [4].

## 3. OWENS VALLEY RIFT

Contrasting basement occurs on either side of the rift. To the west are Mesozoic granitic rocks with minor Paleozoic and Mesozoic roof pendants [5, 6]. The Sierra Nevada block was gently tipped to the west with a series of sharp boundary faults along Owens Valley. To the east are the White and Inyo Mountains composed of Precambrian to Mesozoic sediments of low grade metamorphism [5, 7] with only minor plutons. Although these rocks are complexly faulted and folded, there is a general eastward dip which places Owens Valley along the crest of a gentle rise [8].

The relatively straight boundary fault along the east follows a 160 km scarp that is only interrupted at the Wacobi embayment occupied by a thick section of Pliocene lake beds [7, 9]. In contrast, the western graben fault zone is composed of a series of en echelon scarps stepped to the left with prominent ramps between segments [5]. The rift terminates to the north in the vicinity of Mono Lake where the main fault segments spread and an intervening horst forms Casa Diablo Mountain. The south end of the rift has a similar termination with the medial horst block of the Coso Range.

The rift is only 10 km wide for most of its length but has a vertical displacement of more than 5 km [8]. The graben contains 1.5 to 2.5 km of sediments: the mountain fronts rise up to 2.75 km above the valley floor to an elevation of about 4,000 m.

Based on physiography [10, 11], relative fault displacements of glacial deposits [12], and paleobotanical evidence [13], the eastern scarp dates from the late Pliocene. The most recent faulting in the rift occurred with the 1872 earthquake [14] that produced a 110 km long break extending up the center of the valley

[15]. The maximum displacement on this fault was 5.5 m with a sub-equal component of normal and right-lateral movement [16, 17]. The rift is still seismically active [18, 19].

## 4. VOLCANISM

Quaternary volcanism is concentrated in three areas: 1) Long Valley caldera and Mono Craters ring structure at the northern end of the rift are rhyolitic complexes with minor basalt. 2) Big Pine field at the mid-section of the rift is a basalt field with negligible rhyolite. 3) Coso volcanic field at the southern end of the rift is a bimodal basalt-rhyolite complex.

### 4.1 Mono Craters ring fracture

Mono Craters ring structure [20] is a near-circular feature 13 km in diameter located north of Long Valley (Fig. 2). Along its eastern margin is an arcuate chain of more than 25 high-silica rhyolite domes and lavas [21] that are less than 30,000 yr old [22, 23]. A chamber with molten magma is considered to underlie this structure [24].

### 4.2 Long Valley caldera

Long Valley resurgent caldera [25] is located at a prominent embayment of the Sierra front near the northern end of Owens Valley where the rift widens from 15 km to 60 km due to a series of left-stepping en echelon segments of the Sierra scarp (Fig. 2). The pre-caldera volcanic rocks mainly consist of basalts and basaltic andesites [26] that have ages of 2.6 to 3.2 m.y. [24, 27, 28]. This older sequence predates major rift faulting.

Long Valley collapsed with the eruption of 600 $km^3$ of Bishop Tuff [24] 0.7 m.y. ago to produce a 17 by 32 km caldera. Resurgent doming commenced immediately followed by ring-fracture volcanism until 0.1 m.y. ago [24]. The progressive change in magma composition from 77% $SiO_2$ down to 68% reflects eruption from deeper levels of a large zoned magma chamber [24]. Although intracaldera basalts and basalts along the Sierra escarpment to the north are contemporaneous with the caldera rim volcanoes, they are not genetically related [24].

### 4.3 Big Pine volcanic field

Big Pine volcanic field lies in the central part of Owens Valley rift. The rocks are principally olivine basalts that have erupted

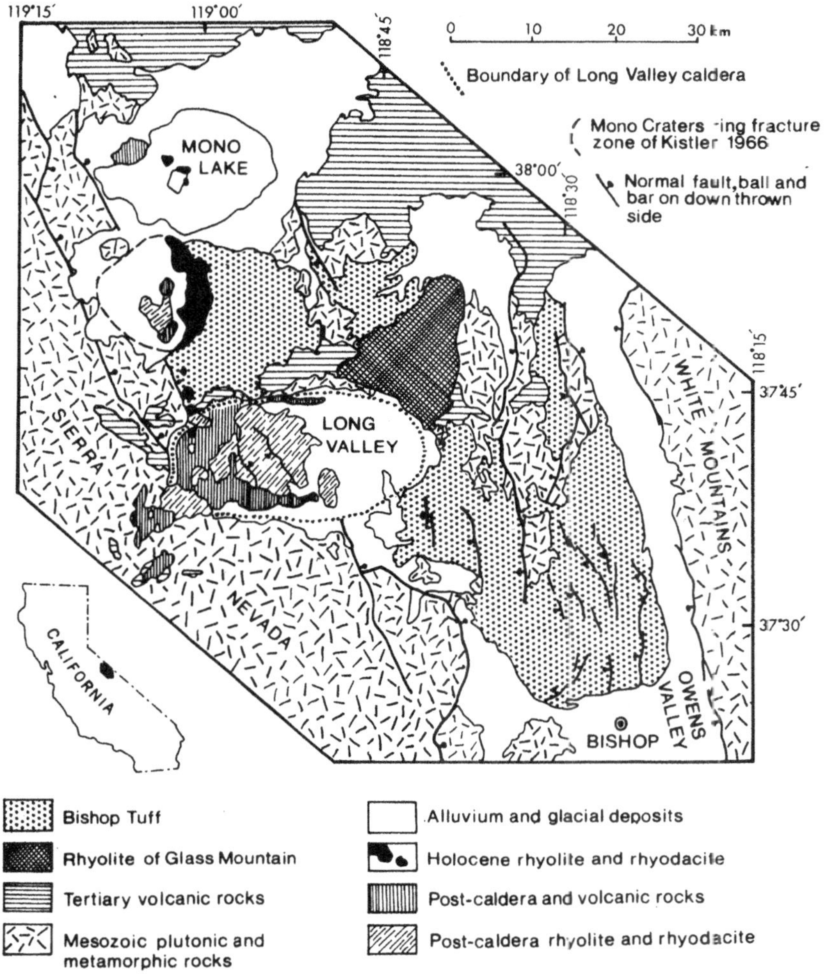


Fig. 2. Volcanic rocks at the north end of Owens Valley. Long Valley caldera and Mono Craters ring structure [20] indicated. Modified from [24].

from vents located along the main graben faults [6]. The single exception is a small rhyolite dome [29].

## 4.4 Coso volcanic field

The Coso volcanic field at the southern end of Owens Valley rift

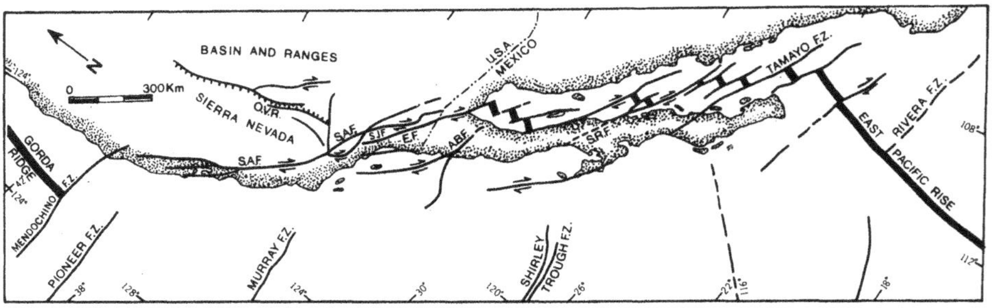


Fig. 3. Present plate configuration affecting tectonics in western U.S.A., after [2, 41].

has a wide-spread sheet of basalt and basaltic-andesite with an age of 3.0 to 3.7 m.y. [30] and a sequence of rhyodacite tuff and lava (Coso Formation) with an age of 2.3 to 3.0 m.y. [30, 31]. A near-circular ring of fractures, 40 to 45 km in diameter, is interpreted as a subsidence system related to a large silicic magma chamber [32]. More than 30 rhyolite volcanoes erupted in the last 1.0 m.y. within the central horst block of the Coso Range [33]: contemporaneous olivine basalts erupted in a peripheral area [30].

## 5. MODEL OF TECTONISM AND VOLCANISM

Most current models of tectonism and volcanism in the southwestern United States follow the scheme of Atwater [1] that assumes early to mid-Tertiary subduction beneath the western edge of the North American plate. After collision with a leg of the Pacific rise a transform system lengthened with time along the contact of the North American and Pacific plates to produce the present plate configuration (Fig. 3). Calc-alkali volcanism in southwestern United States predates the cessation of subduction at 20 to 30 m.y. ago [34]: a basalt-rhyolite association postdates subduction and is coincident with transform motion [35]. Silicic volcanic activity migrated outward from the center of the Great Basin at 30 to 40 m.y. to peripheral areas during 0 to 10 m.y. ago [36].

Owens Valley volcanism is compatible with the above pattern. Although data are sparse, the silicic rocks of Owens Valley rift have low $Sr^{87}/Sr^{86}$ ratios [37] similar to those of the Basin and Range province [38] which precludes an origin by crustal melting or contamination. Volcanic fields with minor rhyolite but major basalt, such as Big Pine or possibly Coso, could be explained by the fractional melting model of Yoder [39] as has been done for the volcanism in the near-by Salton Sea area [40].

The major rhyolitic volcanic centers with several hundred cubic kilometers of silicic magma, such as Long Valley and possibly Mono Craters and Coso present a more difficult problem. Partial melting of mantle is not a likely origin for such large magma bodies. They must represent localization of magmas derived from partial melting of the detached plate that was subducting prior to development of the transform system. A remnant of this detached plate may have become heated and risen to account for the anomalous astheneophere and high heat flow in the Great Basin facilitating crustal distension, thinning, and volcanism. Owens Valley rift may thus be at the leading edge of an expanding astheneophere related to heating and partial melting of a detached plate. This would explain the up-arching of the region prior to rifting and the close association of volcanism and tectonism.

## REFERENCES

1. T. Atwater, Geol. Soc. Am., Bull., 81, 3513, 1970.
2. P.B. King, Tectonic Map of North America, U.S. Geol. Survey Map, 1969.
3. A.H. Lachenbruch, J. Geoph. Res., 73, 6977, 1968.
4. R.F. Roy and D.D. Blackwell, EOS Trans., AGU, 47, 179, 1966.
5. P.C. Bateman, U.S. Geol. Survey Prof. Paper, 470, 1965.
6. J.G. Moore, U.S. Geol. Survey Bull., 1130, 1963.
7. C.A. Nelson, U.S. Geol. Survey Geol. Quad. Map, GQ-528, 1966.
8. L.C. Pakiser, M.F. Kane, and W.H. Jackson, U.S. Geol. Survey Prof. Paper, 438, 1964.
9. C.D. Walcott, J. Geol., 5, 340, 1897.
10. A.C. Lawson, Calif. Univ. Pub. Geol. Sci. Bull., 3, 291, 1904.
11. A. Knopf, U.S. Geol. Survey Prof. Paper, 110, 1918.
12. F.E. Matthes, A Geologist's view, in: The Sierra Nevada, the range of light, ed. by R. Peattie, Vanguard, New York, 1947.
13. D.I. Axelrod, Geol. Soc. Am., Bull., 68, 19, 1957.
14. J.D. Whitney, Overland Monthly, 9, 130 & 266, 1872.
15. W.H. Creighton, Calif. Geol., 25, 51, 1972.
16. P.C. Bateman, Seismol. Soc. Am., Bull., 51, 483, 1961.
17. G.B. Oakeshoot, R.W. Greenfelder, and J.E. Kahle, Calif. Geol., 25, 55, 1972.
18. D.W. Steeples and A.M. Pitt, J. Geophys. Res., 81, 841, 1976.
19. F.J. Gumper and C.H. Scholtz, Bull. Seismol. Soc. Am., 61, 1413, 1971.
20. R.O. Kistler, U.S. Geol. Survey Bull., 1221-E, 1966.
21. W.C. Putnam, Geog. Rev., 28, 68, 1938.
22. G.B. Dalrymple, Earth Planet. Sci. Lett., 3, 298, 1967.
23. I. Friedman, Science, 159, 878, 1968.
24. R.A. Bailey, G.B. Dalrymple, and M.A. Lanphere, J. Geophys. Res., 81, 725, 1976.

25. R.L. Smith and R.A. Bailey, Geol. Soc. Am., Mem., 116, 613, 613, 1968.
26. C.M. Gilbert, Geol. Soc. Am., Bull., 52, 781, 1941.
27. G.B. Dalrymple, Geol. Soc. Am., Bull., 74, 379, 1963.
28. G.B. Dalrymple, Univ. Calif. Pub. Geol. Sci. Bull., 47, 1, 1964.
29. E.B. Mayo, Geol. Soc. Am., Bull., 55, 599, 1944.
30. W.A. Duffield and C.R. Bacon, U.S. Geol. Survey Open File Map, 77-311, 1977.
31. J.F. Evernden, D.E. Savage, G.H. Curtis, and G.T. James, Am. J. Sci., 262, 145, 1964.
32. W.A. Duffield, Geology, 3, 335, 1975.
33. M.A. Lanphere, G.B. Dalrymple, and R.L. Smith, Geology, 3, 339, 1975.
34. P.W. Lipman, H.J. Prostka, and R.L. Christiansen, Roy. Soc. London Philos. Trans., A-271, 217, 1972.
35. R.L. Christiansen and P.W. Lipman, Roy. Soc. London Philos. Trans., A-271, 249, 1972.
36. R.L. Armstrong, E.B. Ekren, E.H. McKee, and D.C. Noble, Am. Jour. Sci., 267, 478, 1969.
37. D.C. Noble and C.E. Hedge, U.S. Geol. Survey Prof. Paper, 650-C, 133, 1969.
38. W.P. Leeman, Geochim. et Cosmochim. Acta, 34, 857, 1970.
39. H.S. Yoder, Am. Mineral., 58, 153, 1973.
40. P.T. Robinson, W.A. Elders, and L.J.P. Muffler, Geol. Soc. Am., Bull., 87, 437, 1976.
41. W.A. Elders, R.W. Rex, T. Meidav, P.T. Robinson, and S. Biehler, Science, 178, 15, 1972.

# RIFT SYSTEMS IN THE WESTERN UNITED STATES

John H. Stewart

U.S. Geological Survey, Menlo Park, CA 94025 U.S.A.

ABSTRACT. Rifting has occurred episodically in the western United States from the Precambrian to the Cenozoic. Speculation concerning Precambrian rifting has focused on the possibility of major continental rifts about 1,450 m.y. and 900? to 650 m.y. ago. The 1,450-m.y. rifting event is seen as cutting across trends of Precambrian crystalline basement rocks. It is considered to have developed along a system of three-armed rifts in which the abandoned arm is preserved as deep epicratonic troughs (aulacogens) extending into the craton at a high angle to the trend of the main rift. The Belt Supergroup, which is locally as much as 20 km thick, was deposited, according to this theory, in one of these abandoned arms.

The 900?- to 650-m.y. rifting event is considered to be related to a major change in the tectonic setting of the western United States that occurred after the deposition of the Belt Supergroup and related rocks. This change is marked locally by voluminous eruptions of mafic lava, in part tholeiitic basalt, and the subsequent development of a well-defined miogeocline in the western United States. Subsidence rates in the miogeocline appear to decrease exponentially, perhaps as a result of the cooling and thermal contraction that followed initial heating and uplift of the lithosphere at the time of rifting.

During the late Cenozoic, extensional faulting produced one of the world's most widespread regions of continental rifting that extends from the United States southward into northern Mexico, a distance of 1,500 km. The region is as much as 800 km wide in the Basin and Range province in Nevada and Utah in the United States and is characterized by a system of basins and

*I. B. Ramberg and E.-R. Neumann (eds.), Tectonics and Geophysics of Continental Rifts, 89-110.*

ranges formed by block faulting. Most estimates of the amount of extension range from 10 to 35 percent of the original width of the basin and range region. Widespread seismic activity, thin crust, low upper-mantle seismic velocities, and high-heat flow are characteristic of the region of extension. Current theories relate extension to one of the following: (1) oblique tensional fragmentation within a broad belt of right-lateral movement and distributed extension related to the development of the right-lateral San Andreas transform fault system, (2) upwelling from the mantle behind an active subduction zone (back-arc spreading), (3) subduction of the East Pacific Rise, or (4) mantle plumes.

## 1. INTRODUCTION

Rifts of several different types and ages have been described in the western United States and include major continental rifts formed in the Precambrian, back-arc rifts formed in the Paleozoic and Mesozoic, and widespread systems of rifts formed in the late Cenozoic. The review concentrates mainly on the well-studied Precambrian and late Cenozoic rifts.

## 2. PRECAMBRIAN RIFTS

The origin of the western margin of the Precambrian North American continent is poorly understood and is the subject of considerable debate. Three concepts that relate the margin to rifting (Fig. 1) have been proposed: (1) the margin was produced by major continental rifting about 1,450 m.y. ago prior to the deposition of the initial deposits of the Belt Supergroup and temporally equivalent rocks [1, 2], (2) the margin was produced by continental rifting about 900? to 650 m.y. ago after deposition of the Belt Supergroup and temporally equivalent rocks [3, 4], and (3) the margin was produced by a combination of rifting, both prior and subsequent to the deposition of the Belt Supergroup [5, 6, 7].

Still another theory, one that is not considered in detail here, does not relate the margin to rifting but rather to a slow outward building of the continent by accretion [8]. The concept is based on the observed irregular outward decrease in age of Precambrian rocks in North America that suggests an accretionary history of the continent. The concept appears to be supported by relations in northwest Canada [9] where 1,700- to 2,300-m.y.-old rocks considered to have been deposited on a continental margin occur directly east of rocks of Belt age and younger, a relationship suggesting that the rocks of Belt age and younger were deposited along a margin only slightly modified from an older one. No evidence of such an older margin, however, has been

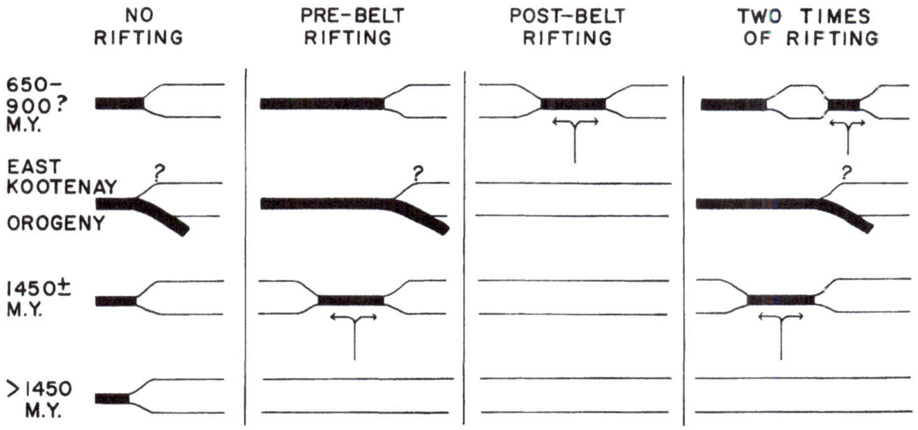


Fig. 1. Models of Precambrian development of the western United States. Black, oceanic crust; unpatterned, continental crust.

described in the western United States.

## 2.1 Pre-Belt rifting (1,450 m.y.)

The Belt Supergroup and temporally equivalent rocks are the oldest relatively unmetamorphosed supracrustal rocks in the western United States (Figs. 2 and 3). They generally are considered to range in age from about 1,450 to 900? m.y. [10, 11], although dating is somewhat inconclusive. The Belt or equivalent rocks rest unconformably on 2,400- to 1,450-m.y.-old crystalline basement rocks.

The Belt Supergroup and temporally equivalent rocks consist mostly of argillite, shale, siltstone, quartzite, and locally abundant dolomite and limestone [10, 12-16]. Conglomerate is locally abundant, particularly near the inferred margins of deposition [17]. Relatively thin mafic lavas and gabbroic or diabase sills occur locally. The Belt Supergroup and the Uinta Mountain Group (Fig. 3) are 20 km and 7 km thick, respectively, and occur in deep epicratonic troughs. The Grand Canyon Supergroup and the Crystal Spring and Beck Spring Formations have also been related to deposition in troughs, but the existence of these troughs is less certain. The Apache Group and Troy Quartzite may be shelf deposits.

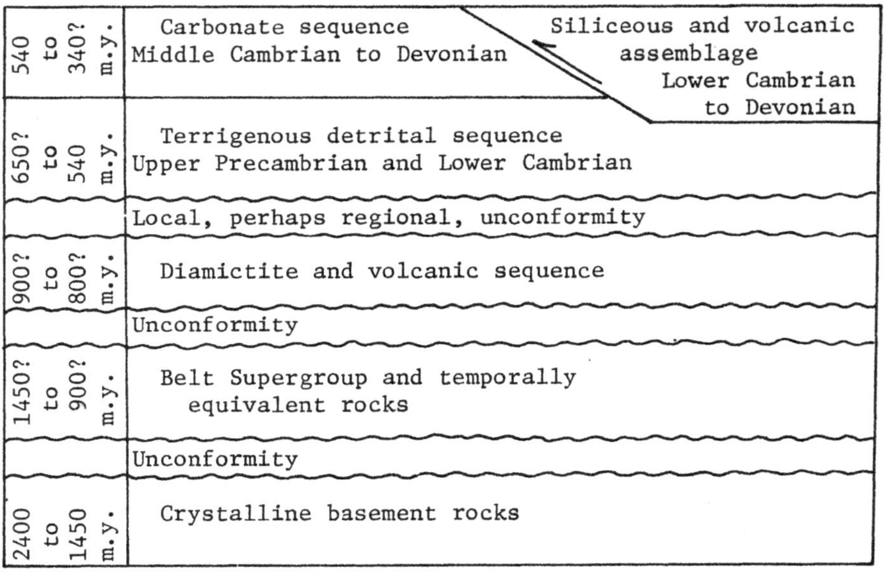


Fig. 2. Major Precambrian and lower Paleozoic sequences in the western United States.

The Belt Supergroup and temporally equivalent rocks in the western United States have been interpreted [18] as deposits formed in embayments extending eastward from an ocean basin along the western margin of the North American Continent, although this concept is not fully accepted by everyone [3]. In Canada, strata temporally equivalent to the Belt Supergroup are more continuous and have been considered by some geologists [19, 20] to represent a miogeoclinal deposit (continental terrace) along the edge of the continent. In terms of plate tectonics theory, Monger and others [1] suggested that the Precambrian margin in Canada was produced by pre-Belt continental rifting that cut across the grain of crystalline basement rocks.

Burke and Dewey [2] have suggested that the deep troughs such as those that contain the Belt Supergroup are failed arms (aulacogens) of three-armed rift systems related to a pre-Belt continental separation. Such a concept could account for the great thickness of strata contained in these troughs (20 km for the Belt Supergroup, 7.3 km for the Uinta Mountain Group, 4 km for the Grand Canyon Supergroup) and for the apparently faulted margins of these basins in some areas.

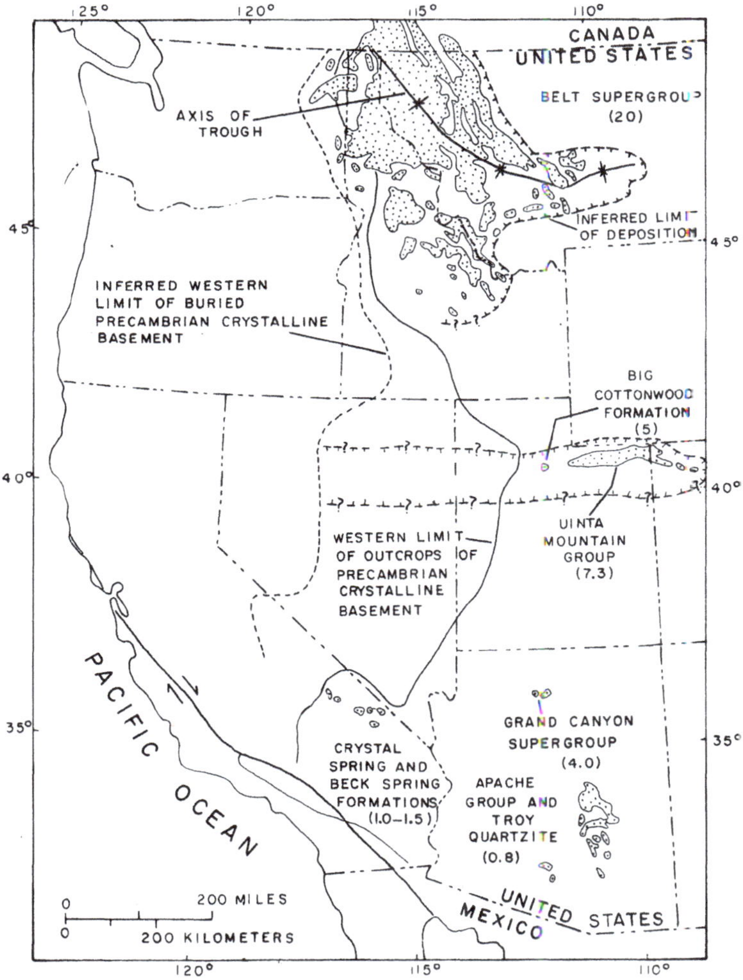


Fig. 3. Distribution and inferred basins of deposition of the Precambrian Belt Supergroup and temporally equivalent rocks (1,450 to 900? m.y.) and known and inferred western limit of Precambrian crystalline basement rocks (2,400 to 1,450 m.y.). Numbers are maximum thicknesses in kilometers.

## 2.2 Post-Belt rifting (900? to 650 m.y.)

In the Northwestern United States and Southern Canada, deposition of post-Belt rocks started after the East Kootenay orogeny [21] in which the Belt Supergroup and the correlative Purcell Supergroup in Canada were slightly metamorphosed, mildly deformed, and unlifted. The Grand Canyon orogeny [22] in Arizona deformed rocks of Belt age and appears to be about the same age as, and perhaps is correlative with, the East Kootenay progeny. Alternatively, as proposed by Elston [22], the Grand Canyon orogeny could be related to the initial stages of rifting that produced rift valleys in the next younger sequence of Precambrian rocks.

Post-Belt upper Precambrian and lower Paleozoic rocks in the western United States can be divided into four major sequences or assemblages (Fig. 2): (1) a diamictite and volcanic sequence of Precambrian age (900? to 800? m.y.), (2) a miogeosynclinal terrigenous detrital sequence (650? to 540 m.y.), (3) a miogeosynclinal carbonate sequence of Middle Cambrian to Late Devonian age (540 to 340? m.y.), and (4) Cambrian to Devonian eugeosynclinal siliceous and volcanic assemblage.

The diamictite and volcanic sequence [4] is a locally thick, but discontinuous, succession characterized by diamictite, mudstone, argillite, sandstone, and conglomerate, and locally thick units of mafic volcanic rocks, which are in part tholeiitic. The most characteristic feature of the sequence is the diamictite, which contains large clasts, many of which are a meter or more across, of locally derived rocks. Dropstones and striated clasts occur but are not abundant, and the diamictite is considered to be, in part at least, of glacial origin.

The depositional setting of diamictite and volcanic sequence is uncertain. In eastern California, the sequence, which is there called the Kingston Peak Formation, apparently accumulated in a fault-bounded trough or rift valley basin [15, 4, 23] that locally contains intact slabs of older rocks as much as 1.5 km in strike length [23]. In places, the thickness of the Kingston Peak changes from a few meters to more than 3,000 m in distances of less than 4 km. Whether the diamictite and volcanic sequence elsewhere in the western United States accumulated in fault-bounded troughs is not known, although the great thickness, the coarseness of the diamictite, abundance of volcanic rock, and the discontinuity of the sequence may indicate that a fault trough origin is likely elsewhere in the western United States.

The miogeosynclinal terrigenous detrital sequence (Fig. 4) [4] lies unconformably on rocks of Belt age or unconformably, and perhaps conformably in places, on the diamictite and volcanic

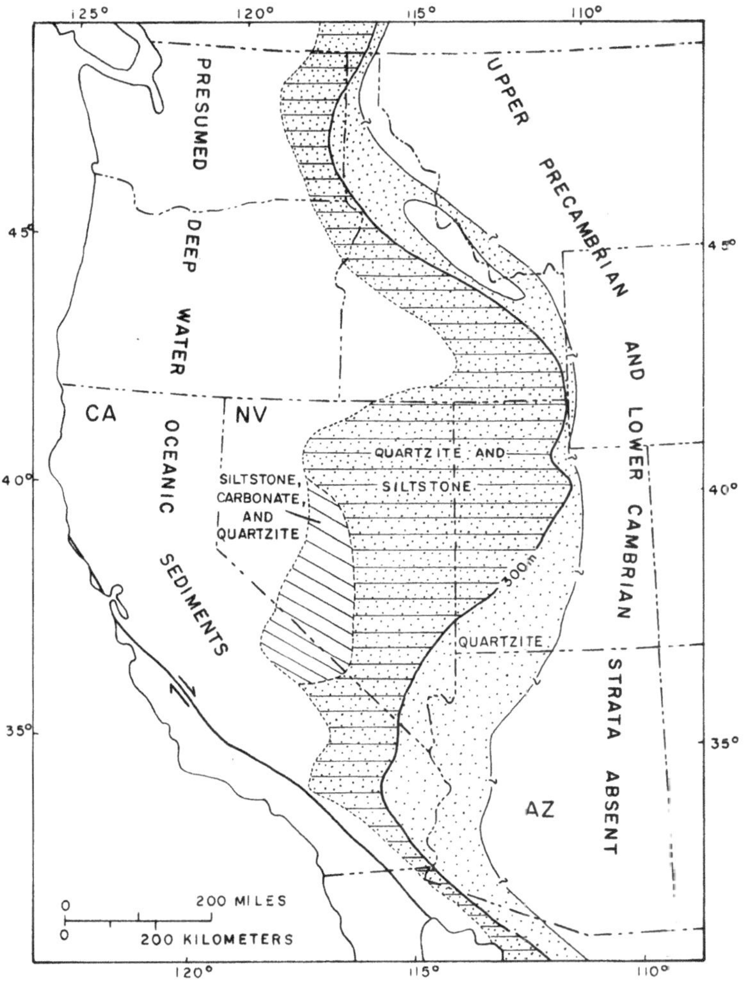


Fig. 4. Distribution, facies, and 300 m isopach of upper Precambrian and Lower Cambrian terrigenous detrital sequence. States: CA, California; NV, Nevada; AZ, Arizona.

sequence. It consists of a westward-thickening wedge of shallow-water quartzite and siltstone and minor conglomerate, limestone, and dolomite. It is generally only a few hundred meters thick in the eastern part of the western United States and thickens to over 6,000 m in the western part. Transport directions are primarily westward, indicating a source to the east. The upper part of the sequence contains Lower Cambrian fossils and the Precambrian-Cambrian boundary lies within the sequence.

A thick miogeosynclinal sequence of carbonate and minor amounts of shale and quartzite conformably overlies the terrigenous detrital sequence. The carbonate sequence contains a great variety of shallow-water and locally moderately deep-water sedimentary rocks ranging in age from Middle Cambrian to Devonian. The carbonate sequence is generally less than 2,000 m thick in the eastern part of the western United States and thickens to locally over 10,000 m in the western part.

Rocks of the Lower Cambrian to Devonian siliceous and volcanic assemblage crop out west of rocks of the miogeosynclinal belt or are tectonically emplaced above rocks of this belt. The eugeosynclinal rocks consist predominantly of deep-water graptolitic shale, radiolarian chert, and mafic lavas. These rocks generally occur in a structural succession of thin thrust plates, and, as a consequence, their thickness is uncertain.

The upper Precambrian and lower Paleozoic miogeosynclinal rocks are predominantly shallow-water deposits, and thus the thick miogeosynclinal accumulation of sediments in the late Precambrian and lower Paleozoic is due to subsidence over a long period of time, rather than to an infilling of an initially deep basin. In general, rates of accumulation appear to be greatest in the Early Cambrian, and presumably in the late Precambrian as well, and decrease in Middle and Late Cambrian and younger Paleozoic time. This relationship is best seen in western part of the miogeosynclinal belt (Fig. 5) where accumulation rates appear to decrease exponentially [4] in a manner similar to that for Cretaceous and Cenozoic shelf deposits along the east coast of North America [24, 25]. If the concept that the subsidence rate decreases exponentially with time is correct, the oldest exposed rocks of the terrigenous detrital sequence in eastern California, and perhaps elsewhere as well, were deposited starting about 650 to 600 m.y. ago [4].

The miogeosynclinal terrigenous detrital and carbonate sequences are considered to represent a continental terrace deposit (a miogeocline) on continental crust along the western margin of the North American Continent [26]. The eugeosynclinal deposits to the west, according to this interpretation, represent deep-water oceanic deposits laid down on oceanic crust. This

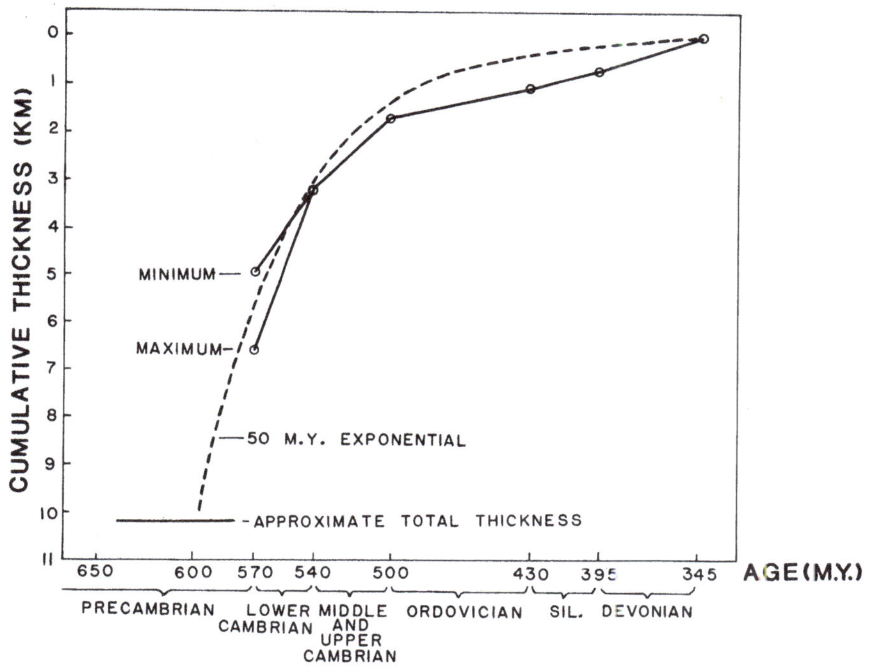


Fig. 5. Cumulative thickness with time for upper Precambrian and lower Paleozoic rocks in eastern California. Detailed description of data given in Stewart and Suczek [4].

pattern of deposition apparently was well developed by late Precambrian time, judging by the lateral continuity of rocks in the upper Precambrian and Lower Cambrian terrigenous detrital sequence. The depositional pattern of the diamictite and volcanic sequence, however, is not clear, although this sequence is not clearly a part of the miogeoclinal succession. It may, in part, represent deposits formed in rift valleys or deep troughs [15, 23, 4] prior to the development of the miogeocline.

Post-Belt rifting (Fig. 6) has been proposed to account for a major change in the tectonic setting of the western United States, from a pattern of deep epicratonic troughs in Belt time to a well-defined post-Belt miogeocline that cuts across the older troughs [3]. This change is marked locally by voluminous eruptions of mafic lava, in part tholeiitic basalt, that presumably are related to extension and the tapping of deep magma sources. The possible exponential decrease in subsidence rates in the post-Belt miogeocline is characteristic of areas of cooling and thermal contraction following initial heating such as

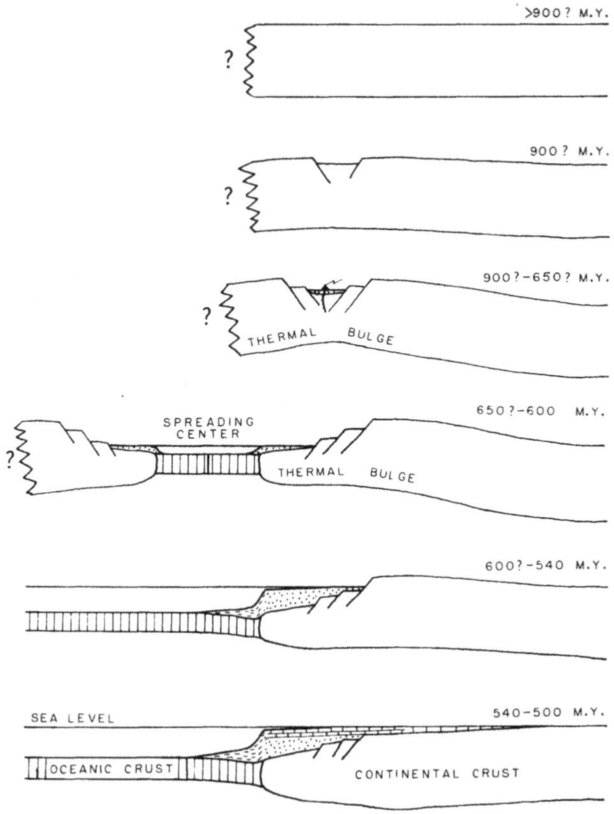


Fig. 6. Diagram showing a model of 900?- to 650-m.y.-old rifting in the western United States [after 4].

might occur at a rift [24, 25]. The concept of post-Belt rift does not preclude the possibility of a complex history for Belt and temporally equivalent rocks, perhaps involving both extensional and compressional events. The main point is that some tectonic event seems to be required to initiate a new cycle of deposition in post-Belt time. The sequence of events including the eruption of mafic lavas and the subsequent development of a miogeocline, is similar to the well-studied events that produced the rifted margin of eastern North America in the Mesozoic.

Initial rifting is considered to have produced the rift valleys in which at least some of the diamictite and volcanic sequence was deposited. Although dating of the diamictite and volcanic sequence is poor, available dates indicate that this rifting may have occurred 900? to 800? m.y. ago. Initial development of the miogeoclinal deposits of the terrigenous detrital sequence may not have started until 650 m.y. ago, however, if the evidence from subsidence rates is correct. Thus, 150 m.y. may have passed from the time that the rift valley deposits formed and the time that the miogeocline formed. This time lapse may indicate that the rift valleys formed long before the main rifting event. Such a lag in rifting has been noted elsewhere, particularly in Norway where the Oslo graben of Permian age developed almost 100 m.y. before the North Atlantic Ocean opened in the Cretaceous.

Thermal expansion during rifting is considered to have produced an uplifted area near the margin of the continent that was the source of the voluminous terrigenous detritus in the miogeocline. After destruction of the uplifted area by erosion and thermal contraction, the sea transgressed eastward, causing detrital material to be trapped in cratonal areas and allowing carbonate deposition to commence on the miogeocline.

## 2.3 Two times of Precambrian rifts (pre-Belt and post-Belt)

The concept of two episodes of rifting relates the shaping of the western margin of North America to rifting in both pre-Belt and post-Belt time [5-7] and reconciles the apparent conflict between the pre-Belt and post-Belt concepts. Three different varieties of the two-rift hypothesis can be considered. First, certain segments of the western margin of North America may have formed during the one time of rifting, whereas other segments did not form until later. Dickinson [5] has made this suggestion and related the Canadian part of western North America to the pre-Belt rift and the United States part to the post-Belt rift. Second, an ocean basin may have formed in pre-Belt time, then closed during the East Kootenay orogeny, and finally opened again in post-Belt time, in much the same way that the Atlantic Ocean opened in the late Precambrian, closed in the middle Paleozoic, and opened again in the Mesozoic. Third, the two-rift hypothesis can be envisioned as consisting of a major continent rift in pre-Belt time and of relatively minor reshaping of the continent during post-Belt time. Microcontinental blocks rifted from the continent in post-Belt could have drifted far away from North America to be emplaced on another continent, or these microcontinental blocks could form the known Precambrian crystalline basement of some borderland terranes [27] west of the presumed Precambrian margin. On the other hand, these borderland blocks have been interpreted by some geologists [1] as having drifted into North America from sites once far removed from North America. Still other possibilities are that the reshaping of the margin in post-Belt time was related to back-arc spreading following the development of a short-lived arc system, perhaps related to the East Kootenay orogeny, or that minor extension occurred along or closely parallel to the continental margin in post-Belt time, related perhaps to differential movement between the continent and the adjacent ocean basin.

## 3. PALEOZOIC, MESOZOIC, AND EARLY AND MIDDLE CENOZOIC RIFTS

Paleozoic and Mesozoic rifting has been related to the inward and outward migration of island arcs in the borderland of the western United States. In particular, the Late Devonian and Early Mississippian Antler orogeny is generally accepted, in term of plate-tectonic theory, to have formed by an island arc-continent collision [28, 6]. Subsequent to this collision, back-arc rifting and spreading [29] appears to be necessary to move the arc system away from the continent and allow for the development of a postorogenic marginal sea because remnants of the arc system are not believed to be preserved east of the marginal sea.

A similar sequence of events may have occurred during the Late Permian and Early Triassic Sonoma orogeny, which is also generally considered to be related to an island arc-continent collision. Possible postorogenic rifting in this case may have produced local block faulting that occurred concurrently with the eruption of Triassic volcanic rocks in central Nevada [30] and may also be related to the development of a deep sedimentary basin in western Nevada, east of a coeval magmatic arc in eastern California. Alternatively, Speed [31] has suggested that the island-arc terrane may floor this basin and that subsidence could be related to thermal contraction after magmatic activity was shut off by the emplacement of the arc on the margin of the continent. Such a concept is also possible for events following the Antler orogeny.

In addition to back-arc rifting and spreading described above, major continental rifting may have significantly modified continental margin in California, according to a concept described by several geologists [32, 28, 33]. These geologists note that the Cordilleran geosyncline is crosscut in southeastern California by the Mesozoic magmatic arc and suggest that the Cordilleran geosyncline was truncated by rifting in late Paleozoic and early Mesozoic time, and that the Mesozoic magmatic arc developed along this truncated margin. Other geologists [34], however, have suggested that the trend of the Cordilleran geosyncline changes abruptly in southeastern California and continues into northern Mexico, a relation that does not require truncation of the continental margin.

## 4. LATE CENOZOIC RIFT SYSTEM

Late Cenozoic extensional faulting extends throughout much of the western Cordillera of North America (Fig. 7). The region is one of the world's most widespread provinces of continental rifting. A review of the geologic and plate tectonic setting, the regional geophysics, characteristics of extensional faulting and basin and range structure, and theories of origin of this region is given below.

### 4.1 Geologic setting

Late Cenozoic extensional faulting in the western United States lies mostly within the region of the late Precambrian and early Mesozoic Cordilleran geosyncline (compare Figs. 4 and 7); in particular, the eastern limit of widespread extensional faulting is along the eastern limit of the miogeosynclinal belt of this geosyncline. In addition, the region of extensional faulting is also the site of Mesozoic tectonic and igneous activity. These relations may occur because the geosynclinal, tectonic, and igneous belts are relatively weak zones along which tensional strain is relieved.

During the early and middle Cenozoic, widespread calc-alkalic volcanism [36] occurred in the western United States. This volcanism is related to magmas generated on a subduction zone along western North America during the early and middle Cenozoic. The distribution of these volcanic rocks is somewhat similar to the distribution of late Cenozoic extensional faults in the western United States and Mexico. Approximately 17 m.y. ago, apparently concurrently with the initiation of extensional faulting, the character of volcanic activity changed markedly [37, 38]. Whereas most of the volcanic rocks prior to 17 m.y. ago are calc-alkalic in character, most rocks younger than 17 m.y. consist of basalt and bimodal assemblages of basalt and rhyolite. Calc-alkalic volcanism during the last 17 m.y. is largely confined in the Cascade Range in California, Oregon, and Washington, and in the Trans-Mexico volcanic belt in Mexico.

### 4.2 Plate tectonics setting

During the early and middle Cenozoic a subduction system existed along the western margin of North America [39, 40]. About 29 m.y. ago, a notable change in this system occurred when a spreading ridge (the East Pacific Rise) intersected the subduction zone, causing a change in plate geometry and the development of a transform fault system along part of the western margin of North America. The length of the transform system increased

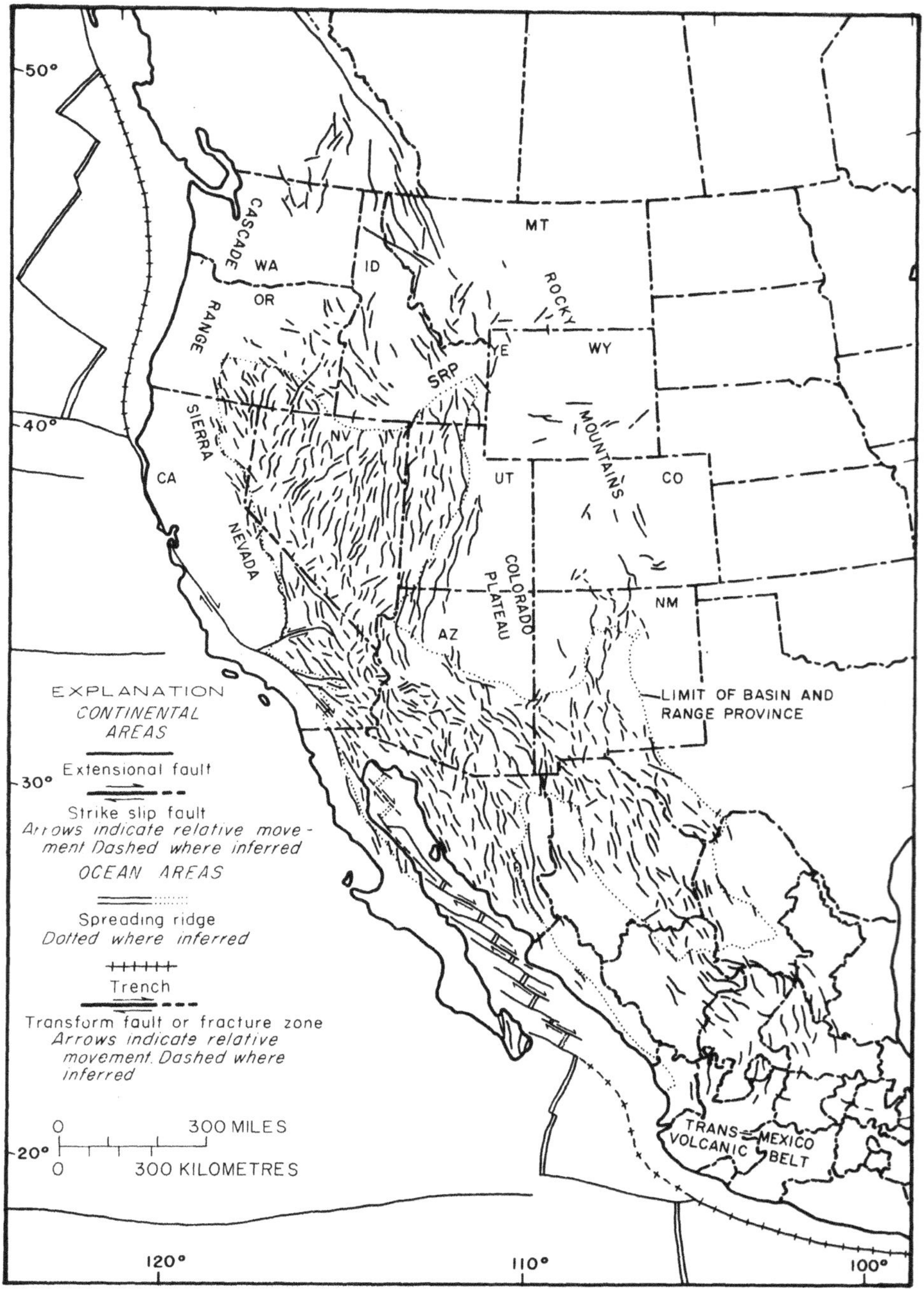
50°
40°
30°
20°
120°
110°
100°
CASCADE RANGE
SIERRA NEVADA
WA
OR
ID
MT
WY
NV
UT
CO
CA
AZ
NM
YE
SRP
ROCKY MOUNTAINS
COLORADO PLATEAU
LIMIT OF BASIN AND RANGE PROVINCE
TRANS-MEXICO VOLCANIC BELT
EXPLANATION
CONTINENTAL AREAS
Extensional fault
Strike slip fault
Arrows indicate relative movement. Dashed where inferred
OCEAN AREAS
Spreading ridge
Dotted where inferred
Trench
Transform fault or fracture zone
Arrows indicate relative movement. Dashed where inferred
0 300 MILES
0 300 KILOMETRES

as the spreading ridge was destroyed along the western margin of the continent. If spreading ridges are passive features, as implied by Atwater [39], and related to the pulling apart of plates rather than to deep convection systems, then mantle upwelling ceased in the area of the transform boundary.

Alternatively, Morgan [41] and Wilson [42] have suggested that mantle plumes are the driving mechanism of lithospheric spreading and that they actively drive plates apart where they occur along spreading ridges. If this is so, plume activity could persist along what was formerly a spreading ridge, even if it is overridden by a continent.

## 4.3 Regional geophysics

The region of extension in the western United States is characterized by high seismicity, although the distribution of earthquakes is uneven [43, 44]. Many areas containing Quaternary fault scarps are seismically inactive, suggesting that the present-day distribution of earthquakes is an inadequate sample of the seismic activity for the entire late Cenozoic. Focal depths of earthquakes are shallow, mostly less than 15 km. First-motion studies indicate general northwest-southeast extension in the Basin and Range province, with local area of strike-slip motion [43, 45].

The region of extension in the western United States is generally characterized by thin crust, anomalously low upper-mantle seismic velocities, and high heat flow [44]. The crust is generally 20 to 35 km thick as compared with 35 to 50 km in the Colorado Plateau and Rocky Mountains. Upper mantle seismic velocities are generally lower than 7.9 km/sec compared with 7.9 to 8.1 km/sec in the Colorado Plateau and Rocky Mountains. Heat flow values greater than 2 $\mu$cal $cm^{-2}$ $sec^{-1}$ characterize the areas of extension as compared with average values of about 1.5.

The Basin and Range province is in nearly perfect isostatic balance. Regional anomalies average no more than 10 mgal [44]. This balance means that any loss of mass by near-surface crustal spreading is almost perfectly matched by lateral backflow in the

---

Fig. 7. Distribution of late Cenozoic extensional faults and a few major strike slip faults in western North America and present-day lithospheric plate boundaries. Faults are generalized and, in part, inferred [35]. States: WA, Washington; OR, Oregon; CA, California; CO, Colorado; ID, Idaho; MT, Montana; WY, Wyoming; NV, Nevada; UT, Utah; AZ, Arizona; NM, New Mexico. Localities: SRP, Snake River Plain; YE, Yellowstone.

mantle [44, 46].

## 4.4 Basin and range structure

Much of the region of extension in the western United States is characterized by a system of elongate mountain ranges and valleys produced by a complex system of normal faults along which movement has resulted in the relative uplift of linear segments to form the mountains and the relative sinking of adjacent segments to form the valleys. The mountains are generally about 15 to 20 km across and the valleys of comparable width. Basin fill is from a few hundred meters to over 3,000 m. The structural relief between the lowest part of bedrock areas under valleys to the highest parts of adjacent mountains is generally from 2,000 to 5,000 m. Strike-slip faults occur locally within the Basin and Range province; the most notable are in a northwest-trending belt (the Walker Lane) in the western part of the Basin and Range province in Nevada [47].

No unanimity of opinion has developed concerning the geometry of faults and structural blocks that underlie areas of basin and range structure. Two basic models of this structure have been proposed [see discussion in 48, 35]. One model relates basin and range structure to a system of structural blocks rotated along curving, downward flattening (listric) faults. The uptilted part of an individual tilted block forms a mountain and the downslope part a valley. The other model relates basin and range structure to a system of horsts and grabens in which individual horsts form mountains and individual grabens form valleys. The horst-and-graben model can be viewed as the result of the fragmentation of a crustal slab into buoyant blocks above a plastically extending substratum.

## 4.5 Amount of extension

The system of normal faults and basin and ranges in the western United States is generally acknowledged as the result of regional extension. A fault that dips 60°, which is perhaps an average figure for the late Cenozoic faults in the western United States, requires 1 km of lateral extension for each 2 km of dip slip. Much more extension is possible if the faults flatten with depth.

Estimates of the amount of extension necessary to produce basin and range structure are generally in the range from 10 to 35 percent of the original width of the region but vary considerably because of disagreements about the geometry of the normal faults at depth. Thus, if the structure is considered to be composed of relatively simple horsts and grabens, estimates of the amount of extension suggests about a 10- to 20-percent increase in the width of the region [49, 48], whereas if the structure is considered to be composed of blocks tilted along downward-flattening faults, estimates have ranged from 30 to 50 percent [50] to as much as 100 percent [51] in local areas.

## 4.6 Regional uplift

The extension province in the western United States lies within a broad region of late Cenozoic uplift. Middle Cenozoic floras in the Cordilleran region indicate elevations of at least a kilometer below present [52]. Uplift of the Sierra Nevada, west of the Basin and Range province, is more than 1,800 m in the last 9 m.y. [53]. Uplift of the Colorado Plateau, east of the Basin and Range province, has also been large and has occurred primarily in the last 5 to 10 m.y. [54]. Within the Basin and Range province in Nevada and Utah, an uplift of 2,000 to 3,000 m in the late Cenozoic time seems likely [35].

## 4.7 Time of extension

Large-scale extension in the western United States is clearly a late Cenozoic event. Extensional faults in many parts of the region cut early and middle Cenozoic tuffs that formed as highly mobile ash flows, which tend to fill troughs much as water does. The occurrence of individual ash-flow units at many different topographic and structural levels can only be explained by faulting.

The volcanic transition from calc-alkalic volcanism in the early and middle Cenozoic to fundamentally basaltic volcanism in the late Cenozoic is considered to mark the change from predominantly compressive tectonics related to a subduction zone to extensional tectonics related to wrench faulting, back-arc spreading, or some other factor. Thus, the time of this transition is considered to date the initiation of extension faulting. In Nevada, the transition was about 17 m.y. ago [38]. It may be somewhat older in southern Arizona and New Mexico, and several authors [37, 51] have suggested a general northward migration of the time of inception of basin and range structure in the western United States.

Direct evidence of the development of basins and ranges is based on the first appearance of fault-controlled sedimentary basins and topographic forms approximately resembling those seen today. In Nevada such basins are well defined approximately 11 to 13 m.y. ago [55, 56], although some of these basins were more widespread than present-day ones. In some areas [56, 57] evidence from sedimentary and volcanic rocks indicates that basins and ranges with a distribution similar to those seen today were not developed until 12 to 7 m.y. ago.

## 4.8 Theories of origin

Theories of origin of late Cenozoic extensional faulting in the western United States can be loosely grouped into four main categories: wrench faulting, back-arc spreading, subduction of the East Pacific Rise, and mantle plumes.

The wrench fault concept relates extension to oblique tensional fragmentation within a broad belt of right-lateral movement along the western side of the North American lithospheric plate [58-63, 39]. According to this view, western North America is within a broad belt of right-lateral movement related to differential motion between the North American and Pacific plates. Some of the right-lateral movement is taken up on the San Andreas fault and related zones of right-lateral shear, such as the Walker Lane in the western part of the Basin and Range province in Nevada. The movement is also thought to produce distributed extension and tensional crustal fragmentation (including basin and range structure) along trends oriented obliquely to the trend of the San Andreas fault.

The theory of back-arc spreading relates basin and range structure to mantle upwelling and associated spreading above a subduction zone [64, 65, 46, 44]. The concept was originally developed to explain the origin of island arcs in the western Pacific that, according to the theory, drifted out from the Asian Continent by spreading (back-arc spreading) on the continent side of the arc at the same time that the arc was being underthrust by the Pacific plate [66-68]. The theory relates spreading to high heat generated along a Benioff zone by friction and the upwelling of mantle material that causes spreading in the crust. The process is slow, and calculation by Hasebe, Fujii, and Uyeda [69] suggests that a subduction system would have to be active for 100 m.y. to reach the heat flow observed in the Japan arc system.

Basin and range structure also has been related to convection currents and lateral spreading on the flanks of the East Pacific high, which, according to this theory, extends under western North America [70, 71, 38, 51]. In support of this view, Menard [70] noted that the East Pacific Rise, if projected northward, would extend under the Great Basin area and that both the Great Basin and the East Pacific Rise are characterized by high heat flow. In addition, Menard pointed out that the ridges and troughs analogous to the basins and ranges also occur on the ocean bottoms on the flanks of the East Pacific Rise.

Present-day plate tectonic theories discount the possibility that the East Pacific Rise extends under western North America. In Atwater's [39] model, the upwelling of mantle material along the East Pacific ridge system is considered to result passively from the separation of plates and is not considered to be the driving mechanism of plate separation. Upwelling should cease, according to this view, once plate separation ceased because of the development of a transform boundary. Thus upwelling related to the East Pacific Rise should not occur beneath western North America.

The concept of deep mantle convection in the form of narrow rising plumes has been advocated as playing a major role in the late Cenozoic history of the western United States [72, 42, 43, 52]. According to Morgan [41], such mantle plumes provide the driving mechanism for continental drift, causing plates to rift and to be driven apart from one another. The Snake River Plain-Yellowstone area in southern Idaho and northwestern Montana has been suggested [52] as the trace of such a mantle plume. According to this theory, the plume lay under the western part of the Snake River Plain about 15 m.y. ago and has traced a path relatively eastward as the North American plate moved westward over it to its present position under the Yellowstone region of northwest Wyoming. According to several geologists [72, 43, 52], this plume may have caused the breakup of a large segment of the western United States. Other geologists [73] relate the volcanism in the Snake River Plain to a major crustal fracture that propagated northeastward, guided by structures in Precambrian rocks.

Most likely, elements of more than one of these theories are involved in producing the late Cenozoic extensional faulting in western North America. The back-arc-spreading model is appealing because of the long history of subduction prior to the time of extension. The spreading, on the other hand, may have been initiated or accelerated by relaxation of confining pressure [64] on western North America by the destruction of the subduction system in the late Cenozoic and its replacement by a transform boundary. In addition, a major component of right-lateral shear related to distributed stress from the transform system can be incorportaed into the back-arc model.

## REFERENCES

1. J.W.H. Monger and others, Am. J. Sci., **272**, 577, 1972.
2. K. Burke and J.F. Dewey, J. Geology, **81**, 406, 1973.
3. J.H. Stewart, Geol. Soc. Am., Bull., **83**, 1345, 1972; Geology, **4**, 11, 1976.

4. J.H. Stewart and C.A. Suczek, Cambrian and latest Precambrian paleogeography and tectonics in the western United States, in: Paleozoic paleogeography of the western United States, ed. by J.H. Stewart and others, SEPM, Pacific Sec., Pacific Coast Paleogeography Symposium 1, 1, 1977.
5. B.C. Burchfield and G.A. Davis, Am. J. Sci., 275-A, 363, 1975.
6. W.R. Dickinson, Paleozoic plate tectonics and evolution of the Cordilleran continental margin, in: Paleozoic paleogeography of the western United States, ed. by J.H. Stewart and others, SEPM, Pacific Sec., Pacific Sec., Pacific Coast Paleogeography Symposium 1, 137, 1977.
7. R.W. Kistler and Z.E. Peterman, unpub. data, 1977.
8. C.K. Seyfert and L.A. Sirkin, Earth history and plate tectonics-An introduction to historical geology, Harper and Row Publishers, New York, 1973.
9. P. Hoffman, Roy. Soc. Lond. Philos., Trans., 273, 547, 1973.
10. J.E. Harrison, Geol. Soc. Am., Bull. 83, 1215, 1972.
11. R.L. Armstrong, Am. J. Sci., 275-A, 437, 1975.
12. M.D. Crittenden, Jr., and C.A. Wallace, Possible equivalents of the Belt Supergroup in Utah, in: Belt Symposium, 1973, Vol. 1, Idaho Bur. Mines and Geology, 116, 1973.
13. T.D. Ford and W.J. Breed, Geol. Soc. Am., Bull., 84, 1243, 1973.
14. D.P. Elston and G.R. Scott, Geol. Soc. Am., Bull., 87, 1763, 1976.
15. L.A. Wright and others, Precambrian sedimentary environments of the Death Valley region, eastern California, in: Guidebook - Death Valley region, California and Nevada Field Trip No. 1, Geol. Soc. Am., 70th Ann. Mtg., Cordilleran Sec., Las Vegas, Nev., 27, 1974.
16. A.F. Shride, U.S. Geol. Surv. Prof. Pap. 566, 1967.
17. W.J. McMannis, Geol. Soc. Am., Bull., 74, 407, 1963.
18. J.E. Harrison and others, U.S. Geol. Sur. Prof. Pap. 866, 1974.
19. R.A. Price, Can. Pet. Geol., Bull., 12, 399, 1964.
20. H. Gabrielse, Am. J. Sci., 272, 521, 1972.
21. W.H. White, Am. Assoc. Pet. Geol., Bull., 43, 60, 1959.
22. D.P. Elston, Precambrian Sixtymile Formation of Chuar Group and 830-m.y.-old Grand Canyon orogeny, northern Arizona, Geol. Soc. Am., Bull. (In press).
23. B.W. Troxel and others, Geol. Soc. Am., Abstr. Programs, 9, 517, 1977.
24. N.H. Sleep, Roy. Astron. Soc. Geophys. J., 24, 325, 1971.
25. N.H. Sleep and N.S. Snell, Roy. Astron. Soc. Geophys. J. 45, 125, 1976.
26. J.H. Stewart and F.G. Poole, Lower Paleozoic and uppermost Precambrian Cordilleran miogeocline, Great Basin, western United States, in: Tectonics and sedimentation, ed. by W.R. Dickinson, SEPM, Spec. Pub. 22, 28, 1974.
27. M. Churkin, Jr., and G.D. Eberlein, Geol. Soc. Am., Bull., 88, 769, 1977.

28. B.C. Burchfiel and G.A. Davis, Am. J. Sci., 272, 97, 1972.
29. M. Churkin, Jr., Paleozoic marginal ocean basin volcanic arc systems in the Cordilleran foldbelt, in: Modern and ancient geosynclinal sedimentation, ed. by R.H. Dott, Jr., and R.H. Shaver, SEPM Spec. Pub. 19, 174, 1974.
30. D.B. Burke, Geol. Soc. Am., Abstr. Programs, 2, 77, 1970.
31. R.C. Speed, Island-arc and other paleographic terranes of late Paleozoic age in the western Great Basin, in: Paleozoic paleogeography of the western United States, ed. by J.H. Stewart and others, SEPM, Pacific Sec., Pacific Coast Paleogeography Symposium 1, 349, 1977.
32. W. Hamilton, Geol. Soc. Am., Bull., 80, 2409, 1969.
33. R.A. Schweickert, Nature, 260, 586, 1976.
34. J.H. Stewart and F.G. Poole, Geol. Soc. Am. Bull., 86, 205, 1975.
35. J.H. Stewart, Basin and range structure in western North America - a review, in: Cenozoic tectonics and regional geophysics of the western Cordillera, ed. by R.B. Smith and G.P. Eaton, Geol. Soc. Am. Spec. Pap. (In press).
36. P.W. Lipman and others, Roy. Soc. Lond. Philos., Trans. 271-A, 217, 1972.
37. R.L. Christiansen and P.W. Lipman, Roy. Soc. Lond. Philos., Trans., 271-A, 249, 1972.
38. E.H. McKee, Geol. Soc. Am., Bull., 82, 3497, 1971.
39. T. Atwater, Geol. Soc. Am., Bull., 81, 3513, 1970.
40. T. Atwater and P. Molnar, Stanford Univ. Pubs. Geol. Sci., 13, 136, 1973.
41. W.J. Morgan, Am. Assoc. Pet. Geol., Bull., 56, 203, 1972.
42. J.T. Wilson, Tectonophysics, 19, 149, 1973.
43. R.B. Smith and M.L. Sbar, Geol. Soc. Am., Bull., 85, 1205, 1974.
44. G.A. Thompson, and D.B. Burke, Earth Planet. Sci. Ann. Rev., 2, 213, 1974.
45. R.B. Smith, Tectonophysics, 37, 323, 1977.
46. G.A. Thompson, Cenozoic basin range tectonism in relation to deep structure, Internat. Geol. Cong., 24th Montreal, Canada, 1972, Proc., 84, 1972.
47. J.H. Stewart and others, Geol. Soc. Am., Bull., 79, 1407, 1968.
48. J.H. Stewart, Geol. Soc. Am., Bull., 82, 1019, 1971.
49. G.A. Thompson, J. Geophys. Res., 64, 217, 1959.
50. L.A. Wright and B.W. Troxel, Shallow-fault interpretation of basin and range structure, southwestern Great Basin, in: Gravity and Tectonics, ed. by K.A. de Jong and R. Scholten, John Wiley and Sons, New York, 307, 1973.
51. J.M. Proffett, Jr., Geol. Soc. Am., Bull., 88, 247, 1977.
52. J. Suppe and others, Am. J. Sci., 275-A, 397, 1975.
53. M.N. Christiansen, Geol. Soc. Am., Bull., 77, 163, 1966.
54. E.D. McKee and E.H. McKee, Geol. Soc. Am., Bull., 83, 1923, 1972.
55. D.I. Axelrod, Geol. Soc. Am., Bull., 79, 713, 1968.
56. C.M. Gilbert and M.W. Reynolds, Geol. Soc. Am., Bull., 84, 2489, 1973.

57. E.B. Ekren and others, Geol. Soc. Am., Mem., 110, 247, 1968.
58. S.W. Carey, The tectonic approach to continental drift, in: Continental drift, a symposium, ed. by S.W. Carey, Hobart, Tasmania, 177, 1958.
59. D.U. Wise, Geol. Soc. Am., Bull., 74, 357, 1963.
60. D.R. Shawe, Geol. Soc. Am., Bull., 76, 1361, 1965.
61. W. Hamilton and W.B. Myers, Rev. Geophys., 5, 509, 1966.
62. J. Sales, Structural analysis of the Basin Range province in terms of wrench faulting, Nevada Univ., Reno, Ph.D. thesis, 178 p., 1966.
63. D.B. Slemmons, Pliocene and Quaternary crustal movements of the Basin-and-Range Province, USA, in: Sea level changes and crustal movements of the Pacific - Pacific Sci. Cong., 11th, Tokyo, 1966, Symposium 19, Osaka City, Univ. J. Geosci., 10, art. 1, 91, 1967.
64. C.H. Scholz and others, Geol. Soc. Am., Bull., 82, 2979, 1971.
65. D.E. Karig and W. Jensky, Earth Planet. Sci. Lett. 17, 169, 197
66. D.E. Karig, J. Geophys. Res., 76, 2542, 1971.
67. D.E. Karig, Earth Planet. Sci. Ann. Rev., 2, 51, 1974.
68. T. Matsuda and S. Uyeda, Tectonophysics, 11, 5, 1971.
69. K. Hasebe and others, Tectonophysics, 10, 335, 1970.
70. H.W. Menard, Jr., Science, 132, 1737, 1960.
71. K.L. Cook, Tectonophysics, 8, 469, 1969.
72. V. Mathews, 3d, and C.E. Anderson, Nature, 243, 158, 1973.
73. G.P. Eaton and others, Science, 188, 787, 1975.

# THE LATE PRECAMBRIAN CENTRAL NORTH AMERICAN RIFT SYSTEM - A SURVEY OF RECENT GEOLOGICAL AND GEOPHYSICAL INVESTIGATIONS

H.C. Halls

Department of Geology, Erindale College,
University of Toronto, Mississauga, Ontario, Canada

ABSTRACT. Geological and geophysical data reveal a partially buried belt of late Precambrian Keweenawan igneous and sedimentary rocks, over 2000 km long, in the heart of the North American continent. The dimensions of the zone, its association with prodigious volumes of flood basalts, and an absence of underlying granitic crust suggest that it is a continental rift, and perhaps a remnant of a global-wide rift system that existed 1.0-1.2 b.y. ago.

## 1. INTRODUCTION

In the central part of the North American continent, Keweenawan mafic igneous rocks, 1.0-1.2 b.y. old, occupy an arcuate belt, over 2000 km long and about 60 km in average width (Fig. 1). These rocks and associated red clastic sediments are considered to be the product of major crustal rifting similar in origin to present-day global rift systems [1-5]. For much of its length the Keweenawan or Central North American Rift System [3] is covered by a mantle of Cambrian and younger Paleozoic strata, to a depth of about 2 km in the west and about 4 km in the east where the rift passes beneath the Michigan Basin (Fig. 1). The rift only emerges from beneath the Paleozoic cover in the Lake Superior region, and it is therefore only here that there is a relatively detailed knowledge of Keweenawan geology. This area thus serves as a model for the interpretation of gravity, magnetic and borehole data further south where the rift is entirely buried.

The purpose of this paper is to briefly summarize our present state of knowledge concerning the nature and origin of the Keweenawan rift system.

*I. B. Ramberg and E.-R. Neumann (eds.), Tectonics and Geophysics of Continental Rifts,* 111-123.

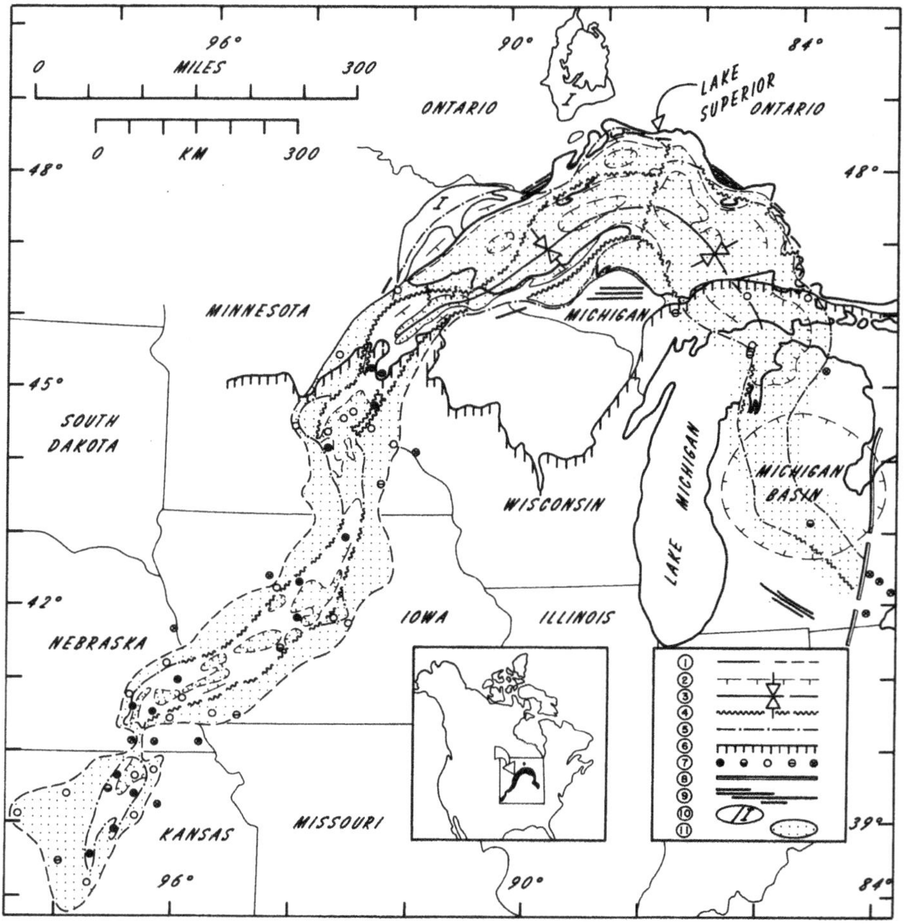


Fig. 1. Geological map of the Central North American Rift System, with Paleozoic and surficial cover removed. Symbols: 1 - geological contact; 2 - structural form lines; 3 - synclinal axis; 4 - fault; 5 - boundary of Keweenawan volcanics; 6 - northern limit of Paleozoic cover; 7 - boreholes which penetrate (i) Keweenawan mafic igneous rocks, (ii) Keweenawan red clastics and underlying mafics, (iii) red clastics only, (iv) red clastics and underlying pre-Keweenawan basement, and (v) basement only; 8 - Grenville Front; 9 - Keweenawan diabase dike swarm; 10 - Keweenawan volcanics, and intrusives (I) within the rift zone; 11 - Keweenawan red clastics. Sources of reference: [6,7,11,31,34,53,54].

## 2. KEWEENAWAN GEOLOGY OF THE LAKE SUPERIOR REGION

Lake Superior lies on the site of a large basin filled with Keweenawan igneous and sedimentary rocks [6-8]. Keweenawan stratigraphy is dominated by a sequence of mafic lava flows and overlying red clastics that have a combined thickness of more than 15 km in the centre of the basin.

Keweenawan volcanics, erupted in a sub-aerial environment, form one of the world's major accumulations of flood basalts with a total volume in excess of 400,000 km$^3$ [9]. Individual flows can be traced over distances of 40 to 90 km [9,10]. Dominant flow types are tholeiitic basalts and olivine tholeiites [11-14]. High Al and alkali basalts, andesites and rhyolites are subordinate but together account for about a quarter of total lava volume. The volcanic sequence may be divided into two parts: a lower, reversely magmetized unit [15,16] with a high proportion of porphyritic and aphanitic flows [11,13,17], and an upper one with normal magnetic polarity [15,16] and a relative abundance of ophitic flows, olivine basalts and interflow sediments [11,13,17]. Locally the two units show slight discordance and may be separated by a clastic sedimentary sequence [13,18]. The volcanics have suffered a low grade burial metamorphism which increases with stratigraphic depth from zeolite to low greenschist facies [14,19]. The apparent simplicity of Keweenawan paleomagnetic reversal stratigraphy, although not without complications [15,20] suggests that reversely magnetized units, whether they be extrusive or intrusive, are older than their normally magnetized counterparts.

Associated with Keweenawan volcanics are numerous intrusive bodies, the largest of which are the Duluth Gabbro sill complex in NE Minnesota and a diabase sill in the Lake Nipigon area (Fig. 1), both of which are intruded at or near the base of the Keweenawan sequence. Of particular relevance to crustal rifting is the nature, age and distribution of lava conduits. In Michigan, volcanics representing the upper part of the normally magnetized sequence appear to have flowed outwards from the basin centre and thus were presumably fed by fissures located along the axis of the basin [9]. Elsewhere around Lake Superior, in older volcanic sequences, flow directions are relatively random [e.g. ref. 11] suggesting that feeders were distributed over a wide area in their vicinity. Parts of these feeder systems are now represented by diabase dike swarms, often reversely magnetized [25], which occur around the periphery of the basin and trend parallel to its axis (Figs. 1 and 4). In a general way, therefore, older feeders occupy more marginal positions than younger ones, an observation consistent with crustal spreading from a central zone of dike emplacement.

The Keweenawan red clastic sequence shows a general upward increase in maturity from arkosic sediments to quartzose, relatively well-sorted sandstones [21,22]. Observations on cross-bedding, lateral facies changes and sediment type [7,23,24] suggest infilling of an irregularly subsiding volcanic basin under semi-arid conditions, where fluvially-transported detritus was derived from uplifted basin margins and later from more distant quartz-rich terrains as the relief of the volcanic uplands became subdued and the basin filled.

The Keweenawan basin has an asymmetric structure with a steeper southerly limb and an axis which gradually changes eastwards from a NE to NW trend (Fig. 1). Highest dips occur adjacent to large high angle reverse faults which follow the strike of the Keweenawan rocks along both lims of the basin in the SW part of Lake Superior region (Fig. 1). Here the central portion of the basin has been relatively raised along these faults by as much as 3 to 4 km [21]. It is not known when these faults were initiated, but their last major displacement occurred after basin infilling, but before deposition of overlying Cambrian rocks. If they were once normal faults formed during the crustal spreading stage of the rift, the sedimentary record indicates that they did not cause significant topographic relief, and thus did not form marginal scarps to a rift valley [26].

## 3. GEOPHYSICAL STUDIES OF THE KEWEENAWAN RIFT SYSTEM

Seismic and gravity studies suggest that the crust beneath the rift is both anomalously thick and dense. Maximum crustal thicknesses of 45 to 50 km occur beneath Lake Superior and further west in Wisconsin and Minnesota [2,27,28]. In these regions and further south in Iowa, rocks with high velocity occur at shallow depths of less than 10 km [2,3,27,29]. These observed velocities (6.5 to 6.9 km/sec) are appropriate to dense gabbro or metamorphosed volcanics such as are exposed around Lake Superior [3,30]. Bouguer and Free Air gravity values support a dense mafic crust and suggest it is isostatically compensated [1,2].

The full subsurface extent of the Keweenawan rift is clearly shown by the gravity and magnetic maps (Figs. 2 and 3). These data together with shallow seismic refraction results [31,32], support an extension of the central horst-like block found in Wisconsin eastwards into central Lake Superior and southwards along the western arm of the rift where it occurs as a series of en echelon segments, offset in a sinistral sense (Fig. 1). Both gravity and magnetic data clearly show the SE continuation of the Lake Superior basin toward eastern Michigan (Figs. 2 and 3), but there is no compelling evidence here that the central horst block also exists.

The gravity expression along the western arm of the rift is a central positive anomaly flanked by minima (Fig. 2). The corresponding magnetic signature is a zone of tight irregular to linear

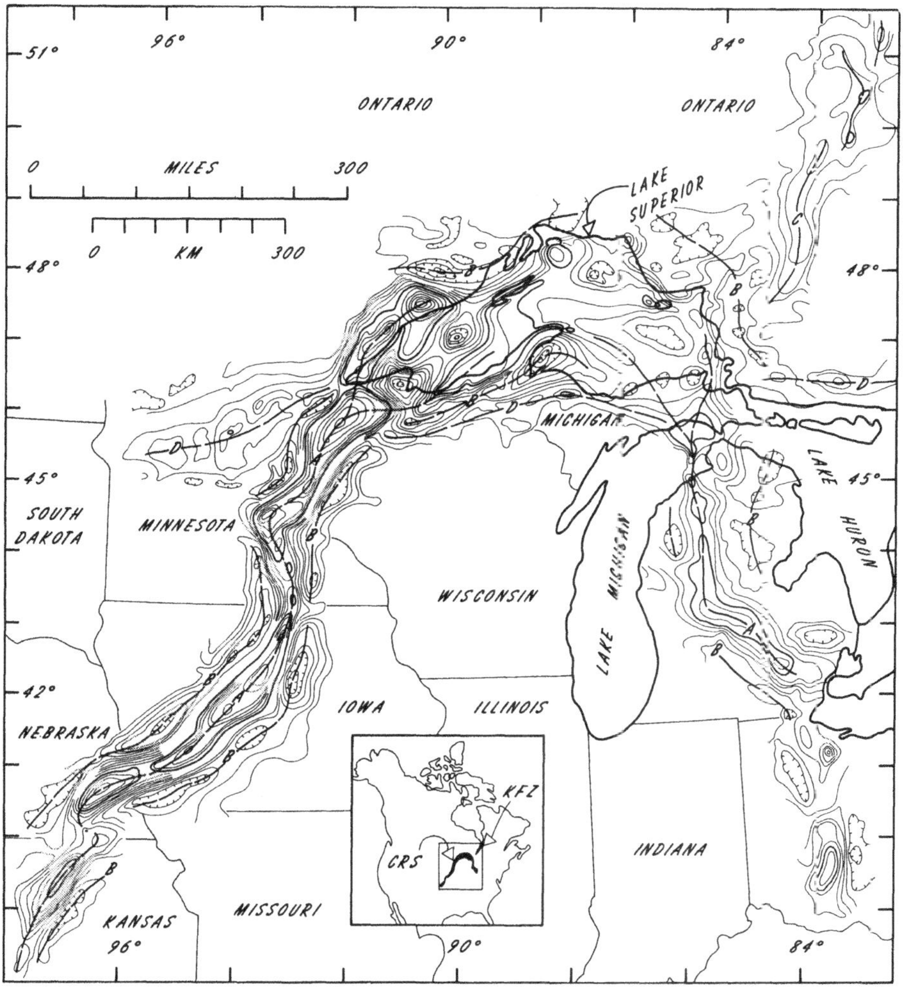


Fig. 2. Bouguer gravity map of the Central North American Rift System (CRS in inset). Heavy contour line = 0 mgals; contour interval = 10 mgals. Interpreted source of anomalies: A - Keweenawan mafics; B - marginal clastic wedges and/or crustal thickening; C - Archean granulites along Kapuskasing fault zone (KFZ in inset); D - Middle Precambrian basins within Archean basement. Source of reference: [44,52,53,55 and Bouguer gravity maps of Canada and USA].

anomalies flanked by smooth zones reflecting a sudden deepening of the source rocks. The gravity and magnetic data, together with well log information, are consistent with a centrally uplifted block of mafic igneous rocks at least 10 and probably more than 30 km thick, which is flanked by marginal wedges of relatively non-magnetic and low density clastics which attain a maximum thickness of at least 2 to 3 km near the faults [1,3]. Crustal thickening may partly account for the flanking gravity minima [1,26], especially as negative anomalies along both northern and southern margins of Lake Superior (Fig. 2) cannot be explained by surface geology alone [26, 31]. The marginal faults of the horst block are inferred from the steep gravity gradients, the abrupt change in wavelength of magnetic anomalies and from small faults and flexures in the overlying Paleozoic cover [1,33].

The continuation of the Keweenawan rift through eastern Michigan is also marked by a linear gravity high and flanking minima (Fig. 2). These anomalies are similar to those along the western rift, both in pattern and in their relation to small Paleozoic folds and faults, suggesting the existence of a locally faultbounded block of Keweenawan mafic igneous rocks [34]. However, in contrast, gravity and magnetic anomalies are relatively subdued owing to deep burial of the horst block beneath the Michigan Basin, and borehole data indicate that marginal clastic wedges are either weakly developed or absent [34]. A recently completed borehole through the Michigan Basin and located on the axis of the gravity high (Fig. 1) confirms the presence of mafic igneous rocks at a depth of about 5.2 km beneath about 1.5 km of Keweenawan red clastics [35]. Further south borehole data and a general NE grain to gravity and magnetic anomalies suggest [34] that the Keweenawan rift is cut off by the Grenville Province (Fig. 1). However, gravity data suggest that a number of small offset segments of the rift continue as far south as Tennessee [36].

Within the central horst of the western rift, magnetic anomalies are conspicuously linear but become more jumbled towards the axial regions [1]. Some linear anomalies can be followed for more than 150 km and may be caused by laterally extensive, upturned flows similar to those around Lake Superior. The irregular anomalies may be due to more gently dipping flows or local intrusive bodies. A number of oval, magnetically smooth and depressed areas occur along the axial zone (Fig. 3). The trend of surrounding magnetic anomalies tends to follow the margins of these quiet areas, indicating either inward dipping flows defining a local sedimentary basin or perhaps mafic rims of small, more felsic intrusions. However, red clastics overlying volcanics occur in a number of drill holes within the horst and an oval quiet area in Minnesota has been drilled to reveal thicker sections of red clastics beneath the Paleosoic cover [1]. All the smooth areas may thus reflect local sedimentary basins developed along the axis of the horst block [1]. The basins tend to occur in areas where the central gravity high is a maximum, although small negative anomalies coincident with the basins remain if the regional field is removed by polynomial surface fitting [1,33]. The basins tend to be elongated in a more east-

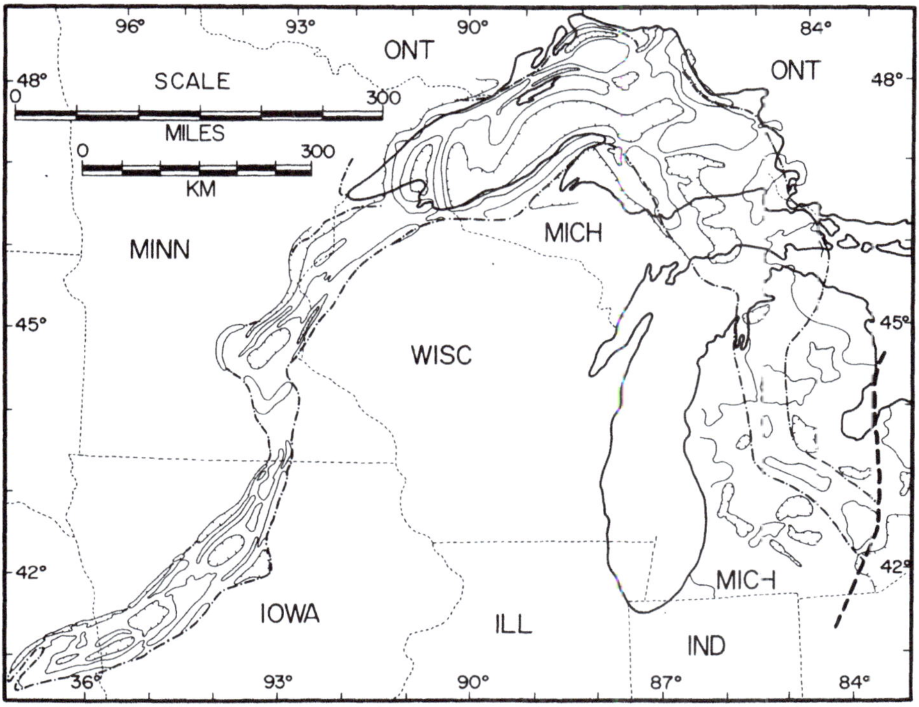


Fig. 3. Magnetic anomaly map of the Central North America Rift System, showing the general shape of local maxima and minima. The margin of the Keweenawan volcanic belt is also shown. Sources of reference: [1,34,51,52,56].

erly direction than the horst block in which they occur and like the rift segments themselves, are offset sinistrally from one another in an en echelon fashion [1] (Fig. 4). The elongate shape of the basins and their coincidence with areas of large Bouguer gravity suggest that they represent sites of major fissure eruption where outpouring of large volumes of magma was accompanied by isostatic subsidence [1]. Local basins have also been postulated within the main Lake Superior basin on the basis of gravity, magnetic and shallow seismic data [5,7,31,37], and may have a similar origin.

A distinctive regional magnetic pattern is associated with the mafic rocks forming the western half of the Keweenawan rift; positive and negative anomalies occur respectively along the SE and NW margins of the horst block (Fig. 3). Model studies [1,37,38] show that this pattern is caused by igneous rocks with a remanent magnetization direction similar to that found paleomagnetically for normally magnetized units in the Lake Superior region [15,16]. In Minnesota and Iowa the magnetic pattern is locally reversed in areas adjacent to interpreted sites of major fissure eruption (Fig. 3), and may reflect older, reversely magnetized flows beneath the marginal clastic wedges (Fig. 4).

## 4. REGIONAL SETTING OF THE KEWEENAWAN RIFT SYSTEM

Magnetic data show that the buried arms of the Keweenawan rift sharply transect the prevailing E-W grain of the older Precambrian basement [1,34]. In the Lake Superior region the rift becomes parallel to basement fabric as it follows the boundary between Early and Middle Precambrian rocks. In detail, however, the Keweenawan rift is composed of a number of segments and lines of offset which either have NE or NW trends. In areas of exposed basement around Lake Superior the dominant strike of faults and fractures is also NW or NE (Fig. 4). Much of this fault pattern predates Keweenawan rifting, so that the shape of the rift may be strongly controlled by the orientation of pre-existing lines of weakness. However, a number of geological features within the surrounding basement appear to be temporally related to the rift and are described below.

The most prominant of these features is the Kapuskasing fault zone (KFZ in Fig. 4) which runs for more than 300 km in a SW direction from Hudson Bay to the east shore of Lake Superior [39]. Upfaulted slices of Archean granulite occur discontinuously along the zone and terminate to the south in a gabbro-anorthosite complex. This southerly limit is shown in Fig. 2 by the termination of the associated gravity high about 100 km NE of Lake Superior. The age of the faulting is unknown, but alkaline complexes along the zone are dated at either 1.0 or 1.7 b.y. [40] and attest to a protracted geological history, the final dated phase of which corresponds closely in time with Keweenawan rifting.

Aeromagnetic data show an extensive orthogonal network of NW and NE trending dikes to the north and east of Lake Superior. The dikes yield positive magnetic anomalies and are either parallel or normal to the trend of the Keweenawan rift. Although many of the dikes are Middle Precambrian in age [41], some of them may be part of the 1.2 b.y. old Mackenzie and Sudbury dike swarms which on paleomagnetic evidence appear to slightly predate Keweenawan magmatism [42]. The NE trending dike set assumes a more northerly trend on crossing the Kapuskasing fault zone, and may indicate that a local crustal stress pattern associated with fault motion was in existence during dike emplacement. North of Lake Superior a 300 m thick, reversely magnetized sill underlies Lake Nipigon (Fig. 4) and may have been fed by a N-S dike system. Immediately south of this lake early pre-volcanic Keweenawan sediments are preserved, in part by subsidence along NNW trending faults.

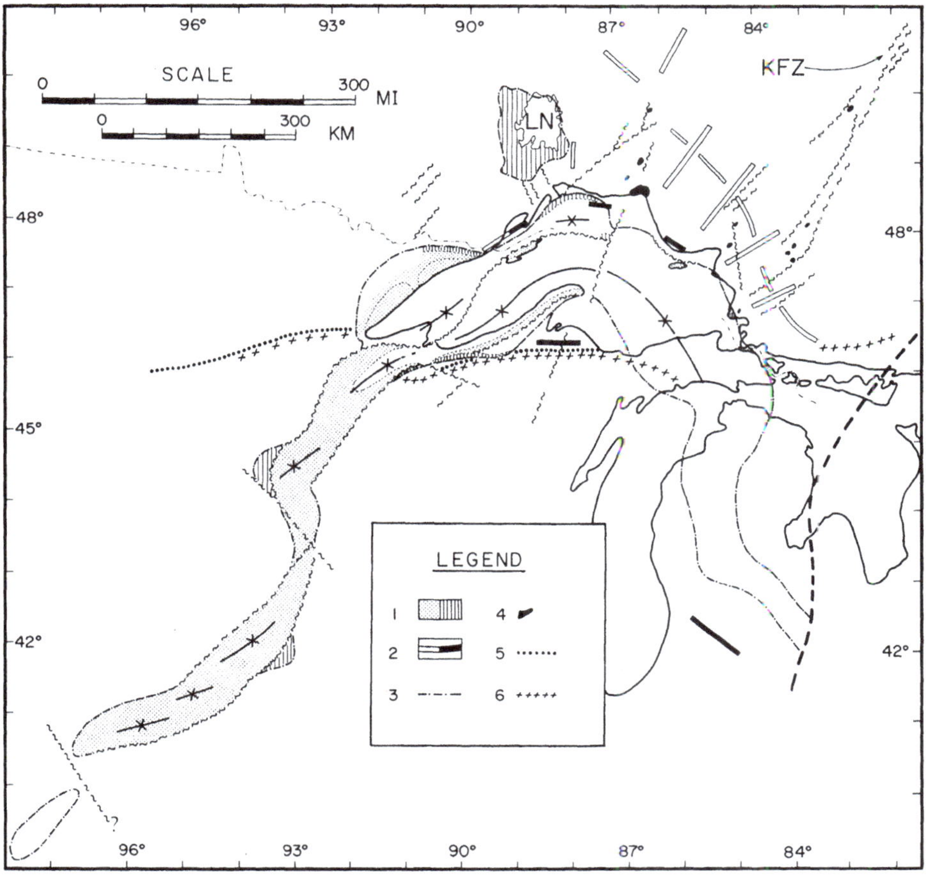


Fig. 4. Map showing the relation of the Central North American Rift to other geological features in surrounding basement. Keweenawan igneous rocks are subdivided according to magnetic polarity. Symbols: 1 - Keweenawan volcanics and sills of normal, reversed polarity; 2 - diabase dike swarms (schematic) of normal, reversed polarity; 3 - boundary of Keweenawan mafic igneous belt; 4 - alkaline intrusives or diatremes of Keweenawan age; 5 - intra-Archean boundary between granite-greenstone to north and gneiss to south; gravity anomaly D of Fig. 2; LN = Lake Nipigon. Sources of references: [1,6,18,25,34,41,43].

A consequence of crustal spreading during Keweenawan volcanism is that offsets in pre-Keweenawan geology should occur across the rift. One such example has been reported from Minnesota and Wisconsin [43,44] in a boundary separating Archean granite-greenstone and gneiss terrains (Fig. 4). A similar offset occurs in a positive gravity anomaly (Fig. 2) which follows this boundary on its south side and which is related to Middle Precambrian basins developed within the Archean basement [45]. The gravity anomaly continues eastwards and trends in a more SE direction toward the Michigan Basin (Fig. 2). A further offset may occur across the rift here as there is a possible correlative anomaly which runs through Middle Precambrian basins along the north shore of Lake Huron (Fig. 2).

## 5. THEORIES OF KEWEENAWAN RIFT DEVELOPMENT

### 5.1 Evidence for rifting

The main lines of evidence which have led to a general acceptance of a rift origin for Keweenawan rocks are: (a) the dimensional similarity of the Keweenawan belt to global rift systems such as those of East Africa and the Red Sea, (b) the occurrence of flood basalts fed by a fissure system more or less parallel to the axis of the belt, and (c) the absence of granitic crust beneath the central part of the belt.

A further indication of rifting, from the distribution and age of feeders, and from apparent offsets in pre-Keweenawan geology, is that a widening of the Keweenawan belt has taken place due to crustal spreading from an axial zone of dike injection. Chase and Gilmer [4] have examined a crustal spreading model for the western rift and find that it fully accords with plate tectonic theory. Their model assumes that the amount of spreading is given by the separation of inflection points between the positive gravity anomaly and flanking minima, and that the offsets of the rift segments represent transform faults [1]. The orientation of the faults and the change in spreading width along the rift suggest a separation of rigid plates about a rotation pole in New Mexico [4]. Rotation about this pole, however, presents difficulties as it does not easily explain the eastern half of the Keweenawan rift, especially the segment through eastern Lake Superior.

An alternative spreading direction, along lines of present longitude, is suggested by the symmetry of the Keweenawan rift about a N-S line and the N-S orientation of the narrowest rift segments [5]. Although reversal of this motion would tend to close Lake Superior and remove the offsets in pre-Keweenawan geology across the rift, it does not satisfy the spatial variation in spreading width required by rigid plate theory, nor account for the en echelon offsets along NW lines seen in the western rift.

### 5.2 Evolution of the Keweenawan rift system

The evolution of the Keweenawan rift system may be summarized in the following sequence of geological events:

(a) Crustal fissuring was initiated with upwelling of reversely magnetized volcanics, particularly in the Lake Superior region.
(b) Local sedimentary basins formed on sites of major fissure eruption.
(c) Crustal spreading with eruption of normally magnetized volcanics resumed along the rift axis, and laterally displaced reversely magnetized dikes and flows and their associated sedimentary basins to more distal positions.
(d) Volcanism gradually waned and sedimentary basins again developed in areas of major fissuring, and also over a wider area in response to a broader isostatic subsidence of the entire Keweenawan mafic assemblage.
(e) During or after the final stages of infilling by red clastic sediments, crustal compression led to local steepening of the basin limbs, formation of the major strike faults and uplift of the central horst block.
(f) During Paleozoic time subsidence continued locally along the rift, especially in eastern Michigan, and Keweenawan boundary faults were locally reactivated.

## 5.3 Cause of Keweenawan rifting

The cause of Keweenawan rifting has been attributed to lines of mantle plumes [46] and to plate collision [47]. In the first theory Lake Superior is regarded as the centre of a plume-generated triple junction whose arms are represented by the eastern and western segments of the Keweenawan rift and by the Kapuskasing fault zone. However, evidence for the third, Kapuskasing arm is rather speculative as the fault zone is weakly developed at its southern end and is only part of a broad NE trending zone of faults and Keweenawan alkaline centres, more than 200 km wide (Fig. 4). An alternative proposal for a third arm [48] is along the NNW lineament through Lake Nipigon where considerable quantities of mafic magma were produced during the early stages of Keweenawan igneous activity.

In the second theory [47] the Keweenawan rift represents a relief of strain energy produced by an obstruction to dextral strike slip motion, following oblique collision, between the Grenville Province and the rest of North America. This model, however, is doubtful because there is presently no convincing evidence for a plate suture close to the Grenville Front, nor anywhere else within the Grenville Province. Radiometric age data, however, show that Keweenawan rifting and Grenville tectonism are closely related if not contemporaneous events, so that a genetic connection between the two is likely. A solution to the problem is presently inhibited by the complexity of Grenville geology, and by the lack of precise ages for the various stages of Keweenawan rift and Grenville Province development. Borehole data in Michigan suggest that the

Grenville event is at least in part younger than the Keweenawan rift. The final compressional phase in Keweenawan history may thus be related to Grenville tectonism, especially as the boundary faults appear to be most strongly developed where they are parallel to the Grenville Front (Fig. 1).

In summary, Keweenawan rocks were formed as a result of crustal rifting and subsequent spreading. However, the cause of the rifting, the details of the spreading process, and the reason why only 60 to 100 km of spreading occurred, remain conjectural. Evidence for contemporary rifting with associated flood basalts is found elsewhere in North America [49], and also in Africa, Australia and India [49,50] so that the Keweenawan rift may represent a preserved remnant of perhaps the oldest global rift system clearly analogous to those which have formed over the last 200 m.y.

## REFERENCES


1. E.R. King and I. Ziets, Geol. Soc. Am. Bull., 82, 2187, 1971.
2. T.J. Smith et al., AGU Mono., 10, 181, 1966.
3. L.C. Ocola and R.P. Meyer, J. Geophys. Res., 78, 5173, 1973.
4. C.G. Chase and T.H. Gilmer, Earth Planet. Sci. Lett., 21, 70, 1973.
5. W.S. White, Geol. Soc. Am. Abstracts, 4, 732, 1972.
6. H.C. Halls, AGU Mono., 10, 3, 1966.
7. W.S. White, U.S. Geol. Surv. Prof. Pap., 524-E, 1966.
8. C. Craddock, Minnesota Geol. Surv. Cent. Vol., 281, 1972.
9. W.S. White, Am. J. Sci., 258-A, 367, 1960.
10. R.N. Annells, Geol. Surv. Can. Pap., 72, 1973.
11. J.C. Green, Minnesota Geol. Surv. Cent. Vol., 294, 1972.
12. H. Wallace, M.Sc. thesis, Dept. Geology, Univ. Toronto, 1972.
13. H.A. Hubbard, J. Res., U.S. Geol. Surv., 3, 529, 1975.
14. R.N. Annells, Geol. Surv. Can. Bull., 218, 1974.
15. K.G. Books, U.S. Geol. Surv. Prof. Pap., 760, 1972.
16. H.C. Palmer, Can. J. Earth Sci., 7, 1410, 1970.
17. L.D. Ayres, Ontario Dept. Mines, Geol. Rept., 69, 1969.
18. H.C. Halls, Can. J. Earth Sci., 11, 1200, 1974.
19. W.T. Jolly and R.E. Smith, J. Petrol., 13, 273, 1972.
20. W.A. Robertson, Can. J. Earth Sci., 10, 1541, 1973.
21. C. Craddock, Minnesota Geol. Surv., Cent. Vol., 416, 1972.
22. W.K. Hamblin, Geol. Soc. Am. Bull., 72, 1, 1961.
23. W.S. White and J.C. Wright, U.S. Geol. Surv. Prof. Pap., 400-B, 1960.
24. R.G. Wolff and N.K. Huber, U.S. Geol. Surv. Prof. Pap., 754-B 1973.
25. L.J. Pesonen, C.J. Hale and H.C. Halls, Unpublished paleomagnetic data.
26. W.S. White, AGU Mono., 10, 28, 1966.
27. M.J. Berry and G.F. West, AGU Mono., 10, 166, 1966.

28. P.N.S. O'Brien, J. Geophys. Res., 73, 2669, 1968.
29. T.J. Cohen and R.P. Meyer, AGU Mono., 10, 141, 1966.
30. H.C. Halls, Can. J. Earth Sci., 6, 555, 1969.
31. H.C. Halls and G.F. West, Can. J. Earth Sci., 8, 610, 1971.
32. H. Mooney et al., J. Geophys. Res., 75, 5056, 1971.
33. R.L. Coons et al., Am. Assoc. Petrol. Geol. Bull., 51, 2381, 1967.
34. W.J. Hinze et al., Am. Assoc. Petrol. Geol. Bull., 59, 1562, 1975.
35. G.D. Ells and B.L. Champion, Am. Assoc. Petrol. Geol. Bull., 60, 1215, 1976.
36. G.R. Keller et al., EOS, 56, 602, 1975.
37. H.C. Halls, Can. J. Earth Sci., 9, 1349, 1972.
38. G.D. Bath, U.S. Geol. Surv. Res. 1960 - short papers, B212, 1960.
39. G.D. Bennett et al., Ontario Dept. Mines, Misc. Pap., 10, 1967.
40. J. Gittins et al., Can. J. Earth Sci., 4, 651, 1967.
41. W.H. Fahrig, Geol. Assoc. Can. Spec. Pap., 11, 571, 1972.
42. E. Irving and J.C. McGlynn, Phil. Trans. R. Soc. Lond. A, 280, 433, 1976.
43. P.K. Sims, Econ. Geol., 71, 1092, 1976.
44. G.B. Morey and P.K. Sims, Geol. Soc. Am. Bull., 87, 141, 1976.
45. J.S. Klasner and D. Bomke, Inst. Lake Superior Geol., 23, 21, 1977.
46. K.C. Burke and J.F. Dewey, J. Geol., 81, 406, 1973.
47. J.A. Donaldson and E. Irving, Nature, 237, 139, 1972.
48. W.H. McIlwaine et al., Geol. Assoc. Can. - Mineral Assoc. Can. Ann. Mtng. Prog. of abstracts, 60, 1974.
49. F.J. Sawkins, Geology, 4, 427, 1976.
50. F. Ahmad, Bull. Volc., 35, 1, 1972.
51. W.J. Hinze et al., AGU Mono., 10, 66, 1966.
52. R.J. Wold et al., AGU Mono., 10, 66, 1966.
53. E. Oray et al., Geol. Soc. Am. Bull., 84, 2763, 1973.
54. R.W. Patenaude, Res. Rept. 64-2, Univ. Wisconsin, Geophys. Polar Res. Centre, 1964.
55. J.R. Weber and A.K. Goodacre, AGU Mono., 10, 56, 1966.
56. R.W. Patenaude, AGU Mono., 10, 111, 1965.

# AN EARLY PALEOZOIC CONTINENTAL RIFT SYSTEM IN GALICIA (NW SPAIN)

P.W.C. van Calsteren and E. den Tex

Geologisch en Mineralogisch Instituut der
Rijksuniversiteit Leiden, the Netherlands

ABSTRACT. A model of mantle-plume diapirism, lower crustal rejuvenation, continental rifting and incipient seafloor spreading is proposed for the Early Paleozoic lithosphere of the northern Hesperian Massif in western Galicia (NW Spain). Geological, geophysical, geochemical and geochronological data are presented that confirm and allow refinement of the model.

## 1. INTRODUCTORY GEOLOGY

The geology and paleogeography of the northern Hesperian Massif in NW Spain indicate the presence of (I) a Precambrian granite crust in western Galicia, (II) an intra-continental oceanic furrow bracketed by shallow seas in the Early Paleozoic of eastern Galicia, Asturias and Cantabria, and (III) a broad Variscan orogenic belt with a geanticlinal axial zone in western Galicia containing the following inliers of upper mantle and lower to middle crustal material arranged in (curvi-) linear belts (Fig. 1).

1.1 The blastomylonitic and polymetamorphic graben is a narrow zone of repeatedly deformed and polymetamorphic paragneisses with occasional amphibolites and some eclogites, interleaved with blastomylonitic and monometamorphic calcalkaline to peralkaline granite gneisses with related hornfelses. A fossil graben structure [2] is inferred from the profusion of pre-Variscan high-level granites, which are closely associated with small gabbro bodies and a mafic dyke swarm, and are intrusive into mainly Upper Precambrian sediments and volcanics. In the Variscan cycle it was subsequently compressed, metamorphosed (but not migmatized) and sparsely intruded by granites.

*I. B. Ramberg and E.-R. Neumann (eds.), Tectonics and Geophysics of Continental Rifts,* 125-132.

1.2 The Ordenes basin is open-ended to the N and contains similar rocks and structures as the fossil graben but to the W, S and E it is surrounded by an inward dipping peripheral belt consisting of at least four meso- to catazonal complexes near Castriz/Agualada, Santiago, Mellid and Sobrado. The latter two have spinel - lherzolite cores and all contain locally retrograded high-pressure

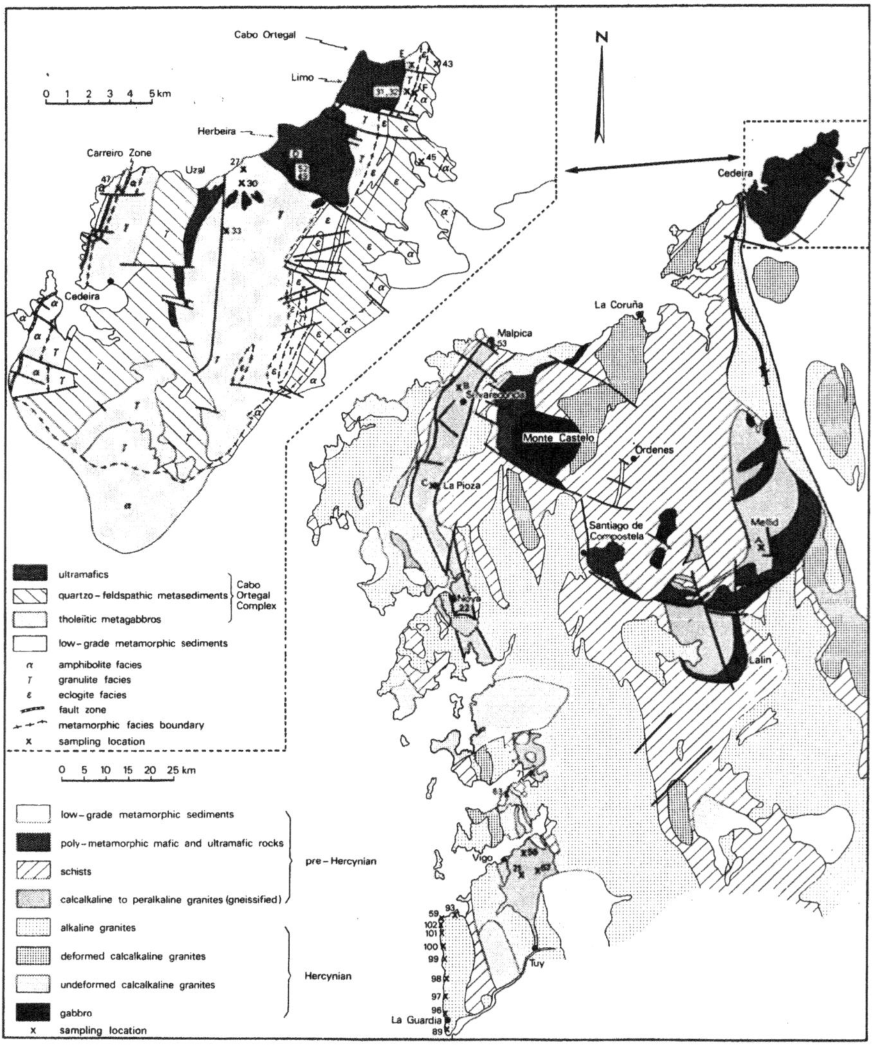


Fig. 1. Simplified geological map, from [6].

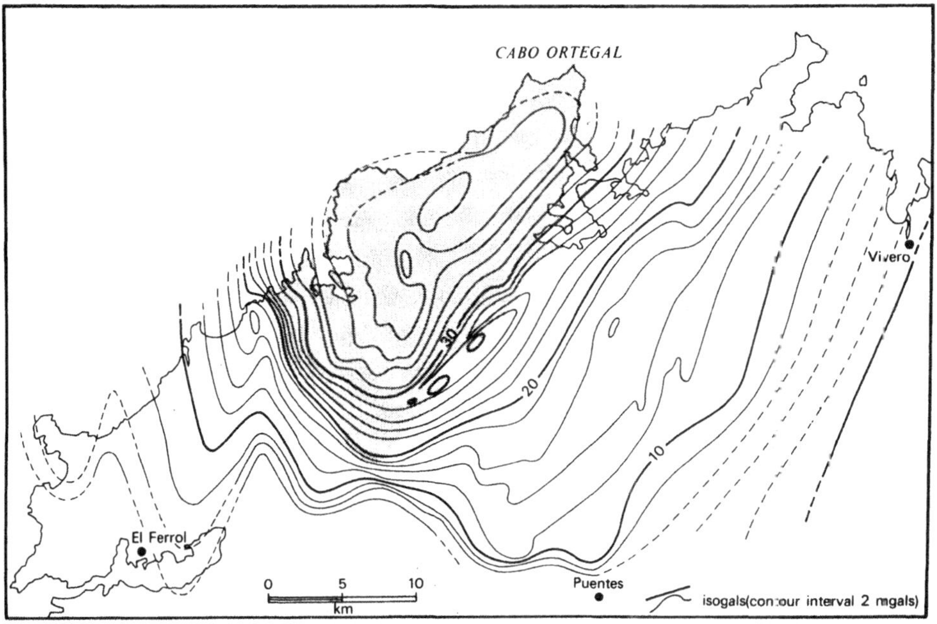


Fig. 2. Bouguer anomaly map of the Cabo Ortegal area, after [3]

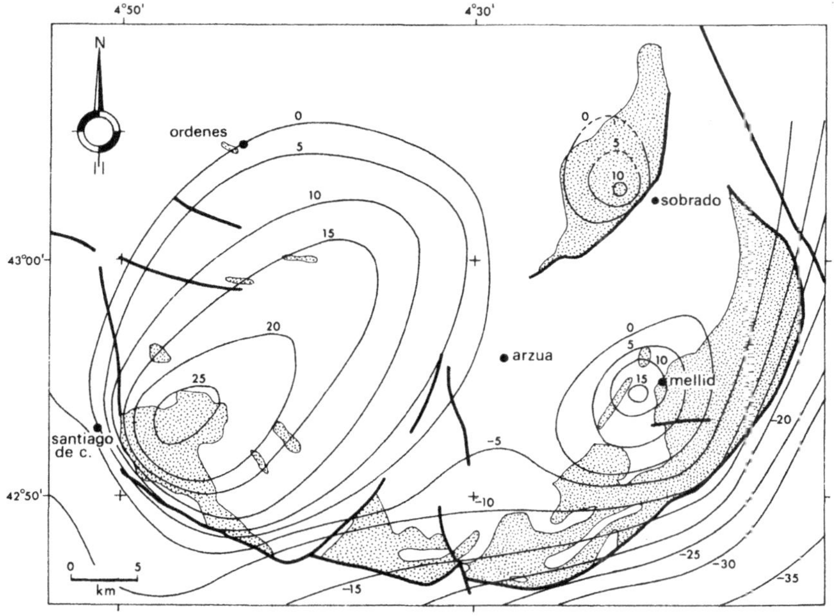


Fig. 3. Bouguer anomaly map of the Ordenes basin, from [4]

granulite zones roughly surrounded by almandine amphibolite facies rocks including garnetiferous augen- and orthogneisses.

1.3 The catazonal complex at Cabo Ortegal is the most deeply exhumed of the central complexes in western Galicia. Its core consists of spinel-lherzolite with pyroxenite and pargasite/phlogopite veins surrounded by and intricately folded with lower crustal ga-cpx-granulites, eclogites, ga-amphibolites and ga-ky-bi-gneisses. In the Variscan cycle this complex was retrograded, refolded and tectonically juxtaposed to a melange of serpentinites, greenschist facies variolites, keratophyres and volcanic breccias together with Siluro-Devonian slates, cherts and limestones (inset, Fig. 1). References to the literature may be found in [1,2].

## 2. GRAVITY INVESTIGATIONS

A gravity survey of the Cabo Ortegal complex and its surroundings [3] has revealed a positive Bouguer anomaly of at least 38 mGal amplitude (Fig. 2). It has been shown that the lower crustal and upper mantle rocks of the core, which are 0.1 - 0.3 $gm.cm.^{-3}$ denser than the associated supracrustal rocks, extend at least some tens of km's S and SE under a gently outward dipping boundary plane, which reverts to a steeply inward dipping contact at the surface. The southwestern boundary plane, however, continues to dip steeply inward to a depth of 3 to 6 km depending on the preferred density contrast. The gravity data are consistent with the interpretation of the subsurface structure as a mushroom-shaped dome tilted to the ESE. Another gravity survey was carried out over parts of the peripheral belt and adjoining segments of the Ordenes basin [4]. This survey has demonstrated the presence of three positive gravity anomalies: an asymmetrical anomaly with an amplitude of 25 mGal, as compared with the regional gravity, located East of Santiago, accompanied by two symmetrical anomalies with amplitudes of 10 and 15 mGal around Mellid and Sobrado, respectively. This indicates the presence of three separate deep-rooted structures of dense material, which may be interpreted as ultramafic diapirs closely associated with mafic igneous and high-pressure metamorphic rocks (Fig. 3). The gravity data do not support the hypothesis that the complexes are the klippen of a basement thrust-nappe rooted to the West [5].

## 3. GEOCHRONOLOGY

Isotope geochronological methods were applied to various rocks from the Cabo Ortegal and Mellid complexes, the blastomylonitic graben and from N of La Guardia (see letters and ciphers in Fig. 1) [6]. The following results emerged: (I) Lherzolites in the Cabo Ortegal complex intruded around 500 Ma ago, during the same times as a calcalkaline granite series in the Mellid complex and in the blastomylonitic graben; (II) granulite facies metamorphic conditions lasted until around 390 Ma in the Cabo Ortegal complex; (III) high-grade metamorphic conditions waned around 350 Ma in the blastomylonitic graben and in the Cabo Ortegal complex shortly before (IV) the generation of an alkaline granite series around 320 Ma ago which was associated with regional migmatization, before the second Variscan deformation phase.

## 4. GEOCHEMISTRY

Major and trace element analyses were carried out [7], on several rock types from the Cabo Ortegal complex i.e. spinel-lherzolites; pyroxenite layers cut out of these; granulite and eclogite facies mafic rocks; amphibolites devoid of higher grade relics and gabbros with incipient high-grade metamorphic imprint. The spinel-lherzolites have major element contents similar to the HT periodotite of the Lizard and compatible trace element contents in accordance with "accepted" averages for ultramafic rocks. Both the $[La/Sm]_n$ and the K/Rb ratios (Figs. 4 and 5), however, are less typical. The low normalized Sm and -less extreme- La contents point to an origin of the spinel-lherzolite as a residual phase in some melting event much like other HT periodotites [8]. The fact that the $[La/Sm]_n$ ratios are greater than unity, i.e. relative La enrichement, leads to the opposite conclusion. The normal to rather high K content compared with the high K/Rb ratios also leads to the conclusion that the Cabo Ortegal spinel-lherzolites were the melt fraction in a first, and the residual portion in a subsequent, melting event. The distribution of the data-points over both diagrams reveals the existence of rather large inhomogeneities with respect to these elements within the lherzolites over the distance of 150 m from which the samples were taken. It has been suggested [9] that the pyroxenite layers that occur in the host-lherzolite originated as its partial melting products. The major and compatible trace element contents of the pyroxenite layers support this suggestion. In Fig. 4 and 5, tie-lines are drawn between pairs of pyroxenite layers and host-lherzolite. The slopes of these lines also confirm the partial melting hypothesis. It is noteworthy that the pyroxenite layer formation did not obliterate the pre-existing heterogeneity.

The amphibolites without high-grade relics and the gabbros with incipient high-grade metamorphic imprint are similar to modern (Iceland) "plume" tholeiites [10] and a similar origin is implied. This conclusion is based on major elements, trace element "fingerprints", the $[La/Sm]_n$ ratio of around unity and the K/Rb ratio of 284.

The granulite and eclogite facies mafic rocks are comparable with continental tholeiites such as the Deccan Trap basalts [11],

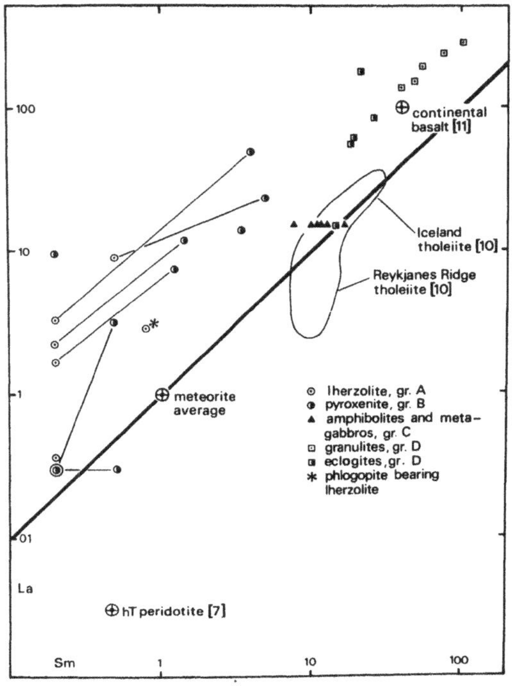


Fig. 4. Normalized La versus Sm diagram, after [8]

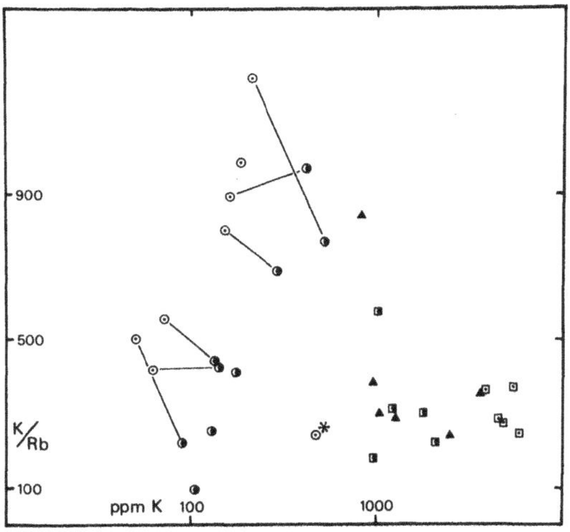


Fig. 5. K/Rb versus K diagram, from [8]

based also on major elements, trace element "fingerprints", the $[La/Sm]_n$ ratio of 3.25 and the K/Rb ratio of 319. Their derivation from a pre-existing continental crust is inferred from the high metamorphic grade they already possessed.

## 5. MANTLE PLUME MODEL

The starting point of our model is a normal continental crust consisting of igneous mafic and sedimentary rocks with a metamorphic grade up into the eclogite facies. It is assumed that in the Early Paleozoic a mantle-plume developed under the crust. Second order diapirs of lherzolitic composition separated from it and intruded the lower crust. The lherzolite of the Cabo Ortegal complex is regarded as representing one of the second order diapirs and its proposed evolution will be discussed. The lherzolites were the melt fraction of the mantle-plume at great depth as indicated by the high $[La/Sm]_n$ and K/Rb ratios. During the diapiric rise the lherzolite solidified. At a depth of about 100 km, in the asthenosphere, parital melting started again, leading to a liquid of tholeiitic composition that was able to leave the diapir and intrude the lower crust. The group of metagabbros and amphibolites in the Cabo Ortegal complex may be regarded as representing this magma. A very small part of the melt-fraction was unable to leave the diapir and crystallized in situ as pyroxenite layers and veins in the lherzolite. Heat given off by the lherzolite diapir induced granulite facies metamorphism in the already medium- to high-grade metamorphic crust. The mafic eclogites and granulites may be regarded as part of this older crust. The regional temperature increase caused by the mantle-plume induced anatexis in the deeper parts of the crust, leading to the generation of calcalkaline granitic magmas. These magmas intruded higher parts of the crust, mainly along deep reaching faults, sometimes causing contact metamorphic aureoles. The oldest representatives of these granite series are now orthogneisses. Some of the calcalkaline magmas were enriched in juvenile material, derived from the mantle-plume or its diapirs, and differentiated to peralkaline granite magmas that also intruded higher crustal levels. Heat from the mantle-plume also caused migmatization in higher levels of the crust and the generation of alkaline two-mica granite series.

## 6. RIFT SYSTEM MODEL

The mantle-plume and its diapiric off-shoots, referred to above, may be fitted in with the concept of an Early Paleozoic continental rift system in W Galicia [2]. A Precambrian orogenic terrain, inferred from the presence of an infra-Cambrian volcano-detrital molasse belt between W and E Galicia, was subjected during the Late Precambrian and Early Paleozoic to taphrogenic movements. In the blastomylonitic and polymetamorphic graben this was accompanied by abundant high-level granitic magmatism with peralkaline tendencies of Ordovico-Silurian ages. At somewhat deeper levels, as in the peripheral belt of the Ordenes basin, these rocks can be traced into garnetiferous granite gneisses, felsic and mafic granulites associated with metagabbros and upper mantle diapirs which caused rejuvenation and mobilization of the lower continental crust. The deepest level of the rift system is exposed at Cabo Ortegal, where the hat of an upper mantle diapir and its lower crustal context are tectonically juxtaposed with a melange of Siluro-Devonian age.

## 7. CONCLUDING REMARKS

In the Early Paleozoic, important thermal and taphrogenic events have affected the upper mantle and continental crust in western Galicia. The evidence of an orogeny of Grenville age in Galicia [12] must be discredited. The relation between the Early Paleozoic events and those of the Variscan cycle remain to be classified.

## REFERENCES


1. P.W.C. van Calsteren, *Proc. Kon. Ned. Akad. Wet.*, **B80**, 156, 1977.
2. E. den Tex, *Krystalinikum*, in press.
3. R.A. van Overmeeren, *Tectonophysics*, **26**, 293, 1975.
4. E.J. Keasberry, P.W.C. van Calsteren and R.P. Kuijper, *Tectonophysics*, **31**, T 61, 1976.
5. A. Ries and R.M. Shackleton, *Nature*, **234**, 65, 1971.
6. P.W.C. van Calsteren, N.A.I.M. Boelrijk, E.H. Hebeda, H.N.A. Priem, E. den Tex, E.A.Th. Verdurmen and R.H. Verschure, *Tectonophysics*, submitted.
7. F.A. Frey, *Geochim. Cosmochim. Acta*, **33**, 1429, 1969.
8. P.W.C. van Calsteren, *Lithos*, **11**, 1, 1978.
9. P. Maaskant, *Leidse Geol. Med.*, **45**, 237, 1970.
10. J.-G. Schilling, *Nature*, **242**, 565, 1973.
11. R.A. Schmitt, R.H. Smith and D.A. Olehy, *Geochim. Cosmochim. Acta*, **28**, 67, 1964.
12. D.E. Vogel and A.A. Abdel Monem, *Geol. Minjb.*, **50**, 749, 1971.

# THE MIDLAND VALLEY AS A RIFT, SEEN IN CONNECTION WITH THE LATE PALAEOZOIC EUROPEAN RIFT SYSTEM

E.H. Francis

Institute of Geological Sciences, Leeds, England*

ABSTRACT. The Midland Valley is first recognisable as a graben after the closure of Iapetus. NE-SW structural lineaments parallel to the closure then controlled deposition of early Devonian intermontane sequences and the siting of some of the calc-alkaline volcanism within the rift. The phase, which included similar volcanism and related plutonism beyond the rift, was ended by a mid-Devonian orogeny. Carboniferous volcanism was alkaline and mid-continental; its location, like the distribution of sediments, indicates continuing NE-SW basement controls. These probably owed less to subduction, hot-spots or plate collision than to adjustment-fracturing and underlying partial melting in an inhomogeneous continental plate drifting northwards.

Late in the Carboniferous new stresses led to emplacement of E-W tholeiite dykes, along extensional fractures transgressing all the earlier structures. They may continue eastwards beneath the North Sea, reappearing along ESE to SE trends in southern Sweden and are possibly comagmatic with the earliest lava flows in the Oslo Graben. This orogenic phase may have been initiated by rotation of a South Europe microplate during the closure of a mid-European ocean. After this brief interruption, continental alkali-basaltic magmatism became re-established and continued into early Permian times. It was more silica-undersaturated than before - a change which may represent a new stress regime, corresponding with NW or NNW basement controls recognised in the western part of the Midland Valley and correlatable with the N-S rifts of the North Sea and the Oslo Graben.

* Dept. of Earth Sciences, University of Leeds, from Oct. 1977.

*I. B. Ramberg and E.-R. Neumann (eds.), Tectonics and Geophysics of Continental Rifts*, 133-147.

## 1. INTRODUCTION

In Scotland, the Midland Valley Graben is about 80 km wide and 200 km long extending from the firths of Tay and Forth south-westward to the Firth of Clyde. The downfaulted rocks within the structure are mainly Devonian and Carboniferous in age. They are separated from folded Ordovician and Silurian rocks to the south by the Southern Uplands Fault and from even older metamorphosed Dalradian rocks to the north by the Highland Boundary Fault (Figs. 1,2). The graben can be traced for a further 300 km or so across Ireland, where it is narrower (c. 50 km) and less clearly defined, partly because the boundary faults are obscured by younger rocks.

Previous syntheses [1-3] have shown that the Midland Valley was not recognisable as a rift until the end of the Caledonian

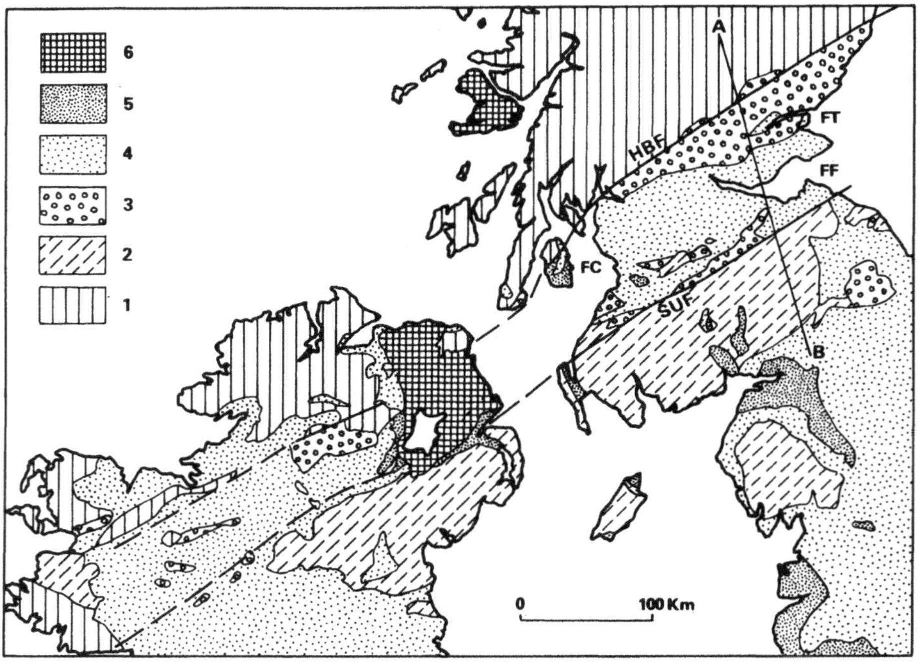


Fig. 1. Geological sketch map of the Midland Valley Graben. 1 Dalradian and older; 2 Ordovician and Silurian; 3 Early Devonian (L.ORS); 4 Late Devonian (U.ORS) and Carboniferous; 5 Mainly Permian; 6 Mesozoic and Tertiary. HBF Highland Boundary Fault; SUF Southern Uplands Fault; FC Firth of Clyde; FF Firth of Forth; FT Firth of Tay. A-B see Figs. 2 and 3.

orogeny and that it was most active as a structure during the Devonian and Carboniferous. However, because it is now believed that the Caledonian orogeny was caused by the closure of a Lower Palaeozoic proto-Atlantic Ocean (Iapetus) some reappraisal of both the timing and foundations of the graben is desirable. Moreover, although the mechanisms governing the main structure and the parallel to sub-parallel internal rifts appear to have operated continuously throughout Upper Palaeozoic times there were profound changes with time in the form taken by penecontemporaneous volcanism. Thus during the early Devonian, the products of activity were extrusive and plutonic calc-alkaline rocks located both within the rift and well to the north and south of it. In early to mid-Carboniferous times, by contrast, volcanism was essentially alkali-basaltic and extrusive and was virtually confined to the rift or along hinge lines of newly formed graben farther south. Finally, the late Carboniferous to early Permian activity, though continuing to be alkali-basaltic except for a brief tholeiitic interruption, seems to be related to a change in structural control: the NE-SW rifting declined in influence and became overprinted by NW or NNW rifting comparable with N-S and NNE structures in the North Sea and Scandinavia.

An attempt is made here to reconcile these progressive changes with some of the current models of lithospheric plate evolution. After oceanic closure by subduction and collision there was continental mid-plate rifting and volcanism modified in late Palaeozoic times, perhaps, by further, more distant oblique collision or by incipient plate separation.

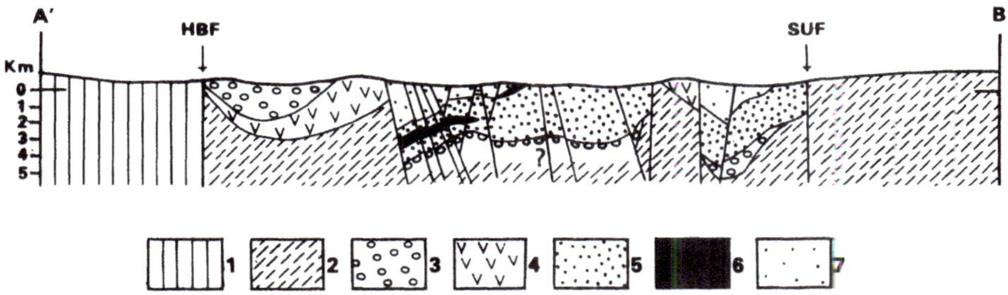


Fig. 2. Horizontal section across eastern part of Midland Valley. Line A′ - B′ approximating to A-B on Fig. 1. 1 Dalradian; 2 Ordovician and Silurian; 3 Early Devonian (L.ORS) sediments; 4 Early Devonian lava and tuffs; 5 Late Devonian (U.ORS) and Lower Carboniferous sediments; 6 Carboniferous basalt lavas; 7 Upper Carboniferous sediments.

## 2. LOWER PALAEOZOIC FRAMEWORK

Stratigraphical evidence [2,3] suggests that the Highland Boundary Fault was a recognisable structure in Arenig times, though then throwing down to the north. Some of the more recent models reconstructing the closure of Iapetus similarly imply that either or both of the boundary faults of the graben were already recognisable lineaments corresponding approximately with trenches above subduction zones. One [4] proposes subduction north-westwards beneath the Midland Valley from the line of the Southern Uplands Fault. Another [5] suggests two Benioff Zones, one dipping north-westwards, the other south-eastwards, with the Midland Valley between them representing the last remaining unsubducted area of Iapetus oceanic crust. Granulitic xenoliths in Lower Carboniferous vents in central Ireland [6] show, however, that this south-western extension of the Southern Uplands is not underlain by oceanic crust; and similar evidence [7] has demonstrated that the Midland Valley, too, is floored by granulites. More specific indications as to the foundations of the structure are now available from the LISPB experiment [8,9] which indicates (Fig. 3) that Lewisian-like continental crust extends at least as far south as the line of the Southern Uplands Fault. Moreover, this profile shows that part of the crust to be thicker beneath the Midland Valley than it is either immediately to north or south, whereas the Ordovician and Silurian rocks are much thinner in the rift than they are in the Southern Uplands.

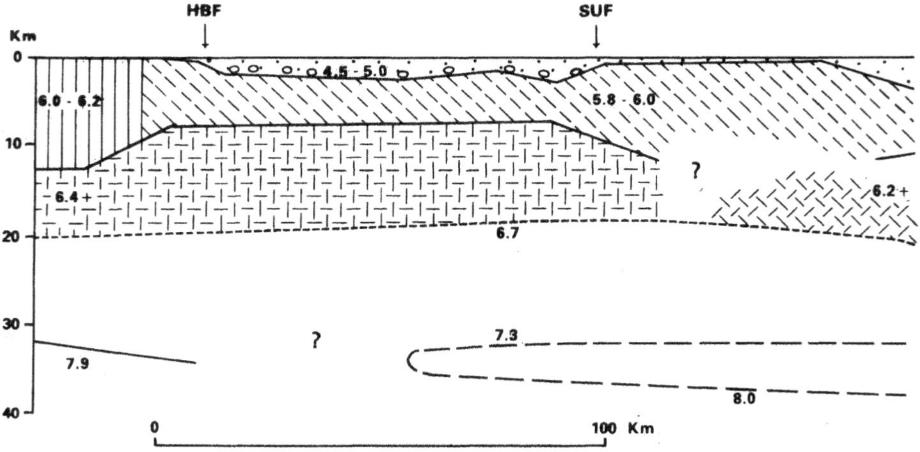


Fig. 3. Part of LISPB profile [8,9] on line A-B shown in Fig. 1: 4.5-5.0 Devonian and Carboniferous; 5.8-6.0 Ordovician and Silurian; 6.0-6.2 Dalradian; 6.2+ Celtic crust; 6.4+ Grampian (Lewisian) crust.

The profile, together with a recent Caledonian plate tectonic model [10] is here used (Fig. 4) to reconstruct a Palaeozoic sequence of events. During the Cambrian, the area of the Midland Valley is assumed to have been a land mass undergoing erosion and providing sediments to the ensialic Dalradian basin to the north. The equivalence of the flanking hinge line of that basin to the

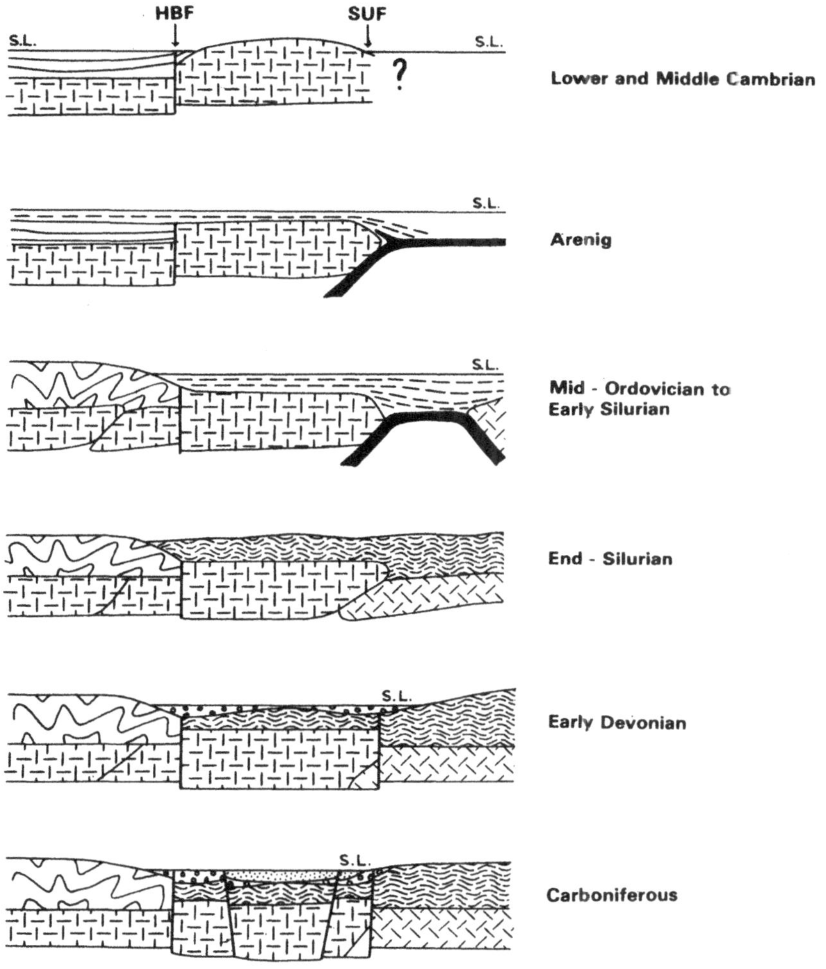


Fig. 4. Palaeozoic plate tectonic model sequence for Midland Valley Graben. Ornament as for Fig. 3. HBF Highland Boundary Fault; SUF Southern Uplands Fault; SL Sea level.

modern Highland Boundary Fault should not be regarded as more than crude. In Connemara, for instance, the Dalradian extends well to the south of the fault and is overlain with successive unconformity by sediments first of Ordovician age with basal Arenig, then of Silurian age [11]. Similarly, the eventual line of the Southern Uplands Fault may not correspond exactly with the still somewhat disputed position of the Iapetus suture [12]. Indeed it may have been determined more by the line of transition from thin Ordovician-Silurian sedimentation on the continental shelf (later to become the Midland Valley) and thicker accumulations on the slopes (later the Southern Uplands). Regardless of any precursor forms taken by the Highland Boundary and Southern Uplands faults, it required specific orogenic forces to have translated them into major fractures bounding the graben through the Upper Palaeozoic. One of the most important of these forces must have been oblique collision between continental plates [10] which brought about transcurrent faulting - also reported from comparable rift systems in the Maritime provinces of Canada [13-15].

## 3. DEVONIAN

The red continental molasse sediments of early Devonian times include fanglomerates which are indicative of the steep slopes of the rising highlands on both sides of the graben [16]. The present distribution of these rocks (Fig. 1) indicates two original main basins adjacent and parallel to the boundary faults [17]. The inner margins of those basins may have been defined by calc-alkaline volcanic rocks which were themselves localised and controlled by major NE-SW lineaments such as the Ochil and Pentland faults. An examination of the larger basins, between the Ochil and Highland Boundary faults supports this view. The volcanics of the Ochil Hills are some 3 km thick and are overlain in the centre of the Strathmore Syncline by 6 km of sediments [18]. It is clear from the LISPB profile (Fig. 3), however, that these are local stratigraphical thicknesses, not continuous in depth. It may follow that there are no great thicknesses of early Devonian volcanics or sediments beneath the Carboniferous basins in the centre of the Scottish part of the graben (Fig. 2), though the distribution in Ireland (Fig. 1) seems to have been different.

In attempting to assess the relationship between the early Devonian calc-alkaline volcanism and the tectonism it is important to note that suites of lavas, virtually identical in petrography and major element chemistry were almost simultaneously erupted outside the rift - in the Glencoe and Lorne Plateau [19] to the north and in the Cheviot Hills to the south. Moreover, related calc-alkaline plutons of comparable radiometric ages were being emplaced even farther to north and south (Fig. 5). It is difficult to relate this spread of activity to the effects of subduction

[19-21] using any of the various plate tectonic models so far proposed. Generation within the newly joined continental plate as a result of continuing transcurrent movements caused by oblique collision seems preferable as an explanation.

It is generally assumed that this (Acadian) orogenic phase ended in mid-Devonian times with uplift and erosion. The late Devonian sediments of the Midland Valley, though also comprising continental red beds, were deposited unconformably on the earlier already-folded rocks and are themselves conformably succeeded by Carboniferous sediments. Volcanic activity had already ceased early in the Devonian. When eruption resumed in the Carboniferous it was very different from the calc-alkaline regime.

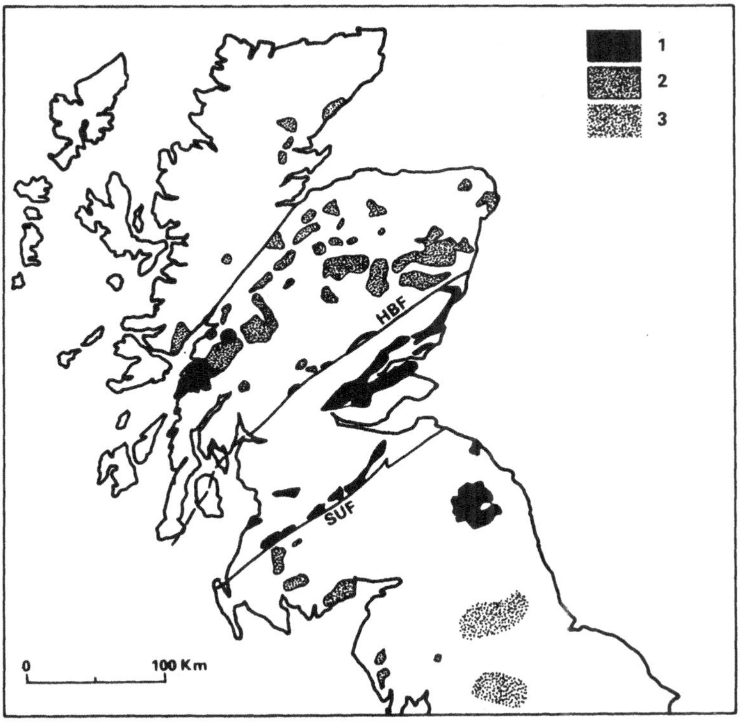


Fig. 5. Map of early Devonian calc-alkaline extrusive rocks and related plutons. 1 Lavas and tuffs; 2 Plutons at outcrop; 3 Plutons entirely concealed.

## 4. CARBONIFEROUS TO EARLY PERMIAN

In a recent summary [22] of the extensive literature on the sediments, volcanic rocks and structures of this period in northern Britain three main phases are recognised. In the first, which continued almost until the end of the Westphalian, the Midland Valley was the largest of several rift structures (others were in England) near the margin of a continental craton covered for much of the time by transgressive and regressive shelf seas. This phase was ended by what seems to have been a brief episode of E-W fracturing and tholeiitic intrusion probably related to the Asturian orogeny. In the final phase, extending throughout the latest Carboniferous into earliest Permian, the uplifted craton was subjected to a new NW or N-S system of rifting which was overprinted on the earlier NE and E-W structures, while the accompanying alkali-basaltic volcanism was markedly more undersaturated than that of the first phase.

### 4.1 Dinantian-Westphalian

Within the Midland Valley NE-trending hinge-line and transcurrent fractures and fold axes controlled the thickness of sedimentation and the location of eruption throughout the period (Fig. 6). There was, however, a change in the mode and volume of activity with time. The earliest and most voluminous outpourings during the Dinantian formed lava piles - but no central complexes - up to 900 m thick. They have been recognised [23] as consisting almost entirely of mildly undersaturated basaltic lavas which can be assigned to several geographically distinct lineages within the rift and with chemical variationswhich may be related to crystal fractionation in high-level magma chambers.

Although the NE-SW basement structures continued to control Namurian and most of the Westphalian sedimentation and volcanism, the volume of alkaline basaltic magma available for eruption decreased. Moreover, because the ascending magma columns had to penetrate increasingly thick accumulations of wet low-density sediments the form of eruption changed from lava effusion to phreatic activity from groups of short-lived centres. The resulting pyroclastic products are closely linked both geographically and chemically with alkali-dolerite sills up to 120 m thick. They are rarely associated with dykes nor are the sills or the pyroclastics commonly found outside the areas of Carboniferous sedimentation. Indeed their absence from the main areas of early Devonian sedimentation - particularly the Strathmore basin - suggests that these flanking parts of the Midland Valley Rift may have been relatively inactive during early and mid Carboniferous times.

From the sporadic stratigraphical and widespread geographical distribution of the Dinantian, Namurian and Westphalian volcanic rocks, a mechanism related to hot spots or plumes or to north-westerly subduction of a mid-European (Rheic) oceanic plate can be discounted [22]. The preferred alternative is a combination of

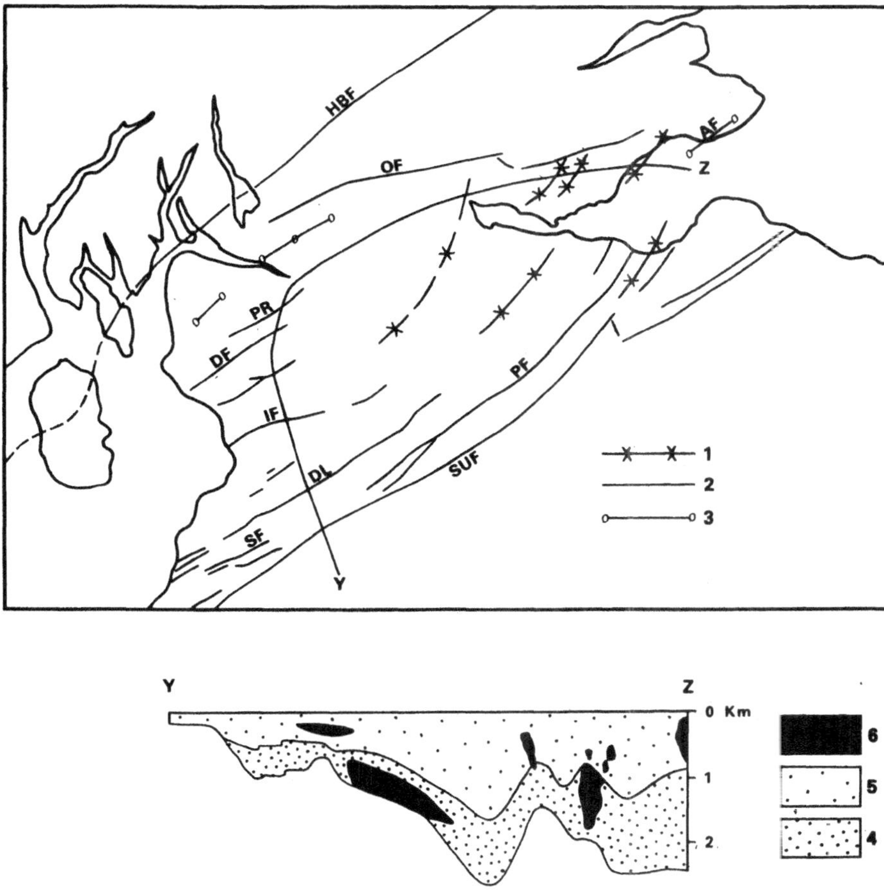


Fig. 6. Map and section showing basement controls of volcanism and sedimentation during the Carboniferous. 1 Axes of syndepositional basins; 2 Syndepositional fractures; 3 Tecto-volcanic lineaments; 4 Lower Carboniferous sediments; 5 Upper Carboniferous sediments; 6 Lavas and tuffs. AF Ardross Fault; DF Dusk Water Fault; DL Dron Line; HBF Highland Boundary Fault; IF Inchgotrick Fault; OF Ochil Fault; PF Pentland Fault; PR Paisley Ruck; SF Straiton Fault; SUF Southern Uplands Fault.

regional epeirogenic movements with local isostatic uplift engendered [24] by the inhomogeneity imparted to the continental crust by the early Devonian emplacement of the granitic plutons. This, added to the northward drift of the continental plate, as established by palaeomagnetic evidence [25] may have sufficed to cause rifting which followed and even reactivated the earlier lines of transcurrent fracture in the Midland Valley. It would also have given rise to lithospheric thinning and increased thermal gradients in the hinge zones so as to propagate repeated interaction between fracturing and magmatic ascent. The application of the membrane tectonic principle [26] may have contributed to these processes.

## 4.2 Late-Westphalian or early Stephanian

A different tectonic and volcanic phase is represented by a group of E-W fractures (Fig. 7) which cuts across all the earlier NE-SW structures and along which tholeiitic dykes were emplaced. They are associated with, and in places feed, a quartz-dolerite sill-complex. In the eastern part of the Midland Valley this complex extends at outcrop and underground over an area of 1500 $km^2$ and is up to 180 m thick. Unlike the earlier alkali-dolerite sills, these rocks have no extrusive equivalents. By way of further contrast, the dykes extend well outside the Carboniferous sedimentary basins. Indeed, including the Whin Sill and dyke complex of north-eastern England, these intrusions cover an E-W belt over 200 km wide (Fig. 8).

The stratigraphical limits for determining the age of this phase indicate merely that they are post mid-Westphalian and pre-Permian. However, the intrusions have consistently given radiometric dates around $295 \pm 5$ m.y. [27] and this not only corresponds well with the $294 \pm 4$ m.y. age recently obtained [28] from WNW-trending dykes in Skåne (i.e. parallel to the Tornquist Line), but also revives interest in the correlation [29] of these intrusions across the North Sea (Fig. 8).

This correlation has further implications. In the Midland Valley and in north-east England the lack of any extrusive equivalents to the tholeiitic intrusions can be explained satisfactorily in terms of mechanical principles [30] suggesting that magma columns ascending through a thick cover of low-density sediments are more likely to spread out as high-level sills than to reach the surface as lavas. In the absence of thick Carboniferous sediments, this explanation would not hold in Norway and it is thus of the greatest interest to note that tholeiites unrelated to the other basaltic rocks of the Oslo Rift are now recognised [31] among the earliest (B1) lavas from the Krokskogen area and that the basalt B1K has given an Rb-Sr age of $292 \pm 8$ m.y. [32].

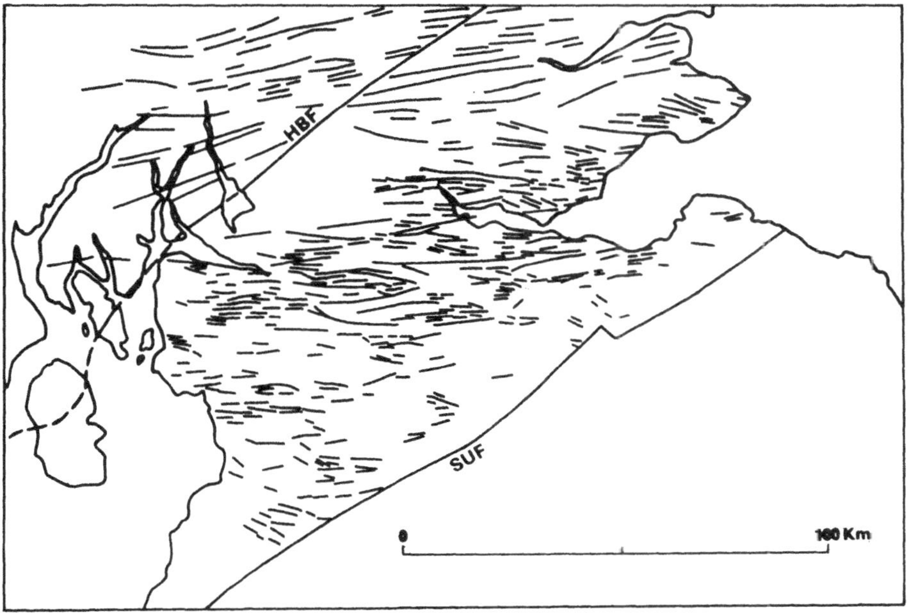


Fig. 7. Late Carboniferous east-west fractures associated with emplacement of tholeiite dykes. The apparently large number of fractures in the Carboniferous basins reflects the volume of structural detail provided by mining for coal and oil-shale.

Fig. 8. Tholeiite dykes and related quartz-dolerite sills in northern Britain and Scandinavia.

The earliest lavas of the Oslo Graben preceded the N-S rifting [33] which was to become the dominant early Permian structural control there and elsewhere. It is therefore tempting to suggest that this easterly to east-south-easterly fracturing and associated tholeiitic volcanism represented a short-lived event about 292-295 m.y. ago. It may have resulted from an (?Asturic) orogeny related to oblique collision of the North America/North Europe plate with a South Europe plate or [34] one of several rotated microplates. Although this suggestion provides a simple tectonic sequence to satisfy the data available, it leaves unexplained the radiometric evidence which indicates that on both sides of the North Sea alkali-basaltic magma, presumably derived from lower levels in the mantle, continued to be available closely before, after and perhaps even simultaneously with, tholeiitic effusion [27,32,35].

### 4.3 Latest Carboniferous-earliest Permian

After having been dated originally as Permian, then as of various late Carboniferous ages, the plants interbedded with the Mauchline lavas of Ayrshire are now reassigned to the earliest Permian [36]. This accords well with radiometric ages obtained from intrusions in Fife and the English Midlands and though these results [27,35] further suggest that alkali-basalt magma was continuously available, it is convenient to regard the post-tholeiitic volcanism as quite separate. This agrees better with the petrochemistry [23,37-39], for these late rocks are more undersaturated than the earlier alkaline basalts. The post-tholeiitic events also seem worth separating because of their different structural associations. Although the Ardross Fault [40] and other NE-trending structures continued to act as controls to isotopically dated early Permian volcanism in the eastern part of the Midland Valley, the western part was overprinted by new NW or NNW rifting with alkaline basaltic volcanism in the Mauchline-Sanquhar rift; there was similar overprinting in England along the west Midlands extension of the Stranraer-Cheshire rift (Fig. 9).

Any mechanism advanced to explain this new volcano-tectonic pattern should also account for the penecontemporaneous N-S rifts and associated igneous activity farther east. The Viking-Central Graben was presumably initiated at this time, judging by its associated early Rotliegende volcanics [41,42] and so must have been the West Norway Trough in view of the 275 m.y. age of related alkaline basaltic dykes in the Sunnhordland region [43]. As the Oslo Graben is younger than the tholeiitic lavas and contains early Permian sediments, it too must be regarded as a member of this conjugate group. Plume-controlled trilite junctions have been proposed [44] as the sources of some of these mid-continental rifts, but if plumes are irreconcilable with Carboniferous evidence

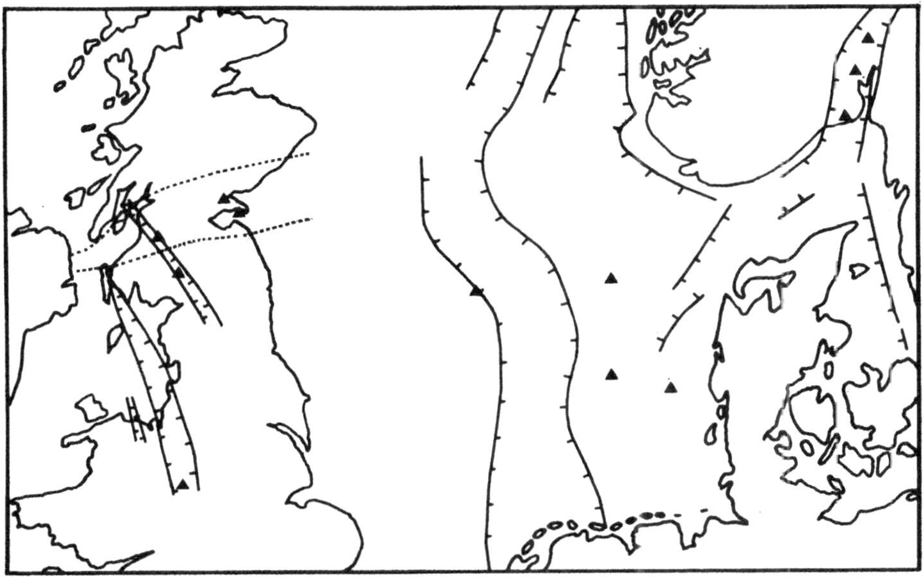

Fig. 9. Map showing Permian rifting. Generalised location of volcanic rocks shown by black triangles.

it would be illogical to accept them in the Permian. A more consistent approach would be to adopt the mechanisms proposed for the Carboniferous rifting and volcanism - that is, drift with eustatic and isostatic movements in an inhomogeneous continental plate. To explain the new orientations of fracturing, some new and extraneous factor would have to be introduced. Collision associated with closure of the Uralides suture is one possibility. Another is the suggestion advanced in respect of the rifts in the western part of the Midland Valley, namely that they were initiated by the opening of the North Atlantic or by prior crustal thinning [45-47]. This, however, seems difficult to accept in the light of the various lines of evidence showing that the opening occurred much later [48,49].

REFERENCES

1. M. Macgregor and A.G. MacGregor, The Midland Valley of Scotland (2nd Edit.), British Regional Geology, 1948.
2. W. Q. Kennedy, Trans. geol. Soc. Glasg., 23, 106, 1958.
3. T. N. George, Trans. geol. Soc. Galsg., 24, 32, 1960.
4. J. F. Dewey, Scott. J. Geol., 7, 219, 1971.
5. P. J. Gunn, Nature, 242, 111, 1973.
6. P. Strogen, Nature, 250, 562, 1974.

7. B.J. Upton, P. Aspen, A. Graham and N.A. Chapman, Nature, 260, 517, 1976.
8. D. Bamford, K. Nunn and C. Prodehl, LISPB-III. Upper crustal structure of Northern Britain, J. geol. Soc. Lond. In press.
9. D. Bamford and C. Prodehl, Explosion seismology and the continental crust-mantle boundary, J. geol. Soc. Lond. In press.
10. W.E.A. Phillips, C.J. Stillman and T. Murphy, J. geol. Soc. Lond., 132, 579, 1976.
11. J.F. Dewey, Geol. Mag., 98, 399, 1961.
12. F. Moseley, Geol. Soc. Am., Bull., 88, 764, 1977.
13. E.S. Belt, Newfoundland Carboniferous stratigraphy and its relation to the Maritimes and Ireland, in: North Atlantic - Geology and Continental Drift, a symposium, ed. by M. Kay, Mem. Am. Ass. petrol. Geol. 12, 1969.
14. G.W. Webb, Paleozoic wrench faults in Canadian Appalachians, in: North Atlantic - Geology and Continental Drift, a symposium, ed. by M. Kay, Mem. Am. Ass. petrol. Geol. 12, 1969.
15. R. White, Can. Inst. Min. Met. Trans., 75, 267, 1972.
16. B.J. Bluck, Old Red Sandstone and other Paleozoic conglomerates of Scotland, in: North Atlantic - Geology and Continental Drift, a symposium, ed. by M. Kay, Mem. Am. Ass. petrol. Geol. 12, 1969.
17. B.J. Bluck, Sedimentation in a late orogenic basin - the Old Red Sandstone of the Midland Valley, in: Crustal Processes and Evolution in NW Britain and Adjacent Regions, Glasgow University Symposium, April 1977, in press.
18. E.H. Francis, I.H. Forsyth, W.A. Read and M. Armstrong, The geology of the Stirling District, Geol. Surv. G.B., Mem., 1970.
19. D.R. Groome and A. Hall, Min. Mag., 39, 621, 1974.
20. M.S. Garson and J. Plant, Nature phys. Sci., 242, 34, 1973.
21. A.E. Wright, Nature, 264, 156, 1976.
22. E.H. Francis, Igneous Activity in a Fractured Craton - Carboniferous Volcanism in N Britain, in: Crustal Processes and Evolution in NW Britain and Adjacent Regions, Glasgow University Symposium, April 1977, in press.
23. R. Macdonald, Scott. J. Geol., 11, 269, 1975.
24. M.H.P. Bott, Proc. Yorks. geol. Soc., 36, 139, 1967.
25. A.G. Smith, J.C. Briden and G.E. Drewry, Phanerozoic world maps, in: Organisms and Continents through time, ed. by N.P. Hughes, Spec. Pap. in Palaeontology, 12, 1973.
26. D.L. Turcotte and E.R. Oxburgh, Nature, 244, 337, 1973.
27. F.J. Fitch, J.A. Miller and S.C. Williams, C.R. 6me Congr. Int. Stratigr. geol. Carbonif. Sheffield 1967, 2, 771, 1970.
28. I. Klingspor, Geol. Foren. Stockh. Forh., 98, 195, 1976.
29. S. Hjelmqvist, Sver. geol. Unders., C. 430, 1939.
30. A.R. McBirney, Bull. volc., 26, 455, 1963.
31. P.W. Weigand, Vid. Akad. Skr. I. Mat. - Naturv. Kl. Ny. serie, 34, 1975.

32. B. Sundvoll, Rb-Sr relationship in the igneous rocks of the Oslo paleorift, 4th European Colloquium of Geology, Cosmochronology and Isotope Geology, Amsterdam (Abstract), 1976.
33. I.B. Ramberg, Norges geol. Unders., 325, 1, 1976.
34. V. Lorenz, Nature, 262, 374, 1976.
35. I.H. Forsyth and C.C. Rundle, Geol. Surv. G.B., Bull., 60, in press.
36. R.H. Wagner, p. 23, in: D.B. Smith, R.G.W. Brunstrom, P.I. Manning, S. Simpson and F.W. Shotton, Spec. Pap. geol. Soc. Lond., 5, 1974.
37. S.I. Tomkeieff, Bull. volc., 2, 59, 1937.
38. A.G. MacGregor, Q. J. geol. Soc. Lond., 104, 133, 1948.
39. R. Macdonald, J.E. Thomas and S.A. Rizzello, Variations in basalt chemistry with time in the Midland Valley province during the Carboniferous and Permian, Scott. J. Geol., in press.
40. E.H. Francis and A.M. Hopgood, Scott. J. Geol., 6, 162, 1970.
41. W.H. Ziegler, Outline of the geological history of the North Sea, in: Petroleum and the Continental Shelf of north-west Europe, 1, Geology, ed. by A.W. Woodland, Applied Science Publishers, London, 1975.
42. J.J. Pennington, The geology of the Argyll Field, in: Petroleum and the Continental Shelf of north-west Europe, 1, Geology, ed. by A.W. Woodland, Applied Science Publishers, London, 1975.
43. R.B. Faerseth, R.M. Macintyre and J. Naterstad, Lithos, 9, 331, 1976.
44. A.J. Whiteman, G. Rees, D. Naylor and R.P. Pegrum, Norges geol. Unders., 316, 137, 1975.
45. J. Hall, Scott. J. Geol., 9, 253, 1974.
46. P.E. Kent, J. geol. Soc. Lond., 131, 435, 1975.
47. M.J. Russell, Scott. J. Geol., 12, 315, 1976.
48. W.C. Pitman and M. Talwani, Geol. Soc. Am., Bull., 83, 619, 1976.
49. A.S. Laughton, Norges geol. Unders., 316, 169, 1975.

# MAIN GEOLOGIC FEATURES OF THE OSLO GRABEN

C. Oftedahl

Geologisk Institutt, Norwegian Institute of Technology, Trondheim.

ABSTRACT. En echelon faults cut a late Carboniferous/early Permian peneplain in South-Eastern Norway to begin subsidence of a complex graben. Volcanic eruptions filled the graben as it gradually subsided and most likely also spilled over the sides. Latitic rhomb porphyries and basalts dominate in preserved areas of volcanic rocks. A lava plateau comprising basalt B1, rhomb porphyries RP1-RP12, and basalts B2 and B3 formed. During the B3 phase at least seven caldera subsidences started. In all 15 to 18 cauldrons formed. Younger flows RP13-26 with several trachytes and additional basalts are now preserved only in the south (Vestfold area) and in some cauldrons. Late in and after the stage of plateau volcanism, monzonitic, syenitic, and granitic magmas stoped upwards, in some cases even penetrating the volcanic rocks to form composite plutonic bodies, in present exposures covering 5000 $km^2$ of the total 8500 $km^2$ of the graben. Especially well-known rocks are nordmarkite (aegerite syenite) and larvikite (augite monzonite), the cryptoperthite of which is a moonstone, in places with a strong blue schiller. Rb-Sr ages range from 294 (early lavas) to 274 (granite) m.y. Gravity anomaly and explosion seismic studies suggest an early Permian basalt magma pillow, 10 km thick at the base of a 20 km crust, as the major source of the magmatism. Trace element contents in lavas indicate that differentiation was far advanced at the start of surface volcanism.

## 1. INTRODUCTION

By 1850 the Oslo region was known among mineral dealers in Central Europe for its rare and beautiful minerals in Langesundsfjorden, at the southern end of the region. Professor Waldemar Christoffer

*I. B. Ramberg and E.-R. Neumann (eds.), Tectonics and Geophysics of Continental Rifts,* 149-165.

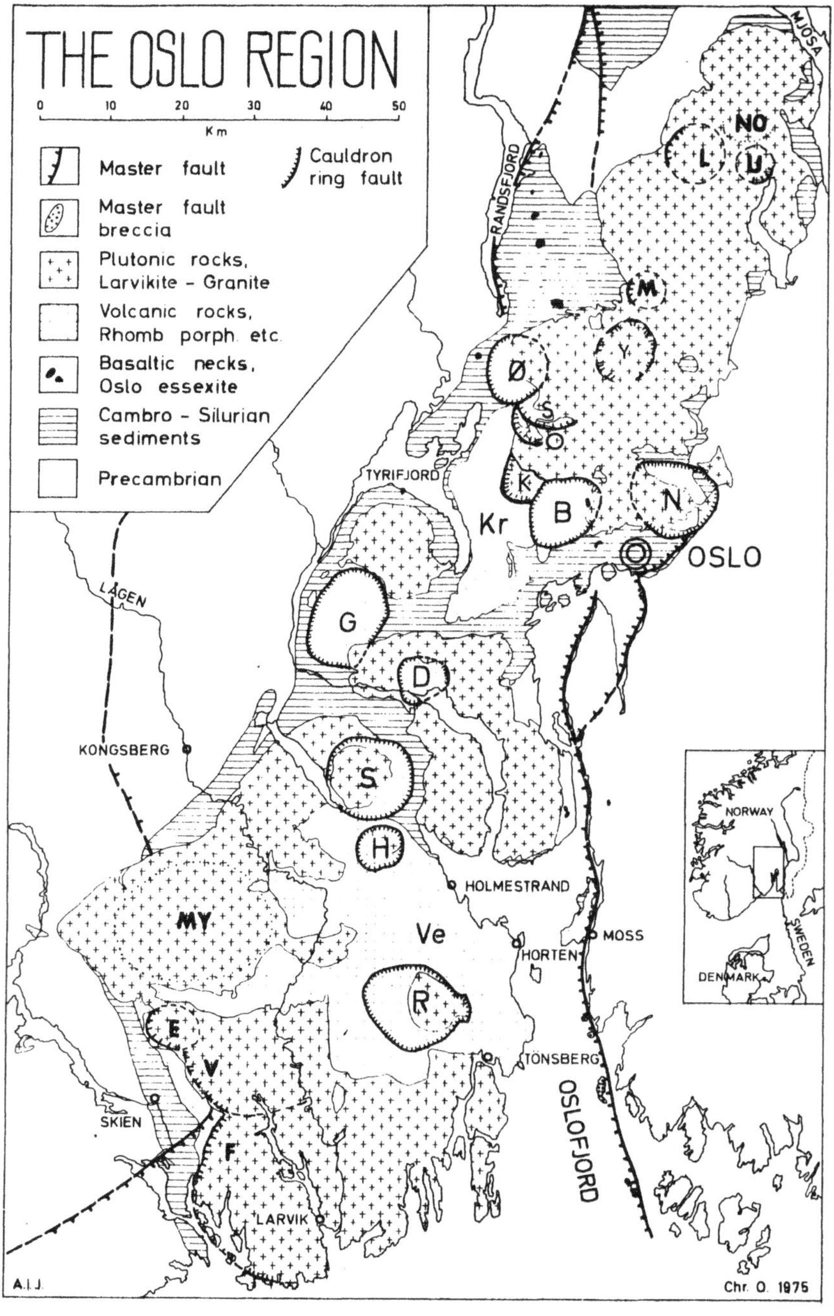


Fig.1 Lava plateaus: Krokskogen (Kr) and Vestfold (Ve). See Table 3 for other map symbols.

Brøgger made the region world famous as a petrographic province by his papers, published during 1882-1933. Most well-known is a comprehensive description of the nepheline syenite pegmatite minerals of the Langesundsfjorden area, together with a general description of the whole region. Then followed seven volumes in the series "Die Eruptivgesteine des Kristiania (Oslo)-Gebietes". The post-Brøgger era started in 1943 when Professor Olaf Holtedahl published the first volume in the new series "Studies on the Igneous Rock Complex of the Oslo Region", now constituting 26 volumes, but few geologists really worked in the region in the period from the early 1940's to the mid 1960's. Fortunately, a number of younger geologists have started to work in the region during the preceding 10 years. The resulting broad range of research is one of the major reasons for the present NATO Study Institute. Geophysical methods have also been utilized; a big monograph on all gravity anomalies and their interpretation by I. Ramberg [1] deserves special mention.

A complete reference list to all older literature was published by O. Holtedahl [2]. A general survey of the area was presented by Oftedahl [3], and a shorter and modern one with references to all of the latest literature by Ramberg [1].

Geologically speaking, the Oslo region was famous as a petrographic province in the Brøgger days (Fig. 1, cauldrons etc., see Table 3) with all sorts of plutonic and dike rocks, as well as a variety of volcanic rocks. With the new publication series

Table 1. Rocks of the Oslo Region. Figures from I. Ramberg [1].

| Rock type | | Areal distribution $km^2$ | | % | |
|---|---|---|---|---|---|
| Rhyolite, trachyte | | 31 | | 0,4 | |
| Rhomb porphyry | | 1153 | | 13,4 | |
| Basalt | | 220 | 1404 | 2,6 | 16,4 |
| Basaltic volc. necks, "Oslo essenxite" | | 15 | | 0,2 | |
| Monzonites: | Larvikite | 1670 | | 19,5 | |
| | Kjelsåsite | 201 | | 2,3 | |
| | Akerite | 52 | | 0,6 | |
| Nepheline rocks: | Lardalite, foyaite | 65 | | 0,8 | |
| Nordmarkite, syenite | | 1425 | | 16,6 | |
| Granite: | Biotite granite | 840 | | 9,8 | |
| | Ekerite | 821 | 5098 | 9,6 | 59,4 |
| Cambro-Silurian: | Shale and limestone | | 1545 | | 19,1 |
| | Sandstone | | 95 | | |
| Precambrian within graben | | | 440 | | 5,1 |
| | Total overall | | 8573 | | 100,0 |

in English, the region became known among other things for well developed cauldron complexes [4] and cauldrons are being discovered up to this day. In the context of plate tectonics, the down-faulted Oslo region is seen as a paleorift, and interest focuses on relations of this paleorift to the continental rifting systems of the Eruasian continent as well as on relations of genesis of its magmas to crustal and upper mantle tectonics.

The main geologic history of the Oslo paleorift is briefly as follows:

1. Subsidence of the area started in latest Carboniferous or earliest Permian with subsidence on some master faults, and on complicated small-scale marginal faulting with an en-echelon fault pattern. A thin red bed series was overlaid by fossiliferous shale, sandstone, and quartz conglomerate.

2. Soon after the start of the subsidence, lava flows covered the floor of the graben and built a level surface. During apparently continuous subsidence the lava flows filled up the graben nearly completely and on occasion spilled over the sides. A lava plateau with thickness within the graben between 1000-2000 m was formed, locally perhaps of more than 3000 m thickness.

3. The monzonitic magmas which rapidly traversed through the continental crust to form rhomb porphyry lava flows also invaded the crust in large masses. Gradually the monzonitic magmas and subsequent syenitic and granitic magmas stoped out larger chambers and worked their way through the Precambrian and overlying Cambro-Silurian rocks into the Permian volcanic rocks. During the early stages of this process a great number of calderas formed.

4. Later erosion removed 1-3 km of the uppermost rocks, so that in the present surface the plutonic rocks dominate over volcanic rocks, with 5000 $km^2$ contrasted to 1400 $km^2$. All important rocks appear in Table 1.

A geologic survey of the region will be presented below, containing the most important facts about the volcanic rocks, the graben formation, plutonic rocks, the history of the rhomb crystals, and the cauldrons. Finally the formation of the magmas and their life in space and time will be briefly discussed.

## 2. GENERAL SURVEY OF THE OSLO REGION

### 2.1 Volcanic rocks

Obviously volcanic rocks covered the whole region, butting against the boundary faults, but owing to removal by stoping and

Table 2. Stratigraphy of the Krokskogen and the Vestfold lava areas and of 10 cauldrons (names in Table 3). B1 etc. - Basaltic lavas. 1, 2 etc. - Rhomb porphyry types RP1, RP2 etc. T1 etc. - Trachytes.

| Krokskogen | Lu | Øy | Sv | He | Ni | Ka | Bae | Dr | Gl | Sa | Ra | Vestfold |
|---|---|---|---|---|---|---|---|---|---|---|---|---|
| | | | | | | | | | | | Ignim: | 26+B6 |
| | | | | | | | | | | | R11 | T4 |
| | | B5 | | | | | | | | | R1 | 24-25 |
| | | 16Ø | | | | | | | | | 27 | T3 |
| | | 15Ø | | | | | | | | | T5 | 23 |
| | | 14aØ | | | | | | | | | B6 | B5 |
| | | 14cØ | | | | | | | | | 26 | 19-20 |
| | | 14bØ | | | | | | | Trach | | | B4+18+T2 |
| | Br. | 14aØ | 15S | 11eH | | | | | lavas | | | 14-17 |
| | -etc. | | 14S | 11dH | | Mega-Br. | B4 | | 14G | | | |
| | 13 | | 13 | 11cH | | Ign. | 13 | 13 | 13 | | | 13 |
| | Ign. | | | 11bH | | Br. | B3 | | | | | |
| | 3 | | | 11aH | Cgl. | | Ign. | | | | | |
| | Ign. | | | Br+Sed. | S.st. | | 12 | | | | | |
| | | | | Ign. | Ign. | | Br+Sed. | Mudst. | | Br? | | T1 |
| B3 | | | | B3(?) | B3+13 | B3 | B3 | B3 | B3 | | | B3 |
| | | | | | | | | | | | | Many |
| 12 | 12? | | | | | | | Mudst. | (R1?) | | | thin |
| 11 | 11(?) | | | | | | | 11 | 11 | | | local |
| 10+B2 | | | | | | | | B2 | B2 | | | RP |
| 9 | | | | | 9 | | | 9 | 9 | 9 | | flows |
| 8 | | | | | 8 | | | 8 | 8 | - | | |
| 7 | | | | | 7 | | | 7 | 7 | 7 | | |
| 6 | | | | | 6 | | | 6 | - | 6 | | 6 |
| 5 | | | | | 5 | | | 5 | 5 | 5 | | 5 |
| 4 | | | | | 4 | | | 4 | 4 | 4 | | 4 |
| 3 | | | | | - | | | - | - | 3 | | - |
| 2 | | | | | 2 | | | 2 | 2 | 2 | | 2 |
| 1 | | | | | 1 | | | 1 | 1 | 1 | | 1 |
| B1 | | | | | B1 | | | B1 | B1 | B1 | | B1 |
| Perm.sed. | S.st. | | | | S.st. | | | | | | | Sandst. |
| Kb.-Sil. | Kb.-S. | | | | Kb.-S. | | | | S.st. | | | Sandst. |
| Precambr. | | | | | PreC | | | | | | | |

erosion the volcanic rocks now occur only in downfaulted blocks of a great number of cauldrons and in some scattered remnants of the originally continuous lava plateau. The two biggest remaining areas of volcanic rocks are the Krokskogen area west of Oslo and the Vestfold lava plateau in the southern Oslo region. In their quadrangle maps Brøgger and Schetelig [5] mapped the Krokskogen area in great detail, showing that its lava plateau consisted of 12 different types of rhomb porphyries with three interlayered basalt flows (see Table 2). The same stratigraphy appeared in the subsided areas later to be described as cauldrons, and was later corrected for the Sande and Glitrevann cauldrons by Oftedahl [4], for the Drammen cauldron by Halsen [6], and for the Bærum and Nittedal cauldrons by Sæther [7]. Other cauldrons and cauldron remnants are under study. In the Brøgger and Schetelig maps 16 rhomb porphyries were mentioned in the Vestfold area, but only RP1 and RP2 correlated with those of Krokskogen. A first revision [8] showed RP1-RP17 and a more comprehensive revision [9] runs to RP26. The latest investigations of volcanic rocks are a general basalt study by Weigand [10], a Skien basalt study by Segalstad [11], and descriptions of the northernmost cauldron volcanics [12,13].

Basalts started the volcanism in nearly all areas where the basal volcanic rocks are exposed (B1). Generally speaking the thicknesses decrease from south to north, with around 1500 m in the area east of Skien [11], 150-100 m in the Vestfold area (Horten-Holmestrand), and ca. 30 m in the Krokskogen area. A mildly alkaline basalt is the usual rock type, dominating in the Vestfold and Krokskogen areas, although B1 is tholeiitic in the southeast corner (Kolsås district) [10]. Surprisingly, Segalstad [11] discovered that the lower part of the thick Skien basalt sequence consists of ankaramitic flows with olivine and titaniferous augite as phenocrysts in a groundmass of melilite, melanite, magnetite, and apatite, followed by melilititic flows (augite phenocrysts in a matrix of melilite and nepheline). A few exotic flows also occur just south of Holmestrand (leucitite, with meta-leucite microphenocrysts in a matrix of meta-leucite, nepheline, and ore).

The rhomb porphyries are mapped from their field appearance, - the shape, size and packing of their plagioclase phenocrysts. As shown earlier [8], the first 7-8 flows (RP1-8) show striking differences in appearance, but especially one type seems to reappear: the standard RP type RP1 or the Kolsås type. Thus RP4, RP9, and RP11, all look fairly similar to RP1.

A summary of the stratigraphy as it appears now in the two bigger lava areas mentioned above and in 11 cauldrons is presented in Table 2. The Krokskogen sequence exists in the neighbouring Nittedal, Drammen, Glitrevann, and Sande cauldrons. This sequence covered more or less completely the central Oslo region, some 3200 km$^2$ around Oslo from the Sande cauldron in the south to the Stryken cauldron in the north. In addition the Krokskogen flows RP1, 2, 4, are regional in the sense that they run from the northern Oslo region to Langesundsfjorden, a distance of 160 km$^2$ along the graben axis and as much as 40 km width across it. The flows RP5 and RP6 flowed at least over the central area and to the Vestfold area, some 100 km north-south. After RP4b the history of the Vestfold area diverges from that of the middle part of the Oslo region in that a great number of local flows covered small parts of the area. Basically a more general stratigraphy developed in the western part (Table 1), whereas many different flows formed in the northeastern and southeastern corners. In all Mr. Henrik Heyer distinguished nearly 40 local flows below B3. After eruption of RP10 the surface of the graben seems to have been evened out again and the thin flow RP10 in Krokskogen may be a branch of the much thicker Greaker flows in the south. RP11 equivalents may be seen in the Lakjeld or the Rød types in the Vestfold area. More certain is the association B3 and rectangle porphyry RP13, found in several cauldrons around Oslo as well as in the Vestfold lava plateau. Whether any of the upper Vestfold flows appear in any of the cauldrons north of Oslo remains to be seen, and several RP types there are considered to be local flows (e.g. RP16Ø in Øyangen cauldron).

With the very easy-flowing lava flows both of basaltic and latitic compositions, the thicknesses most likely reflect the subsidence of the graben just prior to the outflow of each flow unit, meaning that the flows always made ponds and thicker parts in small depressions. Thus the central area around Oslo subsided much more than the Vestfold area during the period RP1-RP11, because the maximum thickness of the Krokskogen plateau is 700 m but the equivalent flows in the central Vestfold area are only 200-300 m thick. The present day saucer shape may therefore be a structure developed by gradual subsidence on many small faults during the lava eruptions. Thus the Krokskogen flows are thicker than the corresponding flows in the Nittedal cauldron to the east and the Glitrevann cauldron to the west. In the Vestfold area the subsidence was greater after the B3 flows, as testified by an estimated 1000-1500 m thickness for the sequence B3-RP26.

Each rhomb porphyry flowed out on the surface like the most easy-flowing basalt to make up cooling units up to 100 m or more in thickness, but consisting of single lava flows with thicknesses as little as 1.0 or 0.5 m. The latest discussion of the surprising property of latites to flow nearly like water is given by Sæther [7] who finds the fluidity proved by the constant appearance of turbulent flow patterns and who sees the explanation in very hot lavas supplied in huge quantities in a very short time.

Interruptions between each flow unit are often marked by a sedimentary horizon, usually conglomeratic gravel of the underlying flow with thickness in the range 10-50 cm or conglomerate bed of rarely more than two meters thickness. Thus, interruptions between the flow units were not very long and have been estimated as between 1,000 and 10,000 years [8]. If we assume a desert climate with seasonal and very strong showers, we have the medium for transporting the gravel to small depressions in the landscape. Presumably the porous ground and the usual dryness prevented the formation of any soil cover. With development of volcanic conglomerates of 1 m thickness or more, time spans between eruptions may have been up to one order longer, 10,000-100,000 years. With the latter figure, the Krokskogen plateau would take about 1.3 m.y. to form and the Vestfold area about 3.3 m.y. Possibly even this time scale could be multiplied by two or even five. With the starting point for the volcanism around 294 m.y. [14], the time span would in the first case be 290 - 294.7 m.y. This tentative inference accords well with a recent result for the nearly 200,000 $km^3$ of Columbia River Basalts which erupted between 16 and 13.5 m.y. [15], with eruptions every few thousand years, and with maximum hiatus 100,000 years.

## 2.2 Graben formation

Was the Oslo "graben" really a graben at any time? An analysis of the more general geologic features suggests that the eastern side of the "graben" did not have any elevated rim whereas to some degree the western side may have been bounded by a scarp [16]. Two features in particular suggest extensive covers of rhomb porphyries outside the present paleorift area. In the first place the Sevalrud breccia found in a region dominated by Precambrian rocks 11 km west of the Randsfjord-Hundselv northwestern master fault contains a number of RP-types, besides blocks of Cambro-Silurian sedimentary rocks and Precambrian rocks. Thus the region here had a sequence closely similar to what existed in the paleorift at the time of explosive formation of the breccia: Precambrian rock overlaid by a Cambro-Silurian sequence, with rhomb porphyries on the top. The rhomb porphyry dikes outside the paleorift also suggest extensive eruptions outside the depression itself because these dikes with 20 - 40 m width can hardly be anything but feeders for flows. The proof that we really had at least a small graben with a very pronounced eastern fault escarpment at one time is derived from the rhomb porphyry conglomerates on small islands close to the master fault along the east side of the outer Oslofjord. After a study of these islands Størmer [17] concluded that the rhomb porphyry conglomerates must have been formed at the foot of a fault escarpment to the east. Since they contain only rhomb porphyries, some basalts and a few blocks of the uppermost Silurian sandstone, the erosion at least at one time exposed the whole volcanic sequence and a little of its Silurian base. Less-evident fault scarp deposits occur in some peripheral volcanic areas. Thus the westernmost part of the Glitrevann cauldron contains a sandstone bed much thicker than usual within B2. Close to the northwestern boundary of the Krokskogen area Sæther [7] reports the existence of breccias with big fragments of a number of different RP flows. Such a deposit may well have been formed near a fault escarpment.

## 2.3 Plutonic rocks

All the important plutonic rocks have been given a quantitative petrographic description by Barth [18]. Much information on a great variety of plutonic rocks in Nordmarka, the area north of Oslo, is obtained from the general description of the area by Sæther [7]. The main natural groups are: The Oslo-essexites of the basaltic necks, the monzonitic rocks, the syenites, and the granites.

Most contacts demonstrate more or less clearly overhead stoping as the mode of emplacement, as has been emphasized by all post-Brøgger workers. But still these plutons cannot be very thick. The gravity analysis of Ramberg [1] shows that the composite, mostly syenitic pluton north of Oslo is as much as 13 km deep. This value is a maximum, because other syenitic to granitic plutons usually are surprisingly shallow, 2 - 4 km in depth. According to a general magma-volume curve [1] the larvikite should have an average thickness more than 10 km.

The gabbroic rocks of some 20 volcanic necks were called essexites by Brøgger. Barth [18] found that the group contains a number of varieties ranging from olivine gabbro through ordinary gabbro, syenogabbro, and syenodiorite to intermediate rocks as the last veins, with a sack-name termed "Oslo-essexite". These necks must have fed basalt flows, but still it is impossible to say with certainty if these necks were feeders for any of the known units B1, B2, B3, or younger flows.

The monzonites contain larvikite as the major member with more basic kjelsåsite as a subordinate rock according to geologic maps of Brøgger and Schetelig. A name given by Brøgger to a finegrained monzonite, akerite, may now go out of use. Both the larvikite and the kjelsåsite are semi-porphyritic rocks with large feldspar tabletes exhibiting plagioclase cores and alkali feldspar rims. Alkali feldspar occurs as smaller grains in the groundmass together with augite, biotite, and ores. Especially around the city of Larvik, varieties of larvikite display a beautiful schiller which has turned these varieties into the most-sold decorating stones around the earth. In these varieties plagioclase and alkali feldspar are intergrown in coarse mesoperthite. It is the alkali feldspar which has cooled so fast as to develop a cryptoperthite with the right dimensions to scatter the light. This explanation still seems a bit unsatisfactory, because the scattering is of reflection type and the display colors, even if mostly bluish, may in places be yellowish, greenish, and reddish. The major masses of monzonites seem to have been emplaced after the termination of the volcanic period, because rhomb porphyries belonging to RP26 are swimming as roof pendants in larvikite in the southern Oslo region.

The syenitic rocks were all termed nordmarkite by Brøgger in his descriptions and early maps. Later he found that it was necessary to have a designation for syenite with visible plagioclase, and he used the term pulaskite. The present usage is to reserve the name nordmarkite for the alkali syenite with arfvedsonite and/or aegerite. A transition into a rock with only alkali feldspar tablets but with ordinary dark silicates exists as well as transitions into granitic and monzonitic rocks.

The transitional contacts or difficult-to-dectect contacts have been a major obstacle to geologic mapping of the plutonic rocks. Sometimes a sharp contact is discovered between two early different varieties, for instance of syenitic rocks, without any visible indications of age relations such as a chilled border, apophyses, or fragments of one rock in the other. Such a contact between larvikite and lardalite was described by Brøgger in the last century and several others have been found later. In modern terms both plutonic rocks with gradual changes or with local sharp contacts but without contact relations, and ring complexes can be explained as derived from pulses of magma ascending from layered magma chambers undergoing gravitative differentiation.

The granitic rocks occur in two apparently distinct types, in plutons of either the biotite-bearing Drammen granite or the arfvedsonite/aegerite-bearing ekerite. The chemical differences between these two varieties are very slight, and rocks from the two groups plot close to or on top of each other in the normative feldspar triangle.

## 2.4 The rhomb crystals

The essential fact about the rhomb shape is that some rhomb porphyries carry phenocrysts exclusively bounded by the crystal faces (110), $(1\bar{1}0)$, $(\bar{2}01)$, and $(20\bar{1})$. Others carry crystals which are more rounded and boat-shaped, with a tendency to other crystal faces. This tendency may increase so that we have just a few good rhomb crystals in a porphyry, with many boat-shaped and rectangular ones. Similariy, semi-porphyritic larvikites carry rhomb crystals in some areas.

Harnik [19] studied a few selected specimens of rhomb porphyries from the Krokskogen area and the Tyveholmen rhomb porphyry dike in Oslo. The freshest possible areas within rhomb crystal cores, rims, and groundmass feldspars were studied by microprobe and single crystal methods. His conclusions may be summarized as follows:

1. The rhomb crystal cores consist of antiperthites with usually 90% host which is a potash andesine to potash oligoclase. The exsolved feldspar is a more An-poor phase. Obviously all stages of exsolution in the plagiocalse core exist. The rims usually have a normal anorthoclase composition. The structural condition of the host ranges from ordered to intermediate order.

2. After comparison with laboratory results Harnik concluded that the plagioclase core has retained a high potash component due to a long stay in magma at high temperature and high pressure, with a very rapid cooling afterwards.

3. With continued crystallization at depth the crystals became constantly richer in potash.

4. After pressure relief and decrease in termperature the cores got rims richer in soda and poorer in potash.

5. Finally the alkali feldspar groundmass crystallized.

If we assume with Ramberg [1] that the monzonite magma formed on top of the basalt pillow at depth around 20 km, we have a rather high temperature and pressure to facilitate continued crystallization of potassic plagioclases. The rims on the rhomb-shaped cores crystallized during the ascent of the magma through the 20 km continental crust. Many melt batches, however, did not ascend so to say instantaneously through the 20 km crust, but had a stay at various levels in the crust with time for equilibration of composition and adjustment of crystal faces. During this period the typical boatshaped crystals with curved crystal faces developed and other crystals attained a new equilibrium with more or less of a match-box shape. Thus the RP lavas may show feldspar phenocrysts with all sorts of outlines, from ideal rhomb to rounded ovals or sharp splinters. Correspondingly the larvikite magmas could have ascended rapidly with the rhomb crystals retained in the pluton or slower with crystal more or less re-adjusted to standard faces (001), (010), and (110). Possibly the cooling speed during crystallization also played an important role.

Thus, from Harnik's [19] conclusion about the origin of the

K-rich plagioclase or the rhomb crystals in appears that unusually K-rich plagioclases and plagioclases with rhomb crystal faces are formed when a monzonite magma crystallizes at high temperature and pressure (20 km depth), with the unusual crystals and their surrounding magma brought up very suddenly either to the surface for quenching or to near-surface positions for rapid cooling.

## 2.5 The cauldrons

The Oslo graben displays a great variety of cauldrons. Four of the most obvious cauldrons were early described in detail [4]. Two cauldron remnants, caldera blocks which have been more or less consumed by later stoping magmas, were also briefly mentioned. A recent survey of the cauldron situation produced results as shown in Table 3, with 15 cauldrons and cauldron remnants and three probable remnants. The increased number in comparison to earlier general descriptions is mostly due to a very detailed map and description of the area north of Oslo by Sæther [7], who did not attempt at all to make out cauldron remnants. Naterstad [20] discovered the Nittedal cauldron, and Sørensen [21] described parts of the Ramnes cauldron. Segalstad [11] discovered two cauldron remnants and a ring complex in the southwestern Oslo region and made assumptions about existence of cauldrons north of Oslo according to Sæther's map. Larsen [22] published a wholly new

Table 3. Cauldrons, probable cauldrons, and ring complexes of the Oslo Graben.

| Name and map symbol | | Diam. (km) | Subsidence (km) | Vol. of subs. ($km^3$) | % of cald. block | | | Ref. |
|---|---|---|---|---|---|---|---|---|
| | | | | | Volc. | Intr. | Plut | |
| Cauldrons | | | | | | | | |
| 1. Lundbergkollen | L | 10? | 1 | 75? | 2 | – | 98 | 13. |
| 2. Hurdal | U | 5 | 1? | – | 15 | – | 85 | 12. |
| 3. Grua | M | 5? | ? | – | 5 | – | 95 | 22. |
| 4. Øyangen | Ø | 10x7.5 | 2,3-4? | 120-190 | 50 | – | 50 | 7. |
| 5. Svarten | A | 11 | 1.5 | 100? | 20 | ? | 80 | 7. 22. |
| 6. Heggelia | O | 7 | 1-1.5 | 30? | 60 | – | – | 7. 22. |
| 7. Nittedal | N | 11 | 0.8 | 50 | 20 | ? | 80 | 20. |
| 8. Kampen | K | 10? | Ca.1 | 75 | 32 | 8 | 35 | 7. |
| 9. Baerum | B | 8.5x12 | 1-1.5(2.5?) | 120-150 | 85 | 15 | – | 4. 7. |
| 10. Drammen | D | 7 | 1.0 | 15-20 | 100? | – | – | 6. |
| 11. Glitrevann | G | 10x16 | 1.2-1.5 | 120 | 50 | 45 | 5 | 4. |
| 12. Sande | S | 12 | 0.5 | 55 | 15 | 85 | – | 4. |
| 13. Hillestad | H | 5 | 0.5 | 10? | 60 | – | 40 | 9. |
| 14. Ramnes | R | 12-16 | Ca.2.0 | 150-200 | 40 | 60 | – | 2. 9. |
| 15. Skrehelle | E | 7 | 0.5-1.0 | | 10 | | 90 | 11. |
| Probable cauldrons | | | | | | | | |
| 16. Stryken | Y | Ca.10 | 1 | – | 20 | | 80 | 7. |
| 17. Vealøs | V | 10-20 | ? | – | 5? | | 95+ | 11. |
| 18. Langesundsfjord | F | Ca.12 | ? | – | 1 | | 99+ | 23. |
| Ring complexes | | | | | | | | |
| 19. Nordlikampen | No | 5 | | | | | | 12. |
| 20. Gjerdingen | Gj | 4-5 | | | | | | 24. |
| 21. Katnosa | Ka | 9 | | | | | | 7. |
| 22. Kikut | Ki | 3-4 | | | | | | 7. |
| 23. Bjørnsjøen | Bj | 7-9 | | | | | | 7. |
| 24. Tryvasshøgda | Tr | 7 | | | | | | 20. |
| 25. Mykle | My | 18-22 | | | | | | 11. |

tectonic graben map, where the existence of several new cauldrons and ring structures is indicated. Several of his ring structures are here called cauldrons.

A recent re-interpretation by Oftedahl [23] permits a new general description of cauldron formation processes (see also the earlier account [4]). The cauldrons formed as subsided blocks when magma chambers 7-15 km in diameter stoped their way towards the surface. As the magmas came close to the surface, they seem to have lost some overpressure and allowed the caldera blocks to drop down to form subsiding pistons near the tops of the magmas. Largescale crustal movements in a tensional regime are thought to be responsible for this magmatic withdrawal. There seems to be no reason to assume that basaltic shield volcanoes existed on top of the cauldron blocks. Quite the opposite, after a long stage of gradual subsidence had started, the caldera basins were more or less filled by the basalts B3, the rectangle porphyries RP13, and younger rhomb porphyries. The formerly-postulated basalt volcanoes are excluded because sedimentary units of caldera depression type formed before or within the basalts. This is the case with the Bærum cauldron with mega- and mesobreccias in the eastern, firstsubsided part and several hundred meters of gravelly volcanic sandstone in the western part, both rocks formed within the basalts and during their formation. More ordinary caldera lake sediments formed in the Nittedal cauldron on top of a small ignimbrite and in the Drammen cauldron just before an ignimbrite. Extensive caldera escarpment breccias and ignimbrites may be present in the Kampen, Heggelia, and Svarten cauldrons.

Only the Ramnes and the Hillsestad cauldrons show thick sequences of ignimbrites, 1400 m and 300 m thick for the two cauldrons, respectively. For these cauldrons one may assume that the subsidence could have been balanced more or less completely by voluminous ignimbrite eruptions. A newly-found ignimbrite formation on top of RP26 confirms the relatively young age of the Ramnes subsidence.

Gravity anomaly analyses of eight cauldrons by Ramberg [1] show negative anomalies which correspond to small amounts of syenitic to granitic magmas existing under cauldron block. No marked positive anomalies exist. Thus the caldera blocks did not fall down into basaltic magma chambers but into the syenitic top layer of monzonitic chambers. The corresponding monzonite now gives no anomaly, having an average-crust specific gravity (2.72). In most cases more or less of granitic magma was formed on top of the syenitic magma coming out as ignimbrites and some late intercaldera plugs. The syenitic magma filled the ring fault and produced the ring dikes of the Øyangen, Glitrevann, Bærum, Nittedal and Ramnes cauldrons. In the Sande and Ramnes cauldrons a remarkable zoning from granitic or syenitic margins towards a

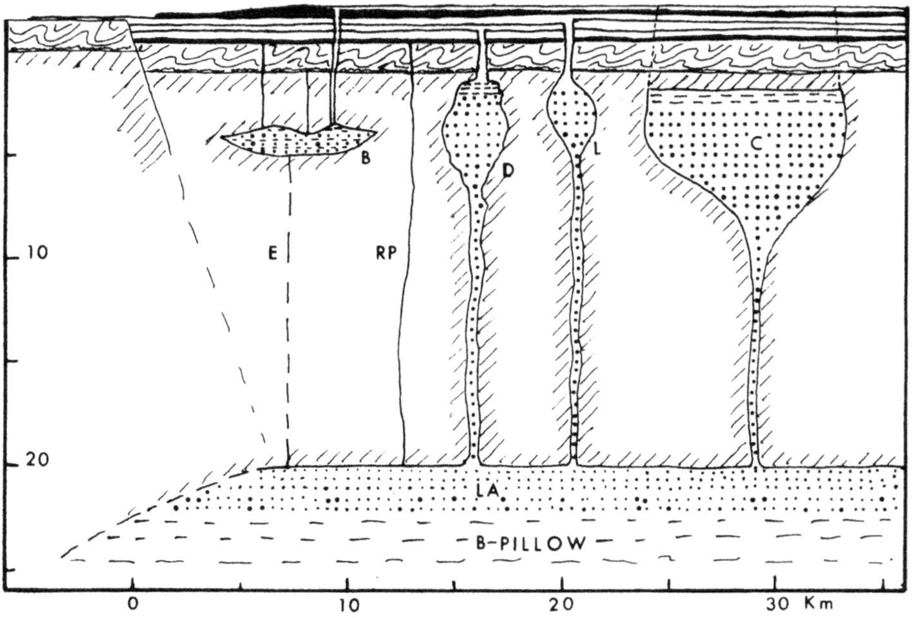


Fig. 2. Idealized section through central part of the Oslo graben block during B3 basalt eruptions and just before the caldera collapse. LA - Latitic magma differentiated on top of basalt pillow. E - Ascent of basalt magma to magma chamber B before latitic magma differentiated. RP - Fissure for latitic magma ascent to form early RP eruptions. RP eruptions also from surface-near latitic magma chamber (L) and similar chamber with uppermost differentiation into trachytic magma (D). C - Layered magma chamber of latitic composition with trachytic and rhyolitic top layers, just before the overlying crust subsides to form a caldera.

monzonitic center occurs [4,21]. This zoning is seen as a reflection of the layered magma chamber invading the block with stoping being most pronounced in the center so that the most-fractional layer was located in the margins.

Resurgent cauldrons are not found in the Oslo region, and the reason may be seen in the composition of the magma chamber below during caldera formation. This magma chamber was syenitic to monzonitic and this magma again was fluid enough to stope its way along ring faults and through caldera blocks without any noticeable doming.

## 3. MAGMATIC HISTORY

Some Rb-Sr ages published by Sundvoll [14] show that the magmatic history covers a much longer time span than earlier supposed, with eruptions of the volcanic rocks starting around 294 m.y. ago, the larvikites and nordmarkites dating to between 284 and 275 m.y., whereas the last little granite pluton is 245 m.y. old. A much better chronology will no doubt appear in the near future. As to relative ages of the plutons, field studies by Nystuen [12] and Naterstad [2] have shown that at least some biotite granites are very early and occur chronologically together with larvikites, thus violating the general rule inherited from Brøgger that on the whole the major age succession was from monzonitic through syenitic to granitic rocks. In this survey comments on ages will be restricted to the ones that rest on the most complete stratigraphy of the volcanic rocks, the Vestfold area. This history of early magmatism rests to a large extent on the general model resulting from Ramberg's gravity study: Magmatism developed from a basaltic magma pillow at the base of the crust, going from 20 km to 30 km in depth. The results are outlined below in form of a stratigraphic table, to be read from bottom up. It is partly based on trace element studies (Oftedahl, this report).

| Stratigraphy | Description |
|---|---|
| RP26 | During the RP26 flows subsidence of the Ramnes caldera block (and Hillestad?) into underlying granitic magma, which vented to the surface as ignimbrite. Remaining top magma (syenite) invaded ring faults to produce a ring dike. Later invasion of caldera block by zoned magma chamber and by a late granite magma to produce a rhyolite plug. |
| RP13-25 | Continued eruptions of RP flows, intercalated with several basalt flows from local magma chambers with syenitic magma uppermost. |
| B3 | Major phase of crustal disturbance. Crustal tearing gave copious B3 flows, distension allowed magma withdrawal in uppermost chambers and caldera collapses (at least seven calderas). |
| B2 | Fracture propagating downward to tap basalt sill. |
| RP1-12 | Fractures puncturing various monzonitic chambers, producing the various RP flows, each with their own phenocryst population. |
| B1 | Strong crustal movement over the whole middle part of the graben, with fissures suddenly penetrating to extensive basaltic, sill-like magma chambers, - the B1 eruptions. In the south some Skien basalts of unusual composition may have originated directly from the mantle. |

Early graben sedimentation

6. Monzonitic magma ascended through the crust to form local magma chambers.
5. The tectonism started with lateral tear faults and graben subsidence on normal faulting.
4. Formation of top layer of monzonitic magma in which rhomb crystals formed.
3. Fractionation of the top of the basaltic pillow, subtraction of olivine, clinopyroxene, and plagioclase.
2. Rise of basaltic magma to form near-surface sill-like magma chambers.
1. Basaltic magma from the mantle ascended and formed a huge sill at the base of a 20 km thick crust on top of a bulge in the Moho. This basalt pillow may have attained its 10 km thickness in the starting phase, or may have gradually built up during the later evolution outlined above.

## ACKNOWLEDGEMENTS

The present work was done while the author worked as a research associate of the Center for Volcanology, University of Oregon. I am indebted to Dr. B. Baker, director of the Center, and to Dr. S. Boggs, head of Department of Geology. The work was further supported by the Norwegian Research Council for Science and the Humanities and by the Fulbright-Hays travel fund. Dr. Gordon G. Goles is thanked for many discussions and for reading and improving the manuscript.

## REFERENCES

1. I.B. Ramberg, Norg. geol. Unders., 325, 1976.
2. O. Holtedahl, Norg. geol. Unders., 164, 1953.
3. C. Oftedahl, Permian rocks and structures of the Oslo Region, in: Geology of Norway, ed. by O. Holtedahl, Norg. geol. Unders., 208, 1960.
4. C. Oftedahl, Nor. Vidensk.-Akad. Oslo, I. Mat.-Natur. Kl., 3, 1953.
5. W.C. Brøgger and J. Schetelig, Geological quadrangle maps Kristiania and Hønefoss, Norg. geol. Unders., 1917.
6. O. Halsen, in: C. Oftedahl, Guide to excursion in the Oslo Region, Norg. geol. Unders., 212, 1960.
7. E. Sæther, Nor. Vidensk.-Akad. Oslo, Mat.-Natur. Kl., N. Ser., 1, 4, 1962.
8. C. Oftedahl, Nor. Vidensk.-Akad. Oslo, I. Mat.-Natur. Kl., 3, 1952.
9. C. Oftedahl, Geol. Rundsch., 57, 203, 1967.
10. P.W. Weigand, Nor. Vidensk.-Akad. Oslo, Mat.-Natur. Kl., N. Ser., 34, 1975.

11. T.V. Segalstad, Nor. geol. Tidsskr., 55, 321, 1975.
12. J.-P. Nystuen, Nor. Vidensk.-Akad. Oslo, Mat.-Natur. Kl., N. Ser., 33, 1975.
13. I.B. Ramberg and B.T. Larsen, Permian supracrustal rocks in Skrukkelia, northern Oslo Region, 1977 (in press).
14. B. Sundvoll, Rb-Sr ages of Oslo igneous rocks, Norw. Geotraverse Meeting, Oslo 1975.
15. E.H. McKee and D.A. Swanson, Geol. Soc. Am. Cordilleran Sect., 73th Annu. Meet. Abstr., 1977.
16. C. Oftedahl, The Oslo Region-northernmost Permian in Europe, NATO Adv. Study Inst., Mainz, 1975.
17. L. Størmer, Nor. geol. Tidsskr., 15, 43, 1935.
18. T.F.W. Barth, Nor. Vidensk.-Akad. Oslo, I. Mat.-Natur. Kl., 9, 1945.
19. A.B. Harnik, Schweiz. Mineral. Petrogr. Mitt., 49, 509, 1969.
20. J. Naterstad, Nytt fra Oslofeltgruppen, 1, 29, 1971.
21. R. Sørensen, Norg. geol. Unders., 321, 67, 1975.
22. B.T. Larsen, Oslo region tectonic map, 1:250.000, Norw. Geotechn. Inst., 1975.
23. C. Oftedahl, Cauldrons of the Oslo Permian Rift, J. Volc. Geotherm. Res. (in press).

# THE TECTONIC HISTORY OF THE OSLO REGION

Ivar B. Ramberg[1] and Nils Spjeldnæs[2]

[1] Department of Geology, University of Oslo, Box 1047, Blindern, Oslo 3, Norway
[2] Palaeocology Dept., Århus University, DK-8000 Århus C, Denmark

## 1. INTRODUCTION

The classical studies of the tectonics of the Oslo Region were by Brögger who in a series of papers towards the end of the last century [1-4] described the structural features in great detail. Later important contributions include those by Cloos [5], Størmer [6,7], Holtedahl [8-11] and Gleditsch [12]. In connection with the extensive studies of the alkaline igneous rock series of the region, numerous observations were made on the tectonics also, and various petrogenetic hypotheses were offered, especially by Barth [13-14] and Oftedahl [15-17]. In recent years special emphasis has been given to geophysics, gravity in particular. These data, and their bearing on our understanding of the geodynamic evolution of the rift, have been analysed by Ramberg [18]. The latter paper also includes references to much of the older literature, as well as appropriate descriptions and maps of the region. Recently, Ramberg and Larsen [19] reviewed the tectonomagnetic evolution of the rift, particularly discussing the lavas and the main, Permo-Carboniferous taphrogenetic event. The present paper is an extension of the latter two papers, and emphasizes the chronology of the development of the region, and succession and significance of tectonic events from Precambrian to Recent.

## 2. TERMINOLOGY AND TECTONIC SETTING

The Oslo Region is a non-genetic geographical term loosely referring to an area of about 11000 km$^2$ in Southern Norway (Fig. 1) in which the rocks are younger than the surrounding Precambrian terrain [18-20]. The Oslo Graben specifically refers to the relatively

*I. B. Ramberg and E.-R. Neumann (eds.), Tectonics and Geophysics of Continental Rifts, 167-194.*

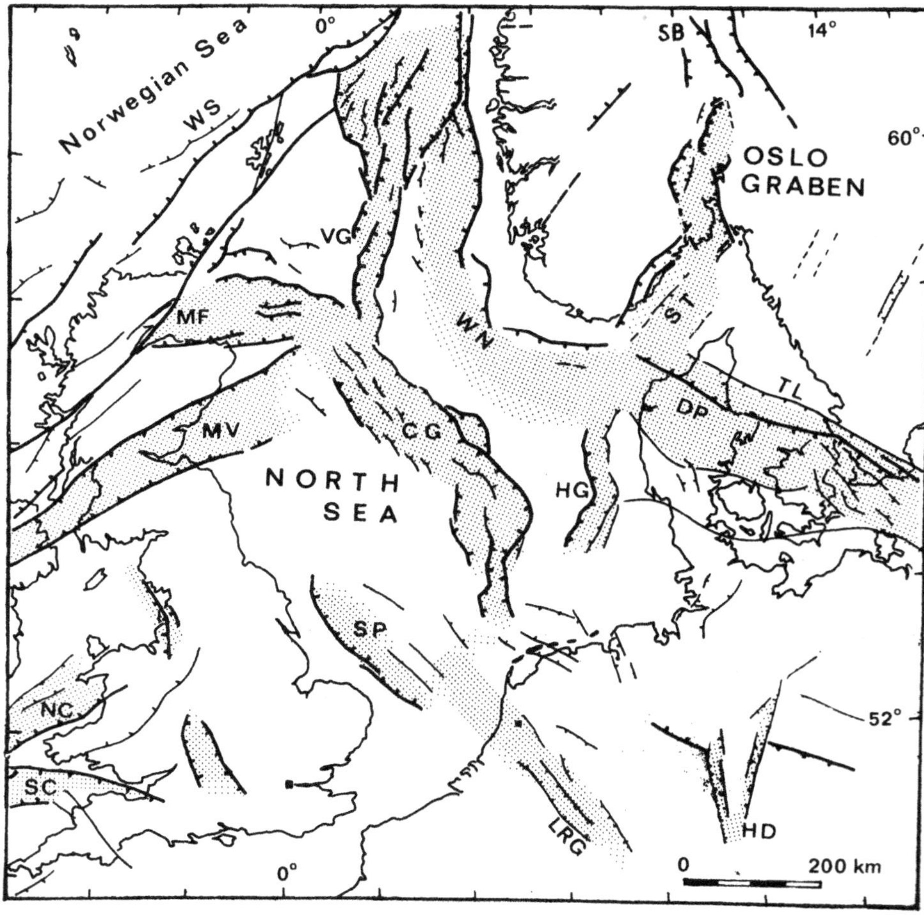


Fig. 1. Tectonic setting of the Oslo Graben. CG-Central Graben, DP-Danish Polish Depression, HD-Hessen Depression, HG-Horn Graben, LRG-Lower Rhine Graben, MF-Moray Firth Trough, MV-Midland Valley, NC-North Celtic Sea Trough, SB-Sparagmite Basin, SC-South Celtic Sea Trough, SP-Sole Pit Basin, ST-Skagerrak Trough, TL-Tornquist Line, VG-Viking Graben, WN-West Norway Trough, WS-West Shetland Basin.

depressed crustal block of about 220 km length on land. The graben itself represents only a part of the Oslo Rift, an extensive fracture zone connecting the Oslo Graben SSW-ward with the North Sea system of grabens, the development of which was also initiated in Late Paleozoic time [21].

The graben consists of two individual, north-trending segments arranged in an en echelon manner relative to each other. These two segments define the overall NNE direction of the rift [19]. Both graben segments are asymmetrically tilted towards the central N-trending Oslofjord-Randsfjord faults, which together have the characteristics of a scissor fault. To the SSE there is evidence of one or more graben segments below the Skagerrak Sea.

In detail, the graben and surrounding areas are criss-crossed by a number of major and minor faults and joints, forming a complex fracture pattern [22]. Whereas north trending fractures predominate in the central graben zone, the Precambrian terrain on both sides of the graben is characterized by NW and NE trends. The diagonal trends have been interpreted as an old regmatic shear pattern. Many of the Permian faults and dikes exhibit a strong influence by older structural trends.

## 3. PRE-PERMIAN DEVELOPMENT

The Pre-Permian sediments of the Oslo Region rest upon a peneplaned basement consisting of metamorphic Precambrian Rocks. This Sveco-Norwegian complex ranges from 1000 to 850 m.y. in age.

North of the Oslo Region is the infaulted Sparagmite basin (Fig. 1), also termed the Østerdal aulacogene [23]. The basin was formed in two phases during the late Precambrian. The first phase produced the smallest, but also deepest basin, which contains fluviatile conglomerate fans along its southern fault-line. The boulder contents and transport directions of these conglomerates indicate that the Oslo Region was a sediment-covered land at that time.

In the second phase, which was contemporaneous with the world-wide late Precambrian glaciation, the basin was wider, and shallower. The infilling of the basin with gradually more mature sediments was not completed until the Lower Cambrian [24].

The predominate structural trends in the Sparagmite Basin are NNW. Evidently the old faults were reactivated during later events, possibly during the formation of the Oslo Rift.

The formation of the Sparagmite basin has been related to the opening of the Protoatlantic [23], but the details of this event are not sufficiently known. Traces of alkaline basaltic volcanism have been observed in the basin [26] and in connection with the late Precambrian faulting [26,27].

During the Lower Cambrian, the Oslo Region was a positive area of low relief. Stratigraphic and lithological studies [28] indicate

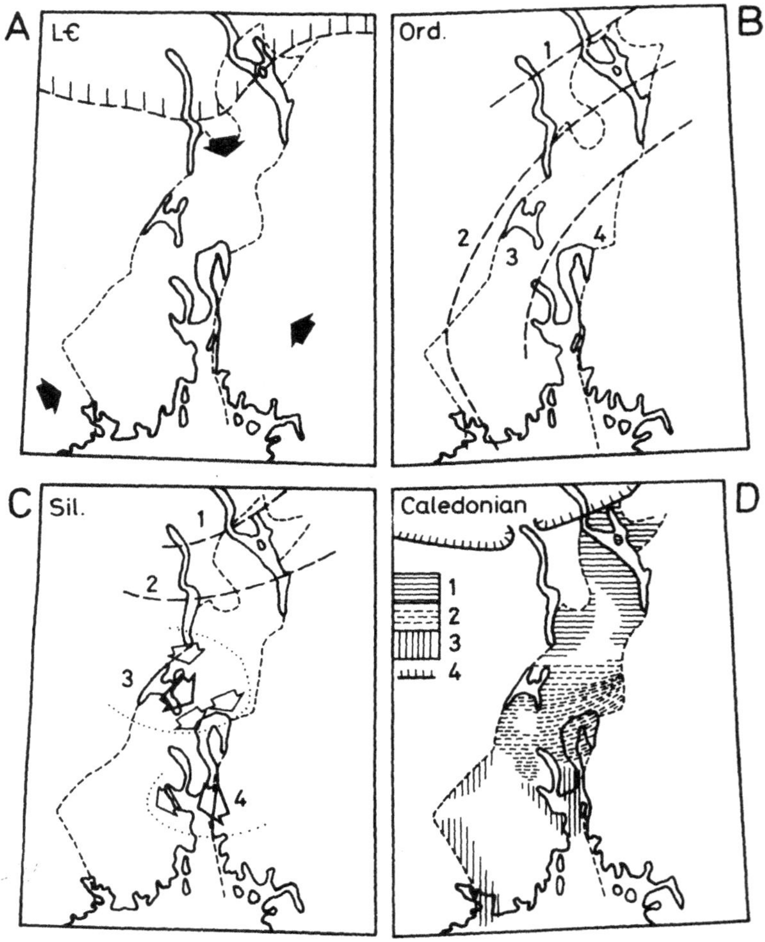


Fig. 2. Palaeogeographic development of the Oslo Region from Cambrian to Devonian. A. The Cambrian situation with the distribution of the Lower Cambrian in the N; arrows indicate possible directions of the Middle Cambrian transgression. B. Facies distribution in the Middle Ordovician: 1 - mainly clastic rocks, 2 - carbonate shale and siltstones (both 1 and 2 are shallow-water sediments with faunal affinities to Estonia), 3 - carbonate-marl zone, 4 - marl-shale zone, both with sediments deposited in deeper water, and faunas of british affinities. C. The Silurian situation. 1 - the zone of the Marienberg beds, where the old Red facies is oldest, 2 - the Bruflat stage, 3 - the Ringerike delta, with transport directions, 4 - the Jeløy-Holmestrand delta, with transport direction. D. Subdivision of the Caledonian folding: 1 - areas with intense folding, 2 - areas with moderate folding, 3 - unfolded areas, 4 - limit of the Caledonian nappes.

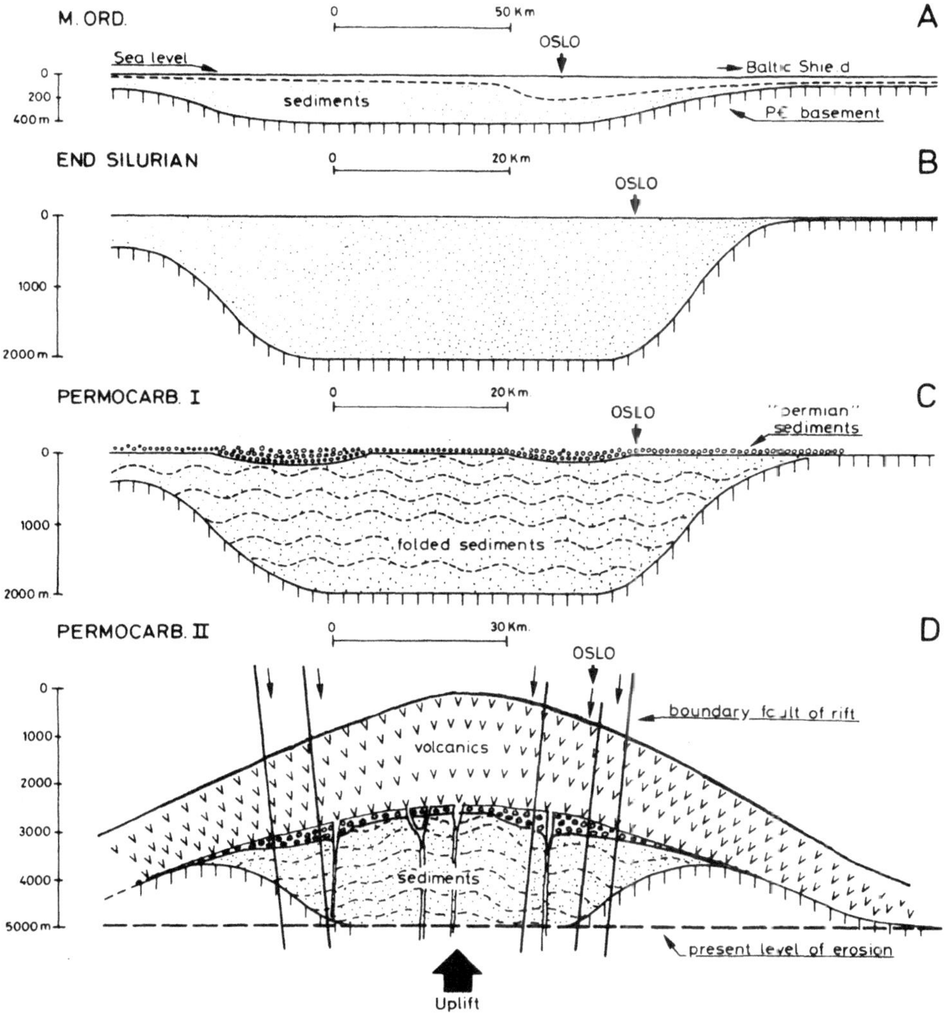


Fig. 3. Diagramatic cross-section of the Oslo Region. A. In Middle Ordovician, the N-S trending basin was with and shallow with the axis of the deposition west of the axis of greatest water depth. B. End of the Silurian; the basin was deeper and narrower, and most of the sediments to the east were removed by erosion. C. After the Caledonian folding, the basin had basically the same configuration, but was now filled with folded Cambrian to Silurian sediments and eroded down to the "pre-Permian" peneplain. On top of this, there was a thin, and irregular veneer of "Permian" sediments. D. In conjunction with the rift formation the area became uplifted (sketchily indicated). The outline of the graben coincided remarkably well with that of the sedimentary basin.

that the peneplain was broken up into a small scale topography along N-S faults, and that a jointing pattern, indistinguishable from that produced by the Late Paleozone (Oslo) rifting, was already in existence during the Middle Cambrian. The region was submerged during the time of the Middle Cambrian zone of Triplagnostus gibbus, which was also the time of widespread transgression all over southern Scandinavia. During the transgression, or earlier, the basement was denuded of older sediments, which are now found only as boulders in the basal conglomerates.

During the reset of the Cambrian and the Tremadoc, the Oslo Region was continuous with the Baltic areas, and even if the sediment thickness was somewhat higher in the Region, those few tens of meters additional sediment cannot be used as an indicator of a separate basin.

In the Arenig, the basin pattern became evident [29]. Along the western and northern margin of the Region shallow-water, mostly carbonate sediment are found, with deeper, and partly thicker ones in the centre. This pattern of basin structure persisted until the end of the Silurian.

During the rest of the Ordovician, the pattern resembled the one shown in Fig. 2B. In the north (zone 1), there are clastic rocks (silt, clay with subordinate carbonates), and in zone 2 carbonates dominate, with some silt and clay. In both zones the sediments were deposited in comparatively shallow water, and the faunas (and floras) show great similarity to those of Estonia. In zone 3 there are mostly marls and some carbonates, and in zone 4 there is a transition from marls in the west to dark shales in the east. These beds have been deposited in somewhat deeper water, and the faunas in zone 3 and 4 show striking similarities with Britain. The water depth was evidently greatest in the southeast, but the axis of highest deposition rate was to the west of the axis of greater water depth (Fig. 3A). During the Silurian, the same basin pattern prevailed, and the rate of sedimentation increased progressively with time. The shallow-water zones are marked with carbonates (reefs and coquinal limestones, with more lagoonal ones in younger beds) and sand-silt rocks, whereas graptolite shales and bituminuous limestones are typical of the eastern, deeper water facies. The rapid facies changes and the extension of the facies belts parallel to the subsequent Permian faults (Bockelie, this vol.), suggest the existence of small scale faulting within the Cambro-Silurian basins.

During the Ordovician, the directions of sediment transport were mostly from the southwest and west in the southern part of the Region, and from north and northwest in the northern part. In the marine Silurian, the transport was mainly from northwest and north. Sediments of Old Red type ended the "Caledonian" sedimentation. Sedimentation commenced in the north, the Brumunddal area, with the Marienberg beds of Lower Wenlock, or even uppermost Llandovery age, and continued southward with time, culminating with the Ringerike Group. In the type region the latter unit, which has been divided into two formations [30], is 1250 m thick. The transport direction in the type area is from northwest and NNW, and these directions are also found at Kolsås, near Oslo. In this northern estuary, the basal, fossiliferous beds seem to be Lower Ludlowian in age. Further south, at Jeløya and Holmestrand, the sandstone reveals transport directions from northeast and ENE [30]. The basal fossiliferous beds [31] seem to be low Downtonian in age. This represents the culimination of the development of the Oslo Basin, which appear to have narrowed and deepened progressively from the Lower Ordovician to the Downtonian. The Jeløy-Holmestrand estuary indicates that a land area existed to the east, and that there was no longer a sea connection to the Baltic area.

The Oslo basin must have been somewhat broader than the present Oslo Region, with a depth of over 2 km (Fig. 3B). There are some indications of volcanic activity, represented by bentonites with maxima in the Lower Caradoc, Upper Llandovery, and in the high Wenlock. The bentonites are widespread, even to Estonia and Bornholm, and do not indicate volcanic activity in the Region itself. However, great thickness (up to 1 m) of the Caradocian bentonites, found in the Oslo Region and in Västergötland, Sweden, may indicate that the centre of volcanic activity was not far away.

The Caledonian folding in the Oslo Region had its main phase after the deposition of the Ringerike Group. There are indications of slight tectonic movements at the end of the Ordovician [32], and perhaps also earlier, and the evolution of the orogen is reflected in the sediment geochemistry [33]. The earlier phases were weak, and are overprinted by the main phase, having the same compressional direction.

The style of folding is that of a Jura-type decollement. There are no indication that the Precambrian basement was involved south of the zone of nappe tectonics (4, in Fig. 2D).

In the north, the folding was intense, and mostly isoclinal, but with some synclines formed by thick competent limestone beds surrounded by strongly folded shales. This is even more marked in the Ringerike Group, where large open synclines demonstrate a clear contrast to the more intensely deformed marine shales and limestones. In front of the synclines (to the south), the soft marine beds are piled up as a result of intense folding, and on the north side older rocks may even be thrusted over the almost horizontal sandstones.

The folding dies out to the south, and it appears (cf. Fig. 2D) to extend farther south in the central part of the basin, and stopped earlier along the margins. There is no direct evidence for the age of the folding in the Oslo Region, as it postdates all the Pre-Permian sediments. Judging from the conditions in the mountain chain, where the main phase was pre-late Lower Devonian (even if there was some deformation reaching into the Late Devonian), it is supposed that the folding took place in the Lower Devonian.

After the folding, the sediments were peneplaned (cfr. [9]). The Permian rests partly on the Ringerike Group (in the synclines), and partly on the older marine sediments (in the anticlines). In the unfolded, southern areas, the Permian invariably rests on the Ringerike Group, and this indicates that the depth from the Pre-Permian peneplain to the crystalline basement is at least 2 km in the central parts of the basin, regardless of which sediments form the base of the Permian (Fig. 3C).

It is observed, therefore, that the Oslo Region was a sedimentary basin, with up to 2 km sediments which were eroded down to the Precambrian on parts of the east side in the Late Silurian. A possible explanation is to assume a dome-like uplift of about 2 km over the basins. Such domes are well known from active rifts, and its supposed absence was regarded as one of the peculiarities of the Oslo graben. Subsequent to rifting, the thermal uplift decreased and the dome seems to have been further removed by peneplanation, down to a level approaching that of the previous Precambrian peneplain (Fig. 3D).

## 4. THE PERMIAN EVENT

The formation of the Oslo Graben (Fig. 4) represents a major event which involved the development of extensional tectonics, volcanism and subsidence. The activity has generally been regarded as Early Permian, but a more extensive period of (repeated?) tectonic and volcanic activity seems now more likely in view of more recent dating results [34], tectonomagmatic [19] and general geological considerations.

### 4.1 Dating of activity

The first dating of the volcanic activity was based on paleontological evidence. The rather rich fossil beds found at Semsvik in 1931 [35], just below the lowest basalt beds, indicated the volcanism commenced in the Late Paleozoic, probably in the Early Permian. Analysis of the original descriptions of fossils (including fishes [36], plants [37-39] and non-marine lamellibranchs [40]), reveals that the evidence from fishes and molluscs give an age between Upper Carboniferous and Permian. The plant evidence

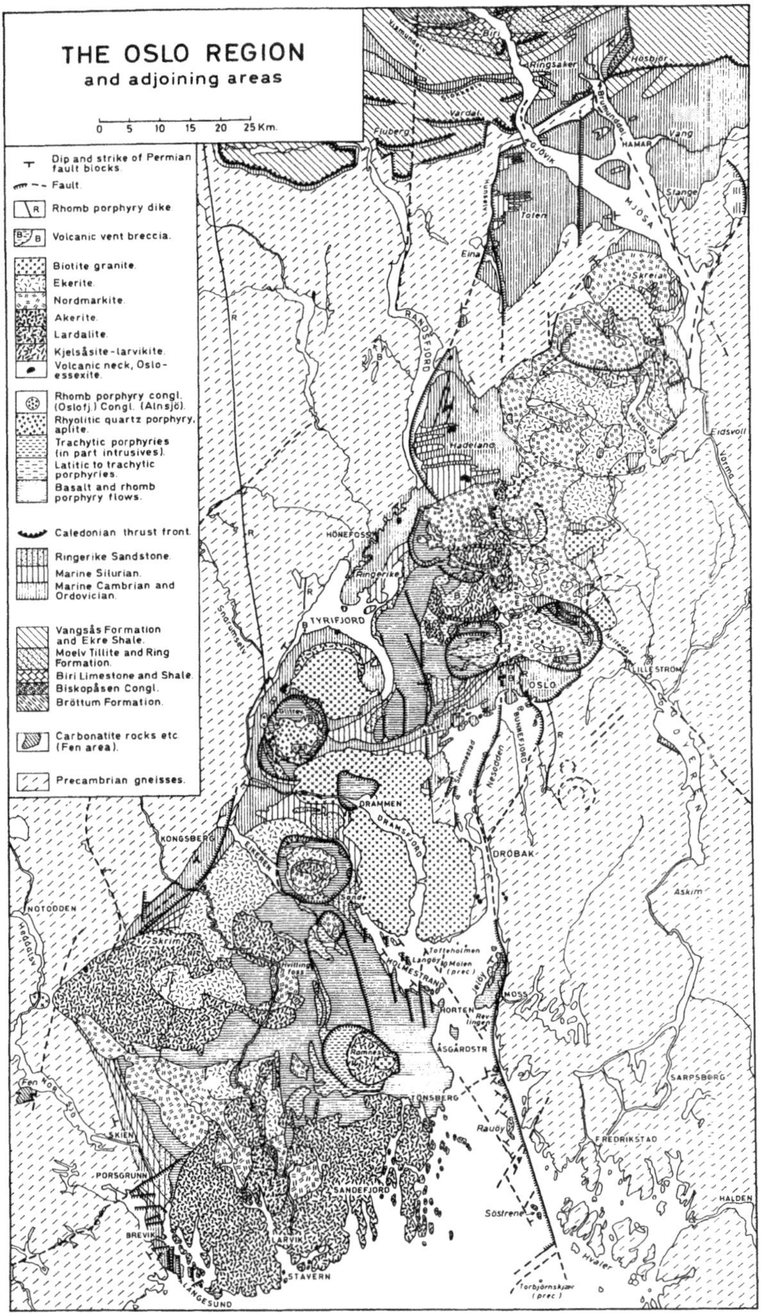


Fig. 4. Geological map of the Oslo Region (after [18]).

narrows the age to Rotliegendes (probably lowest part) and the uppermost Carboniferous, but a more precise age determination for the initial volcanism in the central part of the graben seems extremely difficult on the basis of the Semsvik macro-fossils. If they are of Permian age, they belong to the oldest part of the period.

Since the fossiliferous (and underlying) beds are rich in volcanic material, and since the lowest ($B_1$) basalt in the southern part of the graben is considered to be older than those at Semsvik, the definite possibility remains that the eruptive activity in the Oslo Region started already in the Upper Carboniferous.

The end of the activity is more difficult to define. It has been suggested [41] that the porous Brumunddal sandstone resting on the youngest rhomb porphyry lavas in the northern part of the graben (Fig. 4) is of Triassic age. This is based on a small number of eolian cross-beddings, which conform to the prevailing wind-directions of the British Bunter, and differ from that of the generally E to W Permian direction. The paleomagnetic directions [42] are normal, both in the sandstone and in the underlying, deeply weathered lavas, indicating a post-Permian (or Post-Kiaman) age for the remanent magnetism.

No ash which could have come from the Oslo eruptions has been reported from the Zechstein salt deposits in the North Jutland basin (Fig. 1). Since igneous rocks of the Oslo type are found not only in the Oslo Graben to the north, but also on the Fyn-Ringkøbing Ridge [43] and possibly in the boring C-1 and D-1 west of Jutland [44], the evaporite basin would have been a likely recipient for ash from the copious eruptions. The salt basin also shows a quiet development, indicating that there were no large-scale tectonic disturbances during the deposition of the salt. This seems to imply that the whole eruptive activity, and the rifting, took place during the Rotliegendes ($\sim$ 280 to 250 m.y.) and was finished before salt deposition began during Zechstein time. Alternatively, and more likely, there was a quiet interval in the activity during the Zechstein, with active periods both before and after. Pre $B_1$ faulting, for instance at the Krokskogen lava plateau [45], and downfaulting of the supposedly Triassic Brumunddal sandstone add to the impression of an extended period of repeated tectonic activity.

Radiometric age determinations, mainly by the Rb/Sr and the K/Ar-methods [34,46], generally support the latter hypothesis. The main magmatic activity apparently lasted from about 290 to 245 m.y., while especially K/Ar clay alteration ages on contact and dike-related deposits [46], extend this period to about 220 m.y. The same age (219 $\pm$ 6 m.y.) is also found for a diabase dike near Oslo [47]. A possible Triassic pulse of activity may have surfaced mainly as extensional faulting and emplacement of dikes.

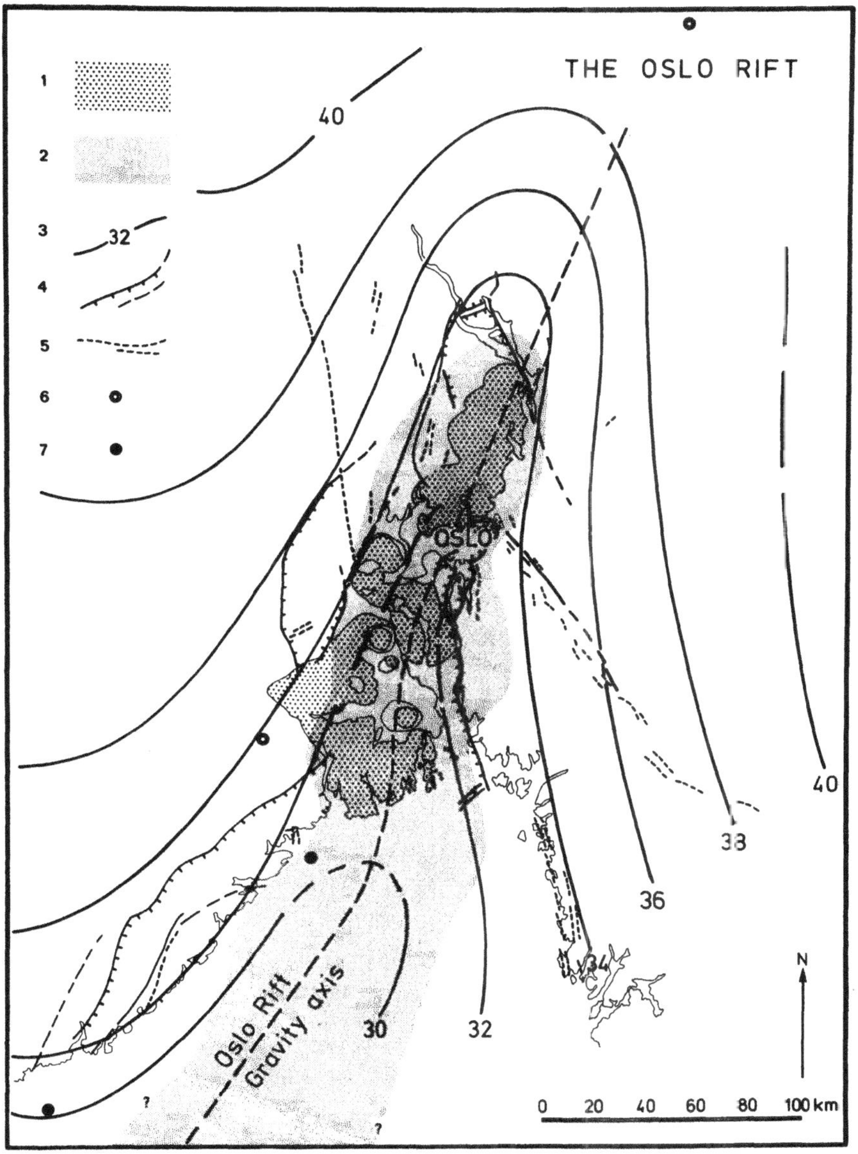


Fig. 5. The Oslo Rift with exposed intrusive rock complexes, outline of deep-seated mafic body, and crustal thickness indicated. 1 - Exposed plutonic rocks, 2 - Outline of deep-seated intrusion (at about 20 km depth), 3 - Moho contours, 4 - Faults, observed and inferred, 5 - Dikes, dike swarms, 6 - Peralkaline complexes (Fen, Särna), 7 - Submarin, mafic plugs or complexes.

## 4.2 Igneous activity

The impressive igneous activity of the Oslo Graben has been dealt with elsewhere [13-19,48], and in several papers in the present Proceedings, Vol. I & II. A short summary of some of the more pertinent facts will be given here.

The volume of surviving volcanic rocks totals about 500 $km^3$. The original lava cover presumably extended far outside the present graben boundaries, judging from the radiating rhomb porphyry and basic dikes (Figs. 4 & 5), the rhomb porphyry fanglomerate along the Oslofjord fault scarp [6,49], as well as a series of volcanic xenoliths in explosion breccias outside the graben area. Conservative estimates [15,18] imply that the original extrusive masses were well in excess of 10 000 $km^3$.

The lava stratigraphy (Oftedahl, this vol., Table II) reveals a sequence of interfingering basaltic lavas (B), rhomb porphyry types (RP), trachytes (T) and rhyolites (R). The abundant basaltic flows clearly predominate in the lower part of the sequence, whereas intermediate to felsic flows are increasingly more common towards the upper part.

Remnants of the lavas presently occur in five distinct regions or volcanic provinces [19]:

(a) The <u>Skien</u> basalt area, is built up of numerous early basalt flows ($B_1$), with an aggregate thickness of 1.5-2 km [18,50]. Situated in the far SW corner of the Oslo Region, the Skien basalts have been linked to early fissure eruptions, forming a stratoid trap series just outside the incipient boundary faults that were later concealed by intrusives. The Skien basalt series of a lower section of ankaramites or melanite-ankaramites (and melanite-nephelinite) and an upper section of basanites [50,51].

(b) The <u>Jeløya</u> area, another marginal province, is situated close to the Oslofjord fault scarp. It consists of several $B_1$ and rhomb porphyry flows intermixed with substantial amounts of agglomerates and other pyroclastic rocks. The Jeløya basalts are alkaline and commonly of mugearite type [19,51].

(c) The <u>Vestfold</u> lava plateau is the most extensive lava area in the Oslo Graben. Original thickness may have exceeded 3 km, and the approximately 130 m thick $B_1$ unit is followed by a large number of rhomb porphyry flows, stratigraphically higher basalts, trachytes and rhyolites [17]. The $B_1$ series consists of 10-15 flows of alkali olivine basaltic to hawaiitic composition [51].

(d) The Krokskogen lava plateau (location see Fig. 2), consists like Vestfold of rhomb porphyry lavas and occasional basalts and trachytes (partly ignimbrites). Original thickness possibly exceeded 1.5. The lower part of the succession in both plateaus probably represents flood volcanics, or products of large shield volcanoes, or both. In Krokskogen the $B_1$ consists of single aphyric flow with composition tending towards the oceanic tholeiites. This is in contrast to all the $B_1$ flows to the south (a-c), showing alkali basalt affiliations.

(e) The cauldrons (about 12), which contain predominantly basaltic rocks (of the central volcanic stage) and trachytes and alkali rhyolite ignimbrites (of the subsidence and caldera-filling stage). The caldera volcanics represent the higher parts of the lava stratigraphy. Basalts and felsic volcanics interfinger with the upper rhomb porphyry flood lavas, thus indicating that the flood (or plateau) lava stage was not finished before the onset of the central volcano/cauldron subsidence stage.

These provinces are composed of different combinations of volcanics that can be related to various geological environments with respect to the graben [19]. Whereas volcanism in the Oslo Graben commenced with basalt eruptions, except in the northernmost part, these early basalts (designated $B_1$) reveal a striking change in number of flows, aggregate thickness and composition, generally from SW towards NE. Chemically the $B_1$ basalts seem to have proceeded from the undersaturated ankaramites (Skien area) through basanites (especially in the central part) to tholeiites (Krokskogen plateau) [19,50,51]. Geological evidence imply that the southern Skien basalts are the oldest and the more evolved Krokskogen basalts (tholeiites) the youngest of the $B_1$ basalts, but all were erupted before the appearance of the first rhomb porphyry flow ($RP_1$). Sr isotope ratios (B. Sundvoll, pers. com.) indicate no direct genetic relationship between the $B_1$ basalt and the rhomb porphyries.

Chemically the rhomb porphyries are strikingly homogeneous both in major and trace element content [19]. Although the various RP units are often uniformly thick and quite extensive, there is again greater tendency for the more copious flows to occur in the southern areas than in the northern ones.

The stratigraphically higher basalt ($B_2$-$B_5$) which erupted mainly from central-vent volcanoes, range in composition from alkali olivine basalts to olivine tholeiites. The $B_3$ basalts are associated with a major phase of crustal distrubance [16] and are typically found interbedded with felsic flows and sedimentary breccias within cauldrons.

The plutonic rocks [e.g. 13,14,16-18], which cover about 60% of the graben area, have been divided into (1) a main series of deep-seated to subvolcanic rocks ranging from dioritic to (alkali-) granitic and nepheline-syenitic composition, and (2) the "Oslo-essexites" (or old volcanic rocks), comprising rocks of generally ultramafic to monzonitic composition. The main plutonic series is found within three subregions (Fig. 4): (a) the northern region of largely (alkali-) syenitic to granitic rocks, (b) a central region predominantly consisting of biotite granites, and (e) the southern region which contains predominantly monzonitic rocks. Thus, the frequency distribution of the exposed Oslo plutonic rocks [14] reveals a striking predominance of intermediate to felsic rocks. Mafic to ultramafic intrusives constitute a negligible amount of only 0.2% of the area. These areal distributions, which in turn led to the hypothesis that the igneous rock series was formed by anatexis and remobilization of the Precambrian crust, are, however, not representative of the volume proportions derived especially from gravity studies and are not supported by isotope compositions [52-54].

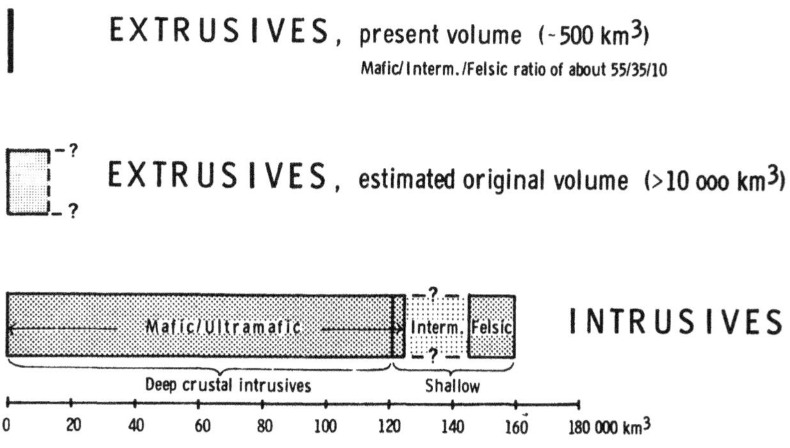


Fig. 6. Volume relations of the Oslo Region extrusive and intrusive rocks (from [18]).

## 4.3 Geophysical evidence

Gravity, magnetic, heat flow and seismic data from the Oslo Region [55] have provided a basis for a series of integrated geophysical models of the deep-seated foundering of the graben, of the lithosphere and the asthenosphere. Here, emphasis will be on gravity and seismic models which have provided estimates of the degree of lithosphere attenuation and on the amount and nature of the intruded masses in the lower crustal part of the graben [18,56]. Inversion of seismological data from the large seismic array, NORSAR, situated over the northern end of the graben [57,58], has given a three-dimensional image of the lithosphere below the array.

In contrast to the rather irregular outline of the graben, a distinct gravity high follows the overall NNE trend of the rift [18]. This gravity anomaly (about + 150 mgal relative to central parts of Norway), although overprinted by local anomalies closely associated with the near-surface felsic intrusives, reveals a substantial mass anomaly at depth. Combined gravity and seismic modelling [18,56,59,60] implies a marked crustal thinning along the entire rift axis of about 6-8 km (Fig. 5). In addition, dense rocks (ca. 3,12 g/cm$^3$) are present in the lower crust between depths of about 20 and 32 km, the latter number being equivalent to the elevated Moho interface below the graben. This body has been interpreted as a paleorift "cushion", originally formed as a result of fractional melting within the decompressed upper portion of a rising asthenolith or "blob".

Minimum mass estimates [18] show that the deep-seated intrusive body of presumed mafic to ultramafic composition along represents a colossal total mass of at least $3.75 \cdot 10^{20}$ g (see Table I). Evidently the felsic rocks that predominate among surface exposures are exceeded in volume by the dense rocks at depth by a ratio of about 10 to 1 (Fig. 6). Thus, volumetrically there is no objection to the idea that the intermediate to felsic rock series originated by magmatic differentiation from a mantle-derived gabbroic source, an interpretation which accords with Rb/Sr and other isotope studies [34,52-54]. The deep seated dense mass is located right below the felsic rocks and closely followed the axis of the graben (Fig. 5). It also follows the zone of crustal attenuation. Although originally emplaced as a basaltic fluid during the active phase of the rift, the paleorift "cushion" (or pillow) is believed to consist presently of a solid, inhomogeneous mixture of pre-existing crust and various meta-intrusives, differentiates and cumulates.

The northern end of the graben is surrounded by the large seismic array NORSAR. A method of direct inversion of travel time data [57] has provided a 3-dimensional model of the lithosphere below the array. Expressed as velocity anomalies in a number of

Table 1: Anomalous mass (Δ M), total mass (M) and volume (V) estimates of Oslo Region plutonic rocks.

| Rock Complex | Δ M (grams) | M (grams) | V ($cm^3$) |
|---|---|---|---|
| Nordmarka- Hurdalen | $11.77 \cdot 10^{17}$ g | $257.0 \cdot 10^{17}$ g | $98.09 \cdot 10^{17}$ $cm^3$ |
| Drammen granite | 2.35 " | 47.3 " | 18.11 " |
| Finnemarka granite | 0.44 " | 8.8 " | 3.36 " |
| Eikern-Skrim granite | 1.79 " | 35.9 " | 13.76 " |
| Siljan syenite | 0.26 " | 6.7 " | 2.56 " |
| Cauldron plutons | 0.61 " | 12.1 " | 4.57 " |
| Sum felsic rocks | $17.22 \cdot 10^{17}$ g | $367.8 \cdot 10^{17}$ g | $140.45 \cdot 10^{17}$ $cm^3$ |
| "Oslo-essexites" | 0.015 " | 0.12 " | 0.047 " |
| Narrefjell mafic intrusive | 0.69 " | 13.1 " | 4.57 " |
| Asker-Lier - " - | 0.38 " | 7.2 " | 2.50 " |
| Horten-Tønsberg - " - | 3.62 " | 69.8 " | 24.15 " |
| Deep crustal intrusive | > 121.0 " | > 3750.0 " | >1210.0 " |
| Sum mafic rocks | > 125 $\cdot 10^{17}$ g | > 3780 $\cdot 10^{17}$ g | >1241 $\cdot 10^{17}$ $cm^3$ |

lithospheric layers from the surface and down to 126 km, the model reveals no clear-cut correlation between the velocity anomaly pattern and gravity anomalies which closely outline the graben features. This apparent lack of correlation is further discussed in Husebye, England and Ramberg (this volume) who points out variations in density are not necessaryly related to the associated P-wave velocities in any simple manner. Another possiblility is that the dense crustal intrusive is compensated by a mass deficiency within a wide area in the lithospheric upper mantle, a situation which suggests that the dense body is more easily picked up by gravity than by teleseismic means. Spectral ratio analysis of long-period P-wave recordings has been applied for detailed Moho mapping in the NORSAR area [58]. The study shows that within the NORSAR area there is an increased crustal thickness from the northernmost part of the graben westwards. This is in accordance with the general picture (Fig. 5) of a thicker crust on both sides of the graben.

## 4.4 Progression of rifting and volcanism

The tectonomagmatic evolution of the Oslo Rift may schematically be represented by a series of stages (Fig. 7). These evolutionary stages did not occur simultaneously along the entire graben, but seem in general to have occurred progressively later from SSW to NNE.

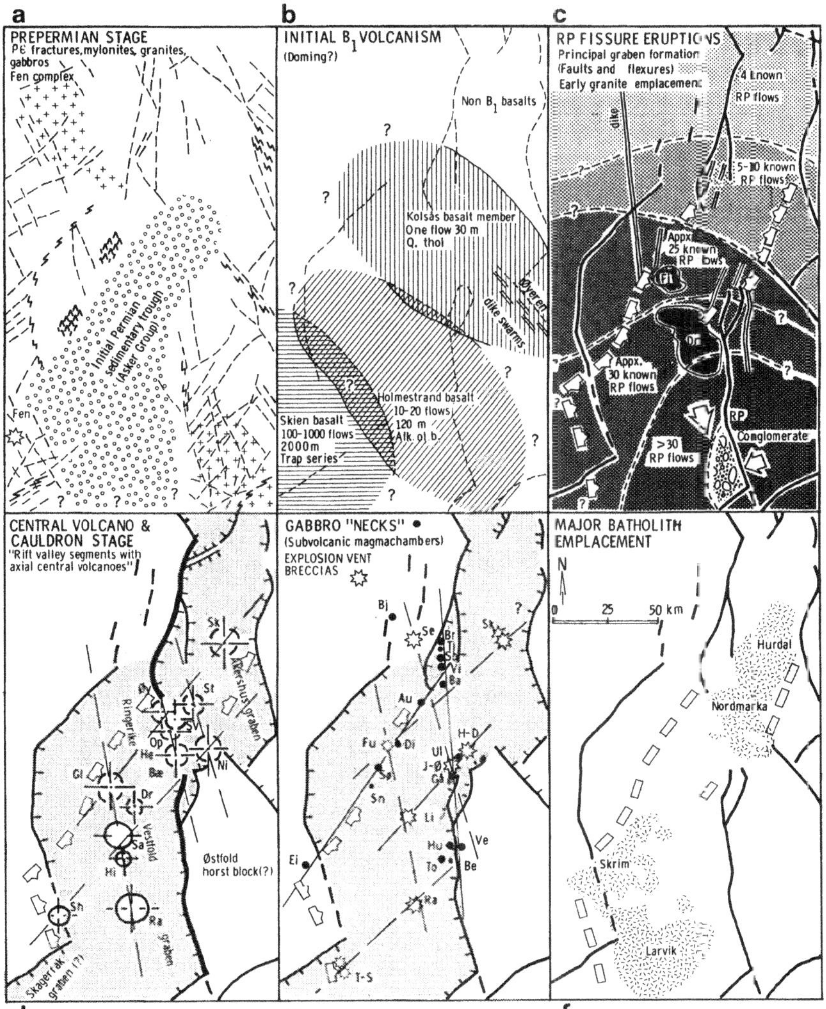


Fig. 7. Igneous and tectonic evolution of the Oslo Graben (schematic). faults, flexures, dikes, sed. transport directions.
Biotite granites: Fi-Finnemarka, Dr-Drammen
Cauldrons: Sk-Skrukkelia, St-Stryken, Ni-Nittedal, Øy-Øyangen, Sv-Svarten, He-Heggelia, Op-Oppkuven, Bæ-Bærum, Gl-Glitrevann, Dr-Drammen, Sa-Sande, Hi-Hillestad, Ra-Ramnes, Sh-Skrehelle
Gabbro necks: Br-Brandbukampen, Ti-Tingelstad, Sb-Sølsberget, Vi-Viksbergene, Ba-Ballangrudhøgda, Ul-Ullernåsen, Gå-Gåsøya, Au-Aurenhøgda, Di-Dignes, Sø-Sønsterbyflakene, Sn-Snaukollen, Ei-Eiangen, Hu-Husebykollen, Ve-Vestby, To-Tofteholmen (and Randvikholmene), Be-Bevøy, Bj-Bjonvika. Explosion vent breccias: Sk-Skrukkelia, Se-Sevaldrud, H-D-Holmen-Dagali, J-Ø-Jar-Øraker, Fu-Furetangen, Li-Lindum, Ra-Ramnes, T-S-Tveitan-Stokkevannet.

Stage 1 (Fig. 7a) represents a general pre-rift stage. Repeated deformational and intrusive events have imprinted a characteristic fracture pattern upon the brittle upper crust. A shallow Late Carboniferous/Early Permian sedimentary basin follows the same trend as the previous Cambro-Silurian basins as well as the overall outline of the future graben. Pyroclastic beds in the upper part of the sedimentary sequence (the Asker Group) indicate that the transition into the next stage has started.

At the onset of rifting, the coincidence between many of the graben features and older tectonic lines reflects the response of the structurally anisotropic continental crust to tensional stress. While the overall position and regional trend of the rift is mantle controlled, the near-surface fracturing was evidently controlled in part by the structural grain of the crust.

Stage 2 (Fig. 7b), initiated about 290 m.y. B.P. [34], is characterized by basalt volcanism ($B_1$). Although direct evidence is lacking (except perhaps in restricted areas in the northern part [18]), a broad domal uplift is envisaged to have occurred at this stage, heralding the subsequent crustal subsidence and intense volcanic activity.

Starting in the Skien area in the SSW, we find a distinct change in number of flows, aggregate thickness and composition of flows. These changes further substantiate a younging of the initial $B_1$ eruptions towards the NNE, and a migration of the volcanism towards more central parts of the rift. The evolution of the basalts, from the primitive undersaturated derivatives of the thick Skien stratoid series, through basanites in the central areas, to the single tholeiitic flow of the Krokskogen plateau, seems to reflect the rise and lateral movement of magma along the developing rift.

Stage 3 (Fig. 7c) is characterized by the continuation of fissure eruptions, but now predominantly of the intermediate rhomb porphyry type. The rhomb porphyry eruptions may have begun at the same time all over the area, or slightly earlier in the south. Again, the number of flows and aggregate thicknesses decreases drastically towards the NNE.

The rhomb porphyry stage coincides with the main phase of normal faulting and the development of a "rift valley" morphology. The principal Oslofjord fault developed at this time, the largest vertical displacements (> 3 km) occurring in the south. The formation of the fault scarp caused the accumulation of rhomb porphyry and other lava debris in the low areas inside the graben.

Stage 4 (Fig. 7d). The transition from the stage of rhomb porphyry fissure eruptions to the central volcano and cauldron stage, which took place somewhere around 280 m.y. B.P., is well documented in the field by rhomb porphyry flows interfingering with basaltic and felsic flows. It is also recognized as a transition from linear to circular tectonics, reflecting the reduction of tensional stress. This stage may prove to be typical of aborted rifts and paleo-rifts in comparison to actively widening rifts. The reduced extension rate (or no extension at all) may also favour the formation of more felsic differentiation products, as seen in many of the continental rifts.

In the Oslo Graben the alignment of the central volcano/cauldron complexes defines a strikingly regular pattern which is parallel to major and minor faults of the two graben segments, which are now fully developed. The somewhat regular spacing of the volcanos is about 28 $\pm$ 10 km [19], which might approximate the thickness of the rigid roof above the intrusive mass [61]. This number compares well with the geophysical indications [18] of about 20 $\pm$ 4 km to the top of the dense intrusive body. The width of the Oslo Graben (40-60 km) is also consistent with a thickness (of rigid layer) no more than about 25 km, assuming boundaries of normal faults with an overall 45$^{\circ}$ dip.

The formation of the central volcanoes throughout time may also have progressed from south to north, as exemplified by the cluster of cauldrons just north of the city of Oslo (Figs. 2 & Fig. 7d). There the cauldron to the north always cuts the nearest one to the south. Stratigraphical evidence [19,45] supports the inferred age relations.

The emplacement of the gabbroic to syeno-dioritic volcanic necks ("Oslo-essexites") and various volcanic vent breccias (Fig. 7e) is believed not to represent a specific stage of its own. The volcanic necks are always found in the periphery of the graben and are generally considered to predate the emplacement of the major felsic batholiths, but otherwise there have been no precise absolute or relative datings of these intrusions. Contrary to previous views, it has been suggested [19,62] that the bulk of the necks formed during the central volcano stage. The volcanic vent breccias which spatially are closely connected with the cauldrons or central volcanoes, probably represent a late, violent phase in the central volcano and (subsequent) batholith stages.

Stage 5 (Fig. 7f) is characterized by the emplacement of the major composite batholiths. The process was certainly already in progress during the previous stages (3 & 4) and especially the biotite granites in the central part of the region (Fig. 2), evidently represent an early intrusive phase (based on radiometric dating [34]). However, the majority of the felsic rock series intruded (or crystallized) some 15 to perhaps 45 m.y. later than the initial B 1 volcanism [34], and gradually obliterated much of the earlier-formed features. To the north the largely syenitic Nordmarka-Hurdalen batholiths belong to this stage, and to the south we find the Larvik and Skien batholiths, predominantly monzonitic in composition, cutting far into the Precambrian (southwestern) shoulder. All these complexes typically contain a number of cauldrons or cauldron remnants.

Although the various plutonic rocks are cut by a number of minor faults and joints, major faults are lacking except in the old granitic complexes. This reinforces the conclusion that most of the plutonic rocks intruded toward the end of a long period of tensional faulting and volcanic activity.

Stage 6. Perhaps after a period of quiescence (around Zechstein time, or abroad 240-220 m.y. B.P.), a terminal stage of faulting and dike intrusion may have occurred in the Early Triassic [34,42, 46,47]. The dikes are mafic in composition, similar to the diabase dikes of the previous stages, and indicate that mafic source rocks persisted throughout the entire period of rifting.

The evolution of the rift may also be illustrated by means of a series of schematic cross-sections (Fig. 8). The anomalous rise of isotherms in pre-Permian time resulted in partial melting at successively more shallow levels and by discharge of basaltic material into the base of the lithosphere (and crust). During the active phase of the rift, the front of this intrusive complex represented the boundary between the rising asthenospheric diapir and the gradually thinned lithosphere. The position of the dense intrusive complex may, therefore, be taken as a measure of the degree of evolution. Whereas in fully developed spreading axes this boundary has reached the surface, completely separating the plates of continental lithosphere, the process never came to a fullfillment in the Oslo Rift, where the interface rose to about 20 km below the present surface. This stage of evolution is perhaps equivalent to the present situation in the Kenya Rift and its northward continuation in the Ethiopia Rift. In the active Rio Grande Rift (USA) we probably see a less mature rift, with a depth to the dense intrusive body of about 28 km [63,64], but with evidence of incipient differentiation and emplacement of near-surface batholiths [65].

The asthenospheric diapir in the Oslo Region may be viewed as a short-lived mantle plume, or blob, similar to the one described from the Baikal rift by Logatchev and coworkers (this volume). Such a blob (or series of blobs) would explain the initiation and subsequent cessation of the rift. A SSW-ward movement of the lithosphere relative to the blob, or to the tear-drop shaped blob, would also explain the observation that the most prolific activity occurred at the "early" (SSW) end of the rift and that the progression of rifting and volcanism, accompanied by increased differentiation, took place towards NNE.

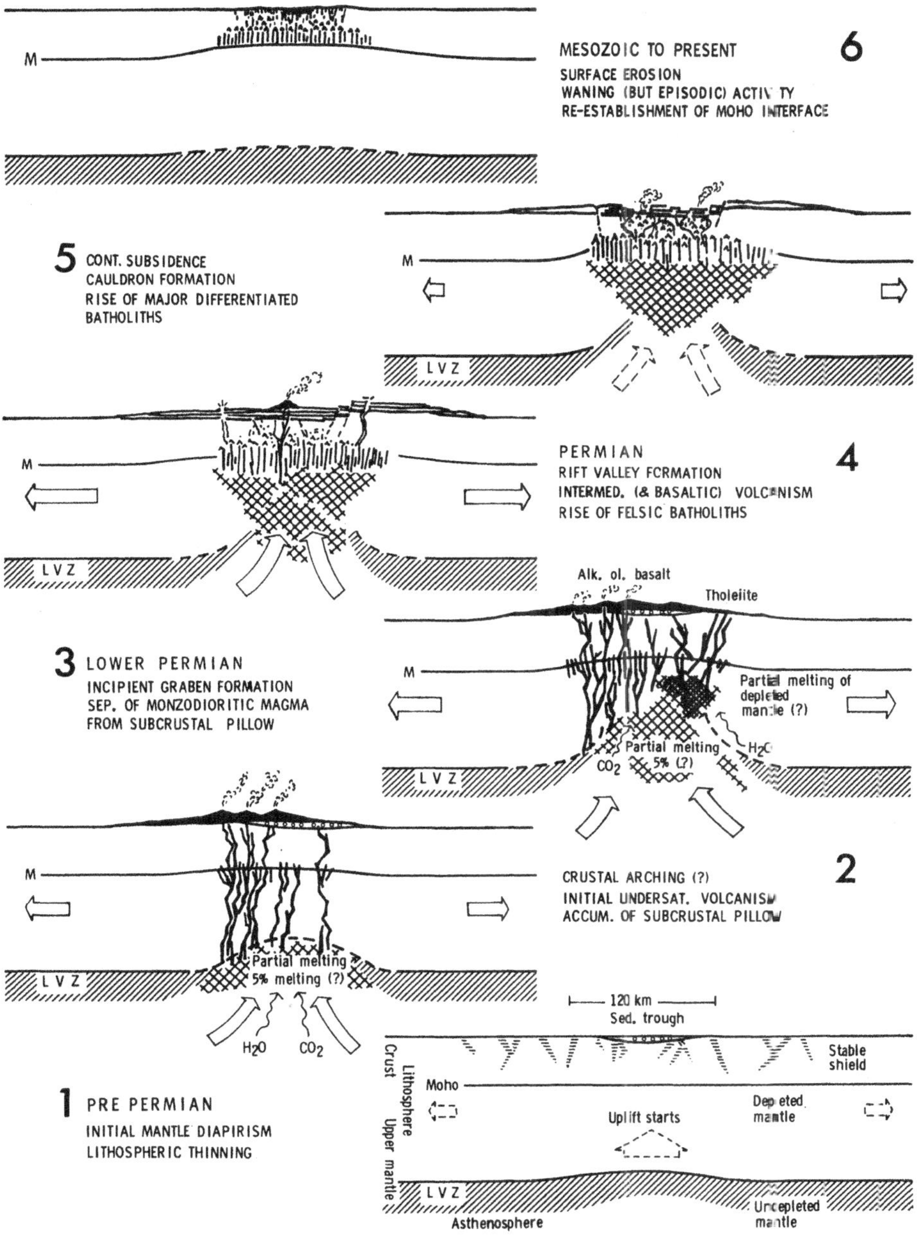


Fig. 8. Evolutionary stages of the Oslo Graben tentatively illustrated by a series of cross-sections (after [19]).

The mantle-derived partial melt provided the material found in the basaltic flows and in the subvolcanic "Oslo-essexite" necks. The same source-magma also generated a monzodioritic magma by crystallization that took place most rapidly near the relatively cold upper part of the body. The monzodioritic magma existed during the middle and final stages of the volcanic period. We deduce this from the frequent rhomb porphyry flows interfingering with both the basalts and the later trachytic to rhyolitic flows. The monzodioritic reservoir itself gave rise to most of the monzonitic, syenitic and granitic bodies that collected in the upper part of the crust.

## 4.4 Structural trends and directions

The fault pattern in the Oslo Region largely consists of a predominant N-S direction combined with two 'diagonal' trends (NN-SE and NE-SW) [22]. All three directions are partly inherited from the Precambrian basement; the faults seem to have been active intermittently during the Late Precambrian, and probably also during the post-Permian.

These features are especially well demonstrated for the great fault SW of the graben [66] (see Fig. 5). On a smaller scale, the parallelism between Permian faults and older Precambrian features can also be observed for instance at Nærsnes (near Slemmestad, Fig. 4) some 20 km SSW of Oslo. Here the fault pattern, with directions parallel to the Permian directions, form characteristic three-sided depressions in the fine-grained Precambrian rocks. One of the three-sided cavities has been found filled with Middle Cambrian fossiliferous sediments [67], indicating the fault pattern was in existence in the Early Paleozoic and that it was reactivated during the subsequent rifting.

Because of the predominance of N-S trending faults, it has generally been supposed that the axis of tension (or least compression) generating the rift was in an E-W direction. An analysis of the intricate fault pattern in the vicinity of the city of Oslo, and especially the trace of the Tyveholmen rhomb porphyry dike (the "type rock" of RPs), suggests that the tensional ($\sigma_1$) axis was probably directed somewhat to the north of west, more in harmony with the overall NNE trend of the graben. The trace of the Tyveholmen dike (Fig. 9) reveals that it has been repeatedly displaced in a direction parallel to the Caledonian fold axes.

It is, however, not unlikely that the dislocation pattern (in Fig. 9) may be derived from different episodes of crustal movement. That such events actually took place and also involved contrasting stress fields, follows from the fact that several zones of Permian (?) thrusting and folding occur with axial planes

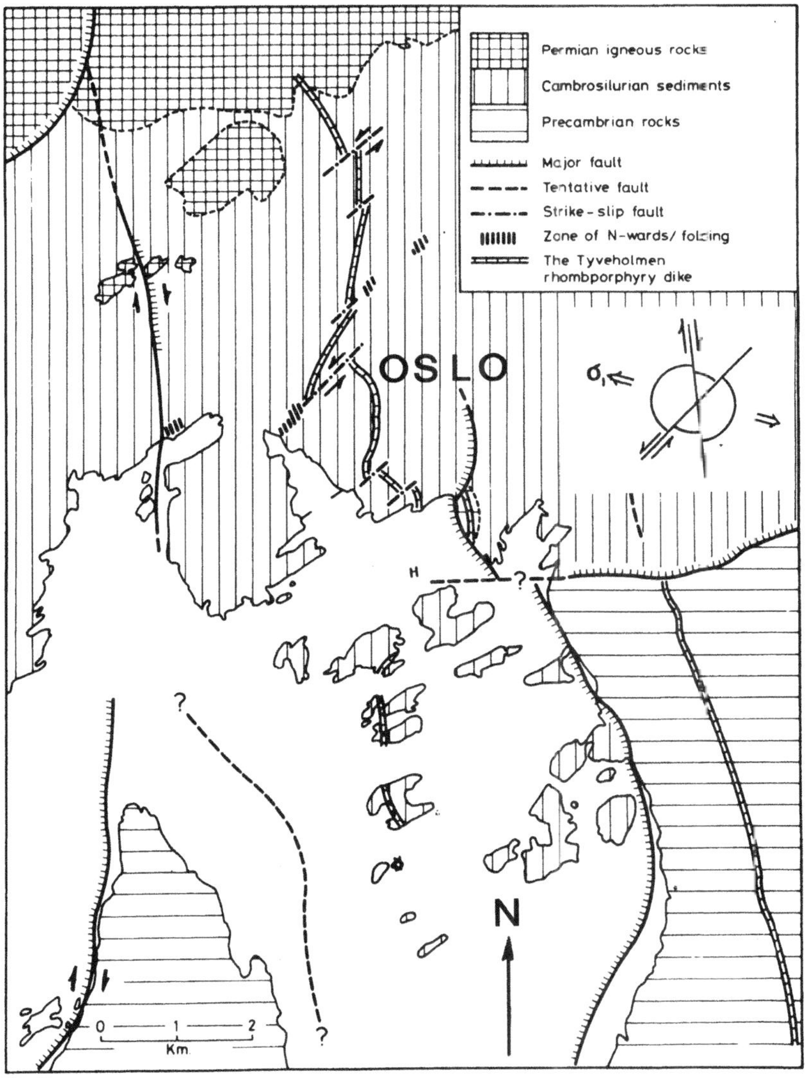


Fig. 9. Tectonic setting of the Tyveholmen rhomb porphyry dike (modified from the Geological Map of the Oslo area, in scale 1:50 000, by Holtedahl & Dons, 1953). Several oblique faults cut the dike; relative movement along major N-S trending faults is also indicated. A tentative stress field solution is shown (insert). Also note that the Tyveholmen dike, which is cut by several faults, was itself injected into the intrusive (nordmarkitic) massif to the north. Its continuation S-ward on the islands is complicated by largescale block-faulting.

of folds generally striking in the NE-SW direction (Fig. 9), i.e. parallel to the oblique faults that offset the dikes of the area. In some cases Permian sills have been folded or thrust indicating a syn- or post-Permian age for the compressional movements [e.g. 10]. Such folds seem to have a NW-ward vergence in contrast to the Caledonian folds which commonly face SE-ward.

## 5. POST-PERMIAN ACTIVITY

As discussed previously, the main rift-forming phase culminated in Permian time. Except for a possible extension of the activity into the Triassic, there are few direct signs of later tectonic and igneous activity within the graben area proper. Along the Norwegian Skagerrak coast, however, i.e. the SSW-ward extension of the rift zone, there are geological and geophysical indications of Cretaceous and Tertiary activity, including Tertiary dolerite dikes [68], post-Jurassic basalt [69] and possible Tertiary volcanic centres in the Skagerrak [70]. Within the graben area, Sæther [71] showed that in contrast to most of the dike rocks in this region, there exists a "late" group of diabase dikes which are conspiceously less altered, and which may be related to Mesozoic and Tertiary events in the North Sea and Skagerrak areas.

The western part of the Fennoscandian Shield was uplifted in connection with the opening of the Norwegian Sea. This uplift can be recognized not only in the topography, but also in the altitude of the Pre-Cambrian peneplain [10]. The exact age of the uplift has been debated, but may be as late as late Tertiary [72] or as early as early Tertiary. The peneplain is almost perfectly flat in the east, but west of the Graben rises with a constant gradient towards the coast. The Oslo Region apparently acted as a type of hinge-line for the uplift in this area, but the geological and geophysical consequences have not been studied properly.

Repeatedly during the Pleistocene and probably late Pliocene, the Oslo Region as part of the Fennoscandian area, was depressed under the load of ice; the area rebounded isostatically when the ice-sheets melted during interglacial times. These movements may have reactivated the old fractures, but direct evidence for this is only fragmentary. Some discrepancies in the recent uplift rates may reflect such movements, or alternatively errors in the measurements. Some of the fault lines are unusually well marked geomorphologically, something which again might indicate late tectonic activity. Finally, Brögger [2, pp. 398-99, text-fig. 29] published a second-hand report about a kettle-hole in the apatite-mine in Bamble (just to the SW of the Oslo Graben) being cut by a fault. The locality is no longer available, and although it is quite probable that small scale movements took place along the fractures even in the Pleistocene, hard evidence is lacking. The historic seismic activity [55] in the Oslo Region has been relatively modest but higher than that of the surrounding Pre-Cambrian terrain; the epicentre patterns seem to follow old fault zones both to the south and to the north of the graben.

## SUMMARY AND CONCLUSION

The main features of spatial/temporal evolution of the Oslo Region are now reasonably well known. Recent geophysical and geological studies have provided a) an estimate of mass and geometry of the various plutonic bodies and b) an improved picture of the gross crustal structure along the rift zone. These results, combined with new petrologic data, have served as a basis for the tectonic model outlined above. This model emphasizes the mantle control on the overall location and evolution of the rift, and the crustal control on the detailed setting and trends of individual graben features. The main taphrogenetic event in Carbo-Triassic time has been separated into several stages including 1) a Pre-Permian stage of fracturing and intrusion, and subsidence of sedimentary basins; 2) an initial (basalt) volcanic stage, possibly associated with crustal doming; 3) normal faulting, continued fissure eruptions and rift valley formation, accompanied by granitic intrusion; 4) a central volcano and cauldron stage accompanied by explosion vents and diatremes; 5) the emplacement of major composite batholiths; and 6) a terminal stage of faulting and dike intrusion.

A number of basic problems, however, remain unsolved: Is it purely accidental that the Late Paleozoic rifting followed the same axial trend as the Lower Paleozoic sedimentary basins? Or does this coincidence violate the idea of rifts being formed by an asthenospheric disturbance whose origin is independent of the overlying lithosphere? In other words, do rift formation and magmatism simply represent the response of pre-existing anisotropy to stresses acting upon the entire plate? Or does intra-continental rifting require the coincidence of a zone of lithosphere weakness with an underlying disturbance? What types of processes cause igneous activity to continue over tens of millions of years? Whether the rifting is caused by processes below or within the lithospheric plates might be analysed by studying surface features such as volcano spacing and frequency, trends and spacing of faults, and age gradients along the rift. Other problems of importance in the Oslo Region include present seismicity, and its possible tectonic implications, and the relation between the Oslo rift and the North Sea rift system.

## ACKNOWLEDGEMENT

Peter Vogt is gratefully acknowledged for critically reviewing the paper and for correcting the style.

REFERENCES

1. W.C. Brögger, Universitetsprogram, 1882.
2. W.C. Brögger, Nyt Mag. Naturvid., 28, 253, 1884.
3. W.C. Brögger, Nyt Mag. Naturvid., 30, 95, 1886.
4. W.C. Brögger, Norg. geol. Unders., 139, 1, 1933.
5. H. Cloos, Fortschr. Geol. Pal., 21, 241, 1928.
6. L. Störmer, Nor. geol. Tidsskr., 15, 43, 1935.
7. L. Störmer, Geol. Soc. London Quart. J., 123, 183, 1967.
8. O. Holtedahl, Am. J. Sci., 29, 504, 1935.
9. O. Holtedahl, Nor. Vidensk.-Akad. Oslo, I. Mat. Natur. Kl., 1943, No. 1, 89 pp., 1943.
10. O. Holtedahl, Norges Geologi, Norg. geol. Unders., 164, 1118 pp., 1953.
11. O. Holtedahl, (ed.) Geology of Norway, Norg. geol. Unders., 208, 540 pp., 1960.
12. Chr. Gleditsch, Nor. geol. Unders., 181 & 182, 1952.
13. T.F.W. Barth, Nor. Vidensk.-Akad. Oslo, I. Mat.-Natur. Kl., 1944, No. 9, 104 pp., 1945.
14. T.F.W. Barth, Nor. Vidensk.-Akad. Oslo, I. Mat.-Natur. Kl., 1954, No. 4, 20 pp., 1954.
15. C. Oftedahl, Nor. Vidensk.-Akad. Oslo, I. Mat.-Natur. Kl., 1952, No. 3, 64 pp., 1952.
16. C. Oftedahl, Permian rocks and structures in the Oslo Region, in: Geology of Norway, ed. by O. Holtedahl, Norg. geol. Unders., 208, 298, 1960.
17. C. Oftedahl, Geol. Rundsch., 57, 203, 1967.
18. I.B. Ramberg, Norg. geol. Unders., 325, 1, 1976.
19. I.B. Ramberg and B.T. Larsen, Tectonomagmatic evolution, in: The Oslo Paleorift, ed. by J.A. Dons and B.T. Larsen, Norg. geol. Unders., 337, 55, 1978.
20. J.A. Dons, Terminology and history of investigations, in: The Oslo Paleorift, ed. by J.A. Dons and B.T. Larsen, Norg. geol. Unders., 337, 3, 1978.
21. P.A. Ziegler, this volume.
22. I.B. Ramberg, R.H. Gabrielsen, B.T. Larsen and A. Solli, Geol. Mijnbouw, 56, 295, 1977.
23. D. Roberts and G.H. Gale, The Caledonian-Appalachian Iapetus Ocean, in: The Evolution of the Earth's Crust, ed. by D.H. Tarling, Acad. Press, London, New York (in press).
24. K. Björlykke, A. Elvsborg and T. Höy, Nor. geol. Tidsskr., 56, 233, 1976.
25. K. Björlykke, Nor. geol. Tidsskr., 49, 313, 1969; K. Björlykke (pers. comm. 1977); J.P. Nystuen (pers. comm. 1977).
26. I.B. Ramberg and T.F.W. Barth, Nor. geol. Tidsskr., 46, 219, 1966.
27. N. Spjeldnæs, Norg. geol. Unders., 251, 53, 1967.
28. N. Spjeldnæs, Nor. geol. Tidsskr., 34, 105, 1955.
29. S. Skjeseth, Nor. geol. Tidsskr., 30, 138, 1952.
30. P. Turner and J.H. Whitaker, Sediment. Geol., 16, 45, 1976.
31. A. Heintz, Nor. geol. Tidsskr., 54, 375, 1974.

32. N. Spjeldnæs, Nor. geol. Tidsskr., 37, 355, 1957.
33. K. Björlykke, Norg. geol. Unders., 305, 1, 1974.
34. B. Sundvoll, Isotope and trace-element chemistry, geochronology, in: The Oslo Paleorift, ed. by J.A. Dons and B.T. Larsen, Norg. geol. Unders., 337, 35, 1978. Also: B. Sundvoll, Proceedings of the NATO ASI in Oslo, Vol. I, 1978.
35. O. Holtedahl, Nor. geol. Tidsskr., 12, 323, 1931.
36. A. Heintz, Nor. geol. Tidsskr., 14, 176, 1934.
37. O.A. Höeg, Nor. geol. Tidsskr., 16, 1, 1936.
38. O.A. Höeg, Nor. geol. Tidsskr., 16, 279, 1937.
39. O.A. Höeg, Nor. geol. Tidsskr., 16, 280, 1937.
40. E. Dix and A.E. Trueman, Nor. geol. Tidsskr., 15, 25, 1935.
41. N. Spjeldnæs, Er Brumundsandstenen Triassisk? X Nordic Geol. Meeting, Oslo, Jan. 6, 1972.
42. K.A. Storetvedt and N. Petersen, Zeitsch. Geophys., 36, 569, 1970.
43. P. Holmsen, Medd. Dansk Geol. Foren., 14, 61, 1958.
44. L.B. Rasmussen, Geol. Surv. Denm., III Ser., 42, 1, 1974.
45. B.T. Larsen, Krokskogen lava area, in: The Oslo Paleorift, ed. by J.A. Dons and B.T. Larsen, Norg. geol. Unders., 337, 143, 1978.
46. P.M. Ihlen, P.R. Ineson and J.G. Mitchell, K/Ar dating of clay-mineral alteration associated with ore deposits in the northern part of the Oslo Region, in: Petrology and Geochemistry of Continental Rifts, ed. by E.-R. Neumann and I.B. Ramberg, Reidel Publ. Comp., 1978.
47. J.A. Dons, News from the Oslo Region Res. Group, 4, 29, 1972.
48. J.A. Dons and B.T. Larsen (eds.), The Oslo Paleorift, Norg. geol. Unders., 337, 1978.
49. B.T. Larsen, I.B. Ramberg and E. Schou-Jensen, Central part of the Oslo Fjord, in: The Oslo Paleorift, Norg. geol. Unders., 337, 105, 1978.
50. T.V. Segalstad, Geology of the Skien basalt, SW Oslo Region. Unpubl. thesis, Univ. Oslo, 1976.
51. P. Weigand, Nor. Vidensk.-Akad. Oslo, I. Mat.-Natur. Kl., Ny Ser., No. 34, 1, 1975.
52. K.S. Heier and W. Compston, Lithos, 2, 133, 1969.
53. E.-R. Neumann, Petrology of the plutonic rocks, in: The Oslo Paleorift, ed. by J.A. Dons and B.T. Larsen, Norg. geol. Unders., 337, 25, 1978.
54. G. Raade, Proceedings of the NATO ASI in Oslo, Vol. I, 1978.
55. E.S. Husebye and I.B. Ramberg, Geophysical investigations, in: The Oslo Paleorift, ed. by J.A. Dons and B.T. Larsen, Norg. geol. Unders., 337, 41, 1978.
56. J. Tryti and M.A. Sellevoll, Report on a reconnaissance seismic crustal study of the Oslo Rift, in: The Norwegian Geotraverse Project, ed. by K.S. Heier, Oslo 1977.
57. K. Aki, A. Christoffersson and E.S. Husebye, J. Geophys. Res., 81, 277, 1977.

58. K.-A. Berteussen, Phys. Earth Planet. Int. (in press).
59. I.B. Ramberg, Nature Phys. Sci., 240, 149, 1972.
60. R. Kaneström, Pure and Applied Geophys., 97, 37, 1973.
61. P.R. Vogt, Earth Planet. Sci. Lett., 21, 235, 1975.
62. J.A. Dons, Nor.Vidensk.-Akad. Oslo, I. Mat.-Natur.Kl., 1952, No. 2, 96 pp., 1952.
63. I.B. Ramberg, F.A. Cook and S.B. Smithson, Geol. Soc. Am., Bull., 89, 107, 1978.
64. D.B. McCullar and S.B. Smithson, EOS, Transactions GSA, 58, 1184, 1977.
65. A.R. Sanford, E.J. Rinehart, P.J. Shuleski and J.A. Johnston, EOS, Transactions GSA, 58, 1185, 1977.
66. R. Selmer-Olsen, Nor. geol. Tidsskr., 28, 171, 1950.
67. N. Spjeldnæs, Nor. geol. Tidsskr., 34, 105, 1955.
68. K.M. Storetvedt, Nor. geol. Tidsskr., 48, 153, 1968.
69. A. Noe-Nygaard, Medd. Dansk Geol. Foren., 17, 285, 1967.
70. K. Åm, Norg. geol. Unders., 287, 1, 1973.
71. E. Sæther, Nor. Vidensk.-Akad. Oslo, I. Mat.-Natur. Kl. 1947, No. 3, 60 pp., 1947.
72. N. Spjeldnæs, Norg. geol. Unders., 316, 116, 1975.

# THE OSLO REGION DURING THE EARLY PALAEOZOIC

J. Fredrik Bockelie

Paleontologisk Museum, Sarsgt. 1,
Oslo 5, Norway.

ABSTRACT. Lithological and biological faciesbelts of Cambro-Silurian age in the Oslo Region coincide in their directions with later Permian fault systems. This together with rhytmic sedimentation, rapid facies changes and other data do not seem to fit existing models for the presence of a shallow epicontinental sea. It is suggested that crustal movement, though on a small scale, played some part in the distribution of sedimentary patterns in the Cambro-Silurian.

## 1. INTRODUCTION

The Cambro-Silurian sediments of the Oslo Region were deposited within a presumed shallow epicontinental sea, which formed the edge of the Baltic Basin [1]. In the western part of the basin (Fig. 1), the central and northern portions of the Oslo Region were folded during the late Silurian. Accompanying this, allochthonous and parautochthonous deposits were also folded and thrust onto the northwestern part of the basin. In addition Permian faulting and subsequent erosion has blurred and removed original basinal shores. Even so, important knowledge of the Baltic Basin can be obtained.

## 2. FACIES BELTS

Since the beginning of this century the presence of often distinct bio- and litho- facies belts both within the Oslo Region and the rest of the Baltic Basin has been known [1,2]. Many of the facies-belts have distinct faunas and lithologies which make correlation

*I. B. Ramberg and E.-R. Neumann (eds.), Tectonics and Geophysics of Continental Rifts,* 195-202.

from one belt to the other difficult or almost impossible. Recently Jaanusson [3] introduced the term confacies to denote the facies-belts defined by common biofacies characteristics and recurrent lithofacies features, particularly pertaining to central and southern Sweden, Estonia and western Russia (Fig. 1). Confacies belts were often constant in geographical distribution, and faunal immigration between belts took place only occasionally. Environmental changes which influenced benthonic faunas were different from those affecting the plankton (graptolites etc.).

An example of immigration to a confacies belt coincides with the slight change of lithology from calcilutite to calcarenite that took place in central Sweden during the early Middle Ordovician. Many taxa previously existing in other parts of the basin, moved into this belt [3].

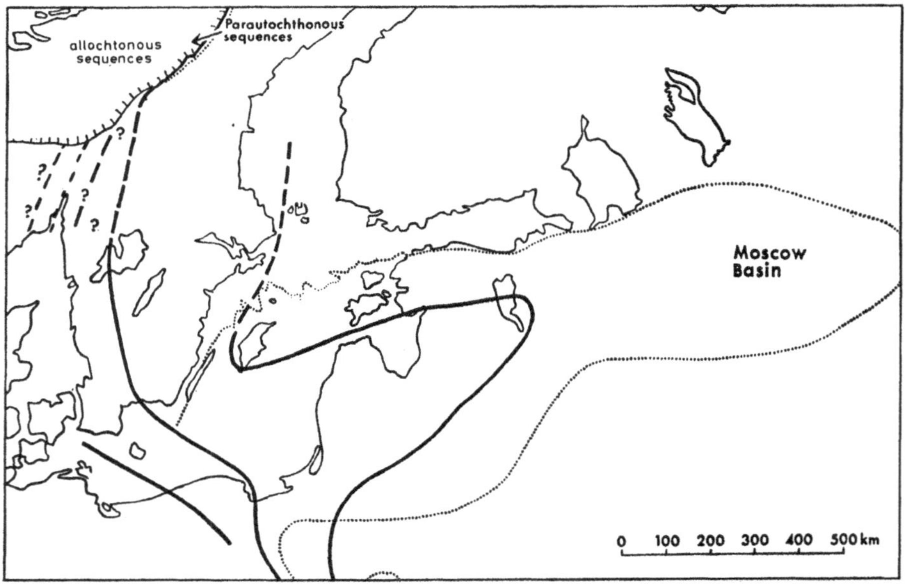


Fig. 1. Confacies belts in the Baltic Basin. Modified from Jaanusson [3]. Dotted: extent of subsurface and submarine Ordovician, chiefly in eastern Europe.

The confacies belts have directions different to those of the later Caledonian NE-SW- foldsystem (Fig. 1). This is evident from the areas in the SW part of Sweden and the SW part of the Oslo Region where the N-S trending belts have not been affected by the Caledonian folding. In central Sweden and the eastern part of the Baltic Basin confacies belts are wide, whereas towards the west and south-west the facies belts are narrower. Preliminary work in the Oslo Region suggests that three, or possibly four confacies belts existed during the Ordovician, some even during the Silurian (Fig. 2).

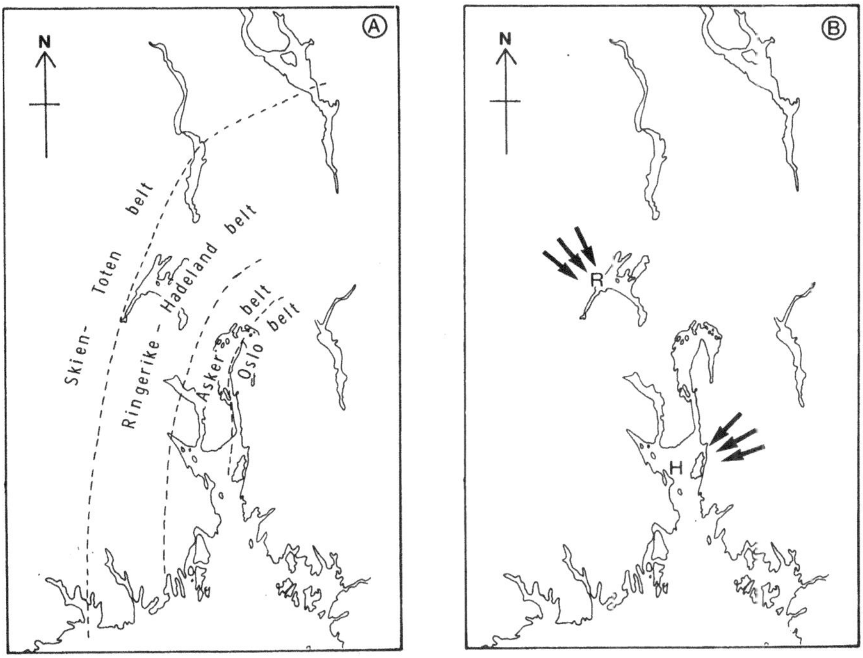


Fig. 2A. Ordovician faciesbelts [1-4], possibly confacies belts. 1: Skien-Toten belt often with reef-building organisms (corals, stromatoporoids etc.), 2: Ringerike-Hadeland belt, commonly containing brachiopods and echinoderms, 3: Asker belt, rich in trilobites, 4: Oslo belt, poor in fossils, often graptolitic.

Fig. 2B. Two directions of sediment-transport to the Oslo Region in the Upper Silurian to Ringerike (R) and Holmestrand (H). Data from Turner [11].

## 3. SEDIMENTOLOGICAL FEATURES OF THE OSLO REGION

The Cambro-Silurian sediments of the Oslo Region differ from those of other areas of the Baltic Basin in having greater thickness (400-500 m shales and limestones in Oslo *versus* 100-200 m dominantly limestones in Estonia) and higher contents of terrigenous components (50-60% versus 5-10%). The high proportion of terrigenous material in the Oslo Region has been assumed to be derived from a westernly source (Figs. 2B, 4A). Skjeseth [4] produced a convincing section across the Oslo Region for the early Ordovician indicating a central graptolitic basin gradually thinning in sediment thickness both eastwards (Sweden-Estonia) and westwards (Snertingdal) with corresponding increase of carbonate and quartz silt in the same directions.

During the Ordovician and Silurian [5] the bio- and lithofacies belts remained remarkably constant at certain intervals (Stages 3c, 4a), but rapidly shifting at other intervals (Stages 5b, 8, 9). Often these changes were linked with shallowing or deepening of the basin, locally or regionally.

The bottom topography was probably uneven, at least during some stages. Such variable bottom conditions are best seen during periods of shallowing. In the Asker facies belt at least four different situations can be demonstrated where NNE-SSW directed shallow water areas occurred (Fig. 3). In the Middle Ordovician Stage 4bβ, distribution of the echinoderm Sphaeronites shows clear eastern and western limits. This echinoderm is also known from shallow water environments of contemporary deposits in parts of Sweden. During the later Middle Ordovician, intraformational conglomerates are found in the same area, but these are lacking in the east and west. During various stages of the Upper Ordovician the same area was occupied by a submarine topographic high which caused some mudflows to change directions (Stage 4cα). In the uppermost Ordovician (Stage 5b), the area was occupied by bars of oolites and coarse sand (Bockelie in prep.).

Net sedimentation rates increased in the Oslo Region during the Cambro-Silurian times [2], but rates and patterns of sedimentation were often rhytmic [6]. The result was in many cases introduction of lime into areas where terrigenous clay was dominant. At other times coarse silt or sand beds were laid down in the same environments. Directions of sediment transport show mostly a western source (Figs. 2B, 4A), but other directions (N, NE, NW) also occur (Fig. 2B).

The Oslo Region can roughly be divided into three areas, a northern "Mjøsa" area, a central "Ringerike-Oslo" area and southern "Langesund-Skien" area (fig. 4). During the Ordovician the northern and southern areas generally have greater thickness of sediment deposited than the central area. In the Silurian, however, this picture was reversed. These two areas also have higher quartz and carbonate content than the "Ringerike-Oslo" area. The latter area has a higher degree of compaction than the other two due to a higher content of terrigenous clay, and hence smaller thickness. In the central area occasional breaks

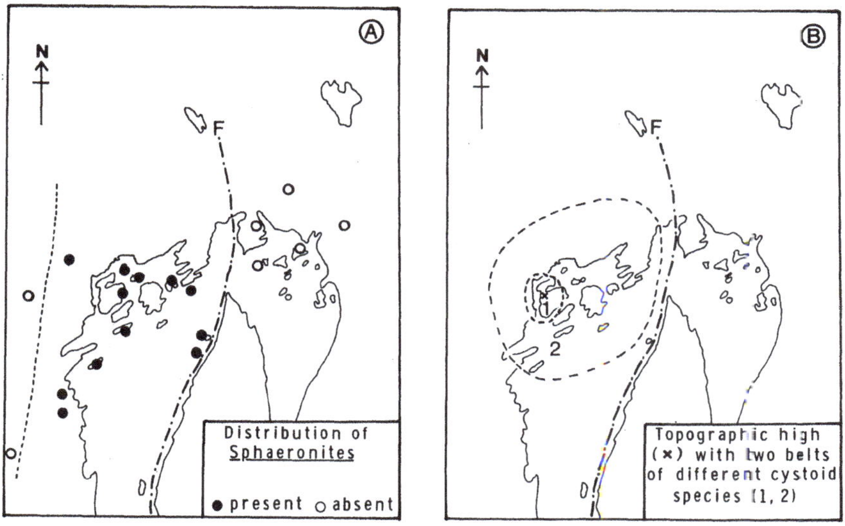


Fig. 3. Distribution of shallow water echinoderms in the Asker confacies belt (Asker-Bærum) in the Middle Ordovician (A) and in the Upper Ordovician (B). X: topographic high, 1 and 2: two facies belts with cystoids. F: direction of the Nesodden fault (Permian).

in sedimentation appeared (Stage 4aα - aβ in Ringerike), whereas such breaks appear to have been more common both in the "Langesund-Skien" area and particularly in the "Mjøsa" area [5].

A survey of fossil echinoderms from the Oslo Region (Bockelie in prep.), has yielded data indicating periods of rapid sedimentation both in the "Skien-Langesund" area and in the "Mjøsa" area, where delicate and often easily broken appendages are present. Such preservation is not unusual in those areas, whereas in the "Ringerike-Oslo" area such fossils with appendages have not yet been found. If a generalization can be made, the "Ringerike-Oslo" area has had a more constant sedimentation rate than the other areas.

## 4. DISCUSSION

The Baltic Basin, and even its western part, the Oslo Region has been considered by many authors to represent a shallow epicontinental sea [1,2].

The presence of channels up to 20 m deep in the Upper Ordovician [7], does not seem to fit such a picture for the Oslo Region. Thus the structure of the Baltic Basin may be more complex than that of a simple, flat shallow sea.

As mentioned above, Caledonian folding and the later Permian

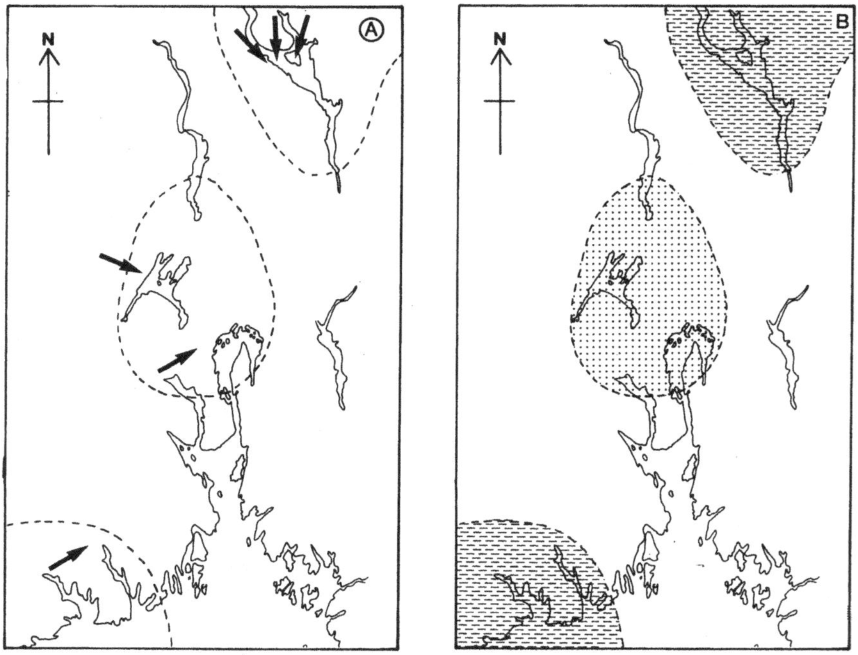


Fig. 4A. The Oslo Region can be divided into three areas, 1: the "Mjøsa" area, 2: "Ringerike-Oslo" area, 3: "Skien-Langesund" area. Arrows indicate common directions of sediment transport during the Cambro-Silurian period.

Fig. 4B. Areas of thickest sediment accumulations in Lower and Middle Ordovician (broken lines) and in the Silurian (dotted).

faulting of the Cambro-Silurian sediments add problems to detailed studies of the Oslo Region. Nevertheless various models have been presented to explain the sedimentation patterns of the Oslo Region syneclise [8] during the Cambro-Silurian. Seilacher and Meischner [9] had two models whereby the sedimentation rates to some extent were controlled by basinal subsidence, either at irregular intervals, or continuously. Kiær [7] and later Spjeldnæs [10] suggested active folding during the late Ordovician to explain certain facies patterns. Turner [11] suggested a basinal structure with sediment transport from two directions (WNW and ENE) in the late Silurian. One of his formations, the Holmestrand Formation, was deposited by powerful braided rivers from ENE (Fig. 2B). Fault scarps or a raised land area may thus have been present. The existence of faunal differences between the Oslo Region and southern Sweden at the end of the Silurian, in spite of similar lithologies [11], may be indicative of a physical barrier between the eastern part of the Oslo Region and Scania.

Changes in facies during the Cambro-Silurian times can be due to eustatic or isostatic changes. Some facies changes appear to be regional, whereas others are not. The local changes of facies may be due to basinal subsidence or uplift. No traces of active faulting can be observed within the Oslo Region during the Cambro-Silurian period, but there is often a correspondence between facies boundaries and later Permian fault directions (exemplified by the Nesodden fault; Fig. 3). This seems to me to be more than just a coincidence and suggests that earlier crustal movements could have provided triggering mechanisms necessary to explain the common occurrence within the Oslo Region of a rhytmic sedimentation.

Facies changes were more rapid and more terrigenous material was deposited in the Oslo Region and its southern continuation (Scania in Sweden), than elsewhere in the Baltic Basin. Great thicknesses of graptolitic shales (1000-1500 m according to Størmer [8], and possible tectonic escarpments between Sweden and Denmark, may be indicative of rapid subsidence in that area during the Silurian. The Oslo Region occurs in the northern continuation of this Scanian depression and continues northwards to the "Mjøsa" area (Fig. 4) where it is linked with the late Precambrian tectonically controlled Sparagmite Basin [12]. It is possible, but not conclusive, that early movements during the Cambro-Silurian times, were precursors to the later Permian rifting.

REFERENCES


1. R. Männil, Evolution of the Baltic Basin during the Ordovician (Russian with English summary), Tallin, 1966.
2. K. Bjørlykke, Norg. geol. Unders., 305, 1, 1964.
3. V. Jaanusson, in: The Ordovician system, ed. by M. Basset, Proc. Pal. Ass. Symp. Birmingham, sept. 1974, 301, 1976.
4. S. Skjeseth, Nor. geol. Tidsskr., 30, 138, 1952.
5. G. Henningsmoen, in: Geology of Norway, ed. by O. Holtedahl, Norg. geol. Unders., 208, 130, 1960.
6. J. Fr. Bockelie and Tove G. Bockelie, Nor. geol. Tidsskr., (in press).

7. J. Kiær, *Norg.geol. Unders.*, **34**, 1, 1902.
8. L. Størmer, *Geol. Soc. London. Quart. J.*, **123**, 183, 1967.
9. A. Seilacher and D. Meischner, *Geol. Rundsch.*, **54**, 596, 1964.
10. N. Spjeldnæs, *Nor. geol. Tidsskr.*, **37**, 355, 1957.
11. P. Turner, *The stratigraphy and sedimentology of the Ringerike group of Norway.* (Univ. of Leicester, Thesis, 1973).
12. S. Skjeseth, *Norg. geol. Unders.*, **220**, 1, 1963.

# THE MARGIN SOUTH OF AUSTRALIA - A CONTINENTAL MARGIN PALEORIFT

M. Talwani[1], J. Mutter[2], R. Houtz[1], M. König[1]

L-DGO CONTRIBUTION NO. 2617
[1]Lamont-Doherty Geological Observatory of Columbia University, Palisades, New York
[2]Bureau of Mineral Resources, Canberra City, Australia

## 1. INTRODUCTION

Several important questions remain unanswered about the details of evolution and crustal structure of passive continental margins. These questions include the location of the boundary between oceanic crust and continental crust, the nature and present location of the continental rift which precedes the opening by sea-floor spreading and the origin of the Magnetic Quiet Zone which often exists over continental margins.

An ideal place to examine these questions appeared to be the southern margin of Australia. Here a well developed Magnetic Quiet Zone exists, and the margin has been regarded as a classic example of a rifted continental margin. In addition, recent studies of deep penetration multichannel seismic reflection data and syntheses of structural and drilling information have led to proposals of detailed evolutionary models for the margin.

A one-month cruise was undertaken on Research Vessel VEMA in 1976 to investigate the crustal structure of the Magnetic Quiet Zone south of Australia and to test some of the proposed evolutionary models. Talwani et al. [1] (in press) have described the detailed results of that investigation. This paper is a summary of their results and conclusions.

## 2. GEOLOGICAL AND GEOPHYSICAL FRAMEWORK OF THE MAGNETIC QUIET ZONE

The physiographic province of the margin south of Australia are sketched in Fig. 1. The continental shelf is narrow in the western

*I. B. Ramberg and E.-R. Neumann (eds.), Tectonics and Geophysics of Continental Rifts, 203-219.*

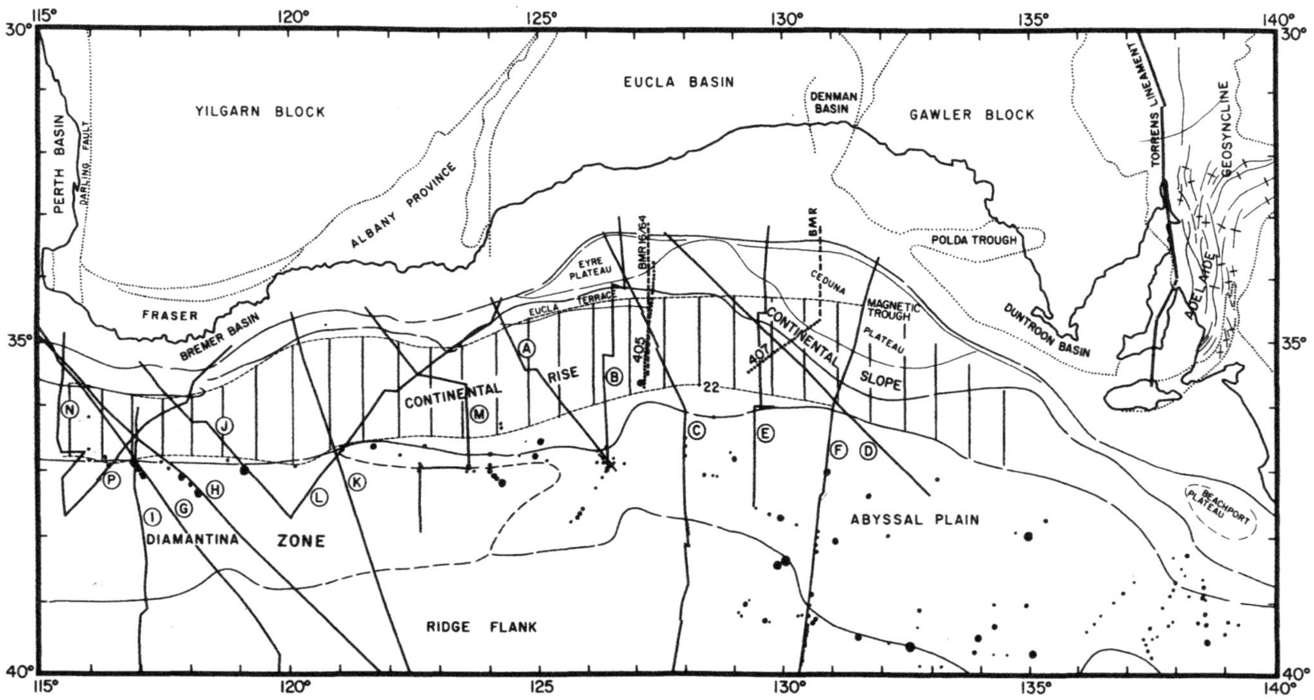


Fig. 1. Physiographic provinces of the southern continental margin of Australia. The profiles used in this paper are shown. The locations of the Magnetic Trough and of magnetic lineation 22 are also shown.

part of the margin but widens to the east and achieves its maximum width in the area of the Great Australian Bight, and is narrow again farther east of it. In the western part the continental slope is steep and narrow. Farther east, south of Eyre Plateau, it is still fairly steep. The Ceduna Plateau is constructional in nature, and here the slope is gentler. The continental rise is very narrow at the foot of Ceduna Plateau. With the exception of this abnormal narrowness of the rise, the continental margin is remarkably uniform in appearance from about 115$^0$E to 130$^0$E.

Weissel and Hayes [2] first outlined the sea-floor spreading history of the southeast Indian Ridge from studies of the magnetic anomaly pattern. König and Talwani [3] extended the anomaly identifications made by Weissel and Hayes. In particular, they noted the presence of the Magnetic Trough. This is the most consistent feature of the magnetic anomaly field south of Australia. Data obtained during the 1976 VEMA cruise have enabled us to establish the continuity of the Magnetic Quiet Zone and to trace the earliest sea-floor spreading anomaly lineations 20, 21, and 22 westward from the area of the Great Australian Bight.

Fig. 2 shows selected projected magnetic profiles and Fig. 3 is a presentation of the newly revised picture of magnetic lineations including the Magnetic Trough, superimposed on the physiographic provinces. (Locations of seismic refraction stations are also given in Fig. 3, but will be discussed later.) Anomaly lineations 19, 20, 21, and 22 have been identified through most of the Diamantina Zone although the basement topography is very rugged and significantly disturbs the magnetic signature thus making the identification of sea-floor spreading anomalies less reliable than in other areas.

The position of the Magnetic Trough is now known in considerable detail. The relation of the trough to physiographic features can be seen in Fig. 3. In the west, the trough lies near the base of the continental slope. This relation is maintained to the east until the Ceduna Plateau is encountered at about 130$^0$E. The trough crosses the Ceduna Plateau about mid-way down its length, and before terminating near 136$^0$E, nearly regains its formerly held relation to the physiography.

Lineation 22 lies along the northern flank of the Diamantina Zone. East of the Diamantina Zone anomaly 22 continues to be associated with rugged basement topography, but the basement is buried under a cover of sediments which renders the bottom topography relatively flat.

In the eastern part of the margin the magnetic field is very smooth between anomaly 22 and the Magnetic Trough, and the

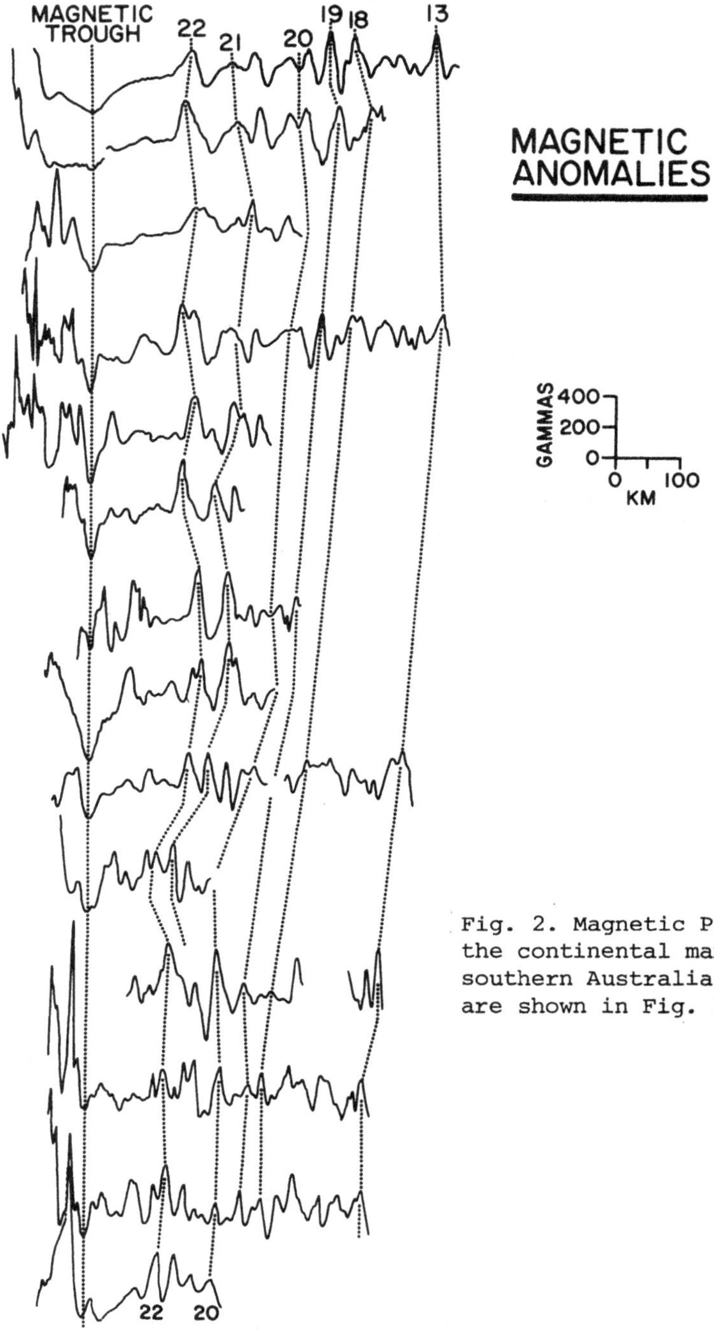


Fig. 2. Magnetic Profiles on the continental margin off southern Australia. Locations are shown in Fig. 1.

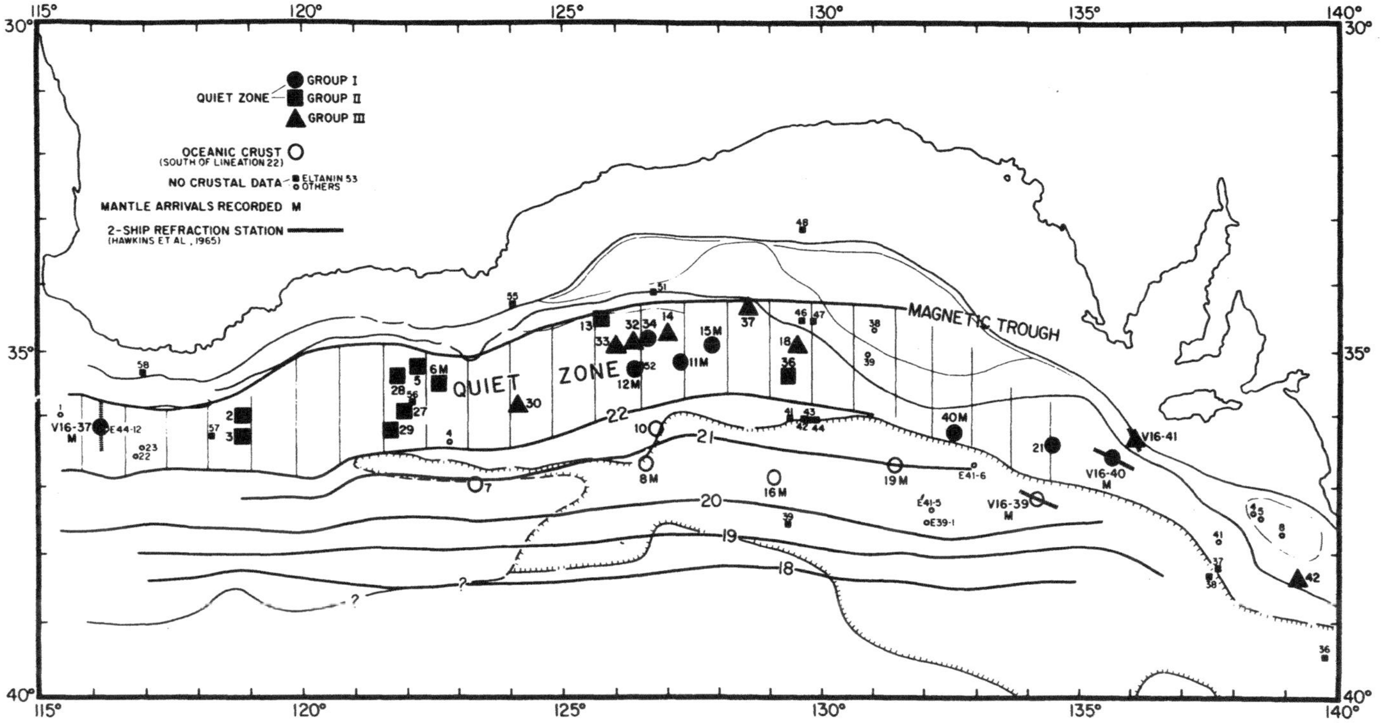


Fig. 3. Magnetic lineations on the continental margin off Southern Australia. The physiographic provinces are demarcated (see names in Fig 1). The locations of seismic refraction stations are also given.

designation Magnetic Quiet Zone has been applied to the area. The magnetic field becomes progressively more disturbed towards the west, in the area lying between anomaly 22 and Magnetic Trough. The magnetic field, although irregular and disturbed in the west, is not strongly lineated, and we refer to the entire area between anomaly 22 and the Magnetic Trough as the Magnetic Quiet Zone.

The depth to basement map shown in Fig. 4 was produced largely from the sonobuoy refraction data discussed later in this paper. The sediments over the Precambrian basement in the shelf area are not thick. The basement descends steeply under the shelf edge thus giving rise to an increased sediment thickness on the slope. The Magnetic Trough lies either within or immediately seaward of the steepest descent in basement depth. The trough is clearly related to an edge effect caused by the step in basement level which occurs from shelf to rise [3]. The sediment thickness varies from about 2 to 4 km on the continental rise which roughly coincides with the Magnetic Quiet Zone. There is a sharp decrease of sediments over basement ridges farther south. These ridges which are exposed in the west as the Diamantina Zone, but are buried to the east, are associated with the earliest sea-floor spreading - type crust, as judged from their coincidence with anomaly 22, the earliest identifiable magnetic lineation.

Seismic reflection profiles together with available well log information on the shelf, have been used to develop geologic sections across the continental margin. These cross sections have been interpreted in terms of the evolutionary history of the margin. The most important study in this connection is that of Boeuf and Doust [4] who interpreted the seismic reflection data obtained aboard R/V PETREL. Boeuf and Doust made an important distinction between basement which appears to be block-faulted in the reflection records and basement characterized by a prominent envelope of diffractions. They interpret the former as block-faulted continental basement and the latter as oceanic basement created by sea-floor spreading. In Fig. 5 the block-faulted continental basement underlies the Magnetic Quiet Zone including the Ceduna Plateau. The basement faults as interpreted in Fig. 5 extend upwards through the Lower and Upper Cretaceous sequence. Faulting commenced in the Lower Cretaceous unit, and died out in the Upper Cretaceous. Beneath the Lower Cretaceous unit, a non-marine sequence of Permain and Jurassic sediments with associated basalts may be locally preserved in faulted inliers. An extrapolation of well log results on land and shallow water, indicates that the Lower Cretaceous unit consists of fluviatile-lacustrine deposits. There is generally an unconformity between the Lower Cretaceous and Upper Cretaceous units although none is seen in the sections such as the one shown in Fig. 5, which is from the deepest part of the basins. The Upper Cretaceous beds constitute a deltaic sequence. Marine conditions had not been fully developed

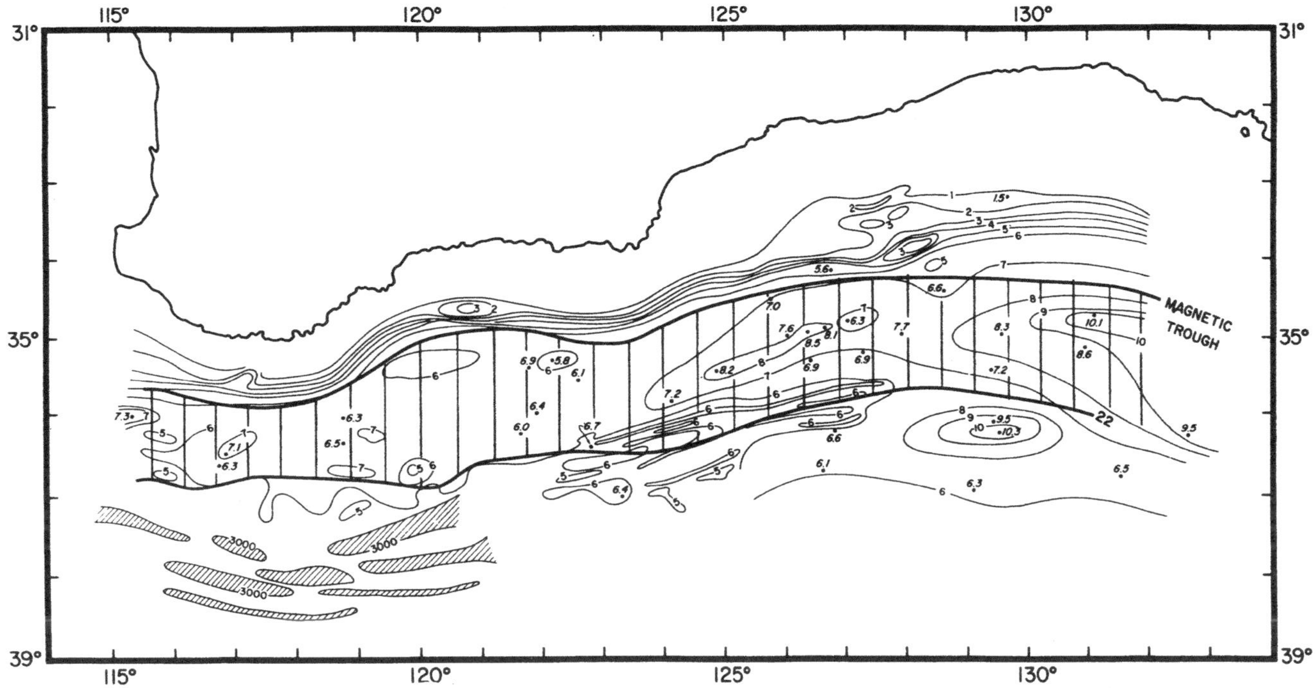


Fig. 4. Depth-to-basement map produced from sonobuoy and reflection data. Note the association of the trough with an escarpment in the basement, and of lineation 22 with ridges in the basement beneath the lower part of the continental rise, and with the northern edge of the Diamantina Zone. Depths are in km. In the southwest area of this map topographic highs of the Diamantina fracture zone as outlined by 3000 fathom contours are also shown.

in the east at this time though they are suspected to have progressed farther in the west. The Upper Cretaceous units thin considerably seawards - in Fig. 5, they are completely absent in the south where the Tertiary beds lie directly over the Lower Cretaceous beds. The unconformity at the base of the Tertiary is generally very well developed and the Tertiary beds are believed to represent open marine conditions for the first time.

Generally there is little faulting in the Tertiary beds although faults in the Lower Tertiary section are seen in some areas, particularly the Ceduna Plateau as shown in Fig. 5.

The bathymetric and magnetic data, the sediment-distribution pattern, and the seismic-reflection profiler results, taken together, demonstrate that the Magnetic Quiet Zone is a tectonic unit with distinct boundaries to the north and south. The northern boundary is indicated by the Magnetic Trough and consists of sharp landward decrease in topographic and basement depth as well as sediment thickness. To the south, the quiet-zone boundary is characterized by a seaward decrease in depth to basement and the appearance of sea-floor spreading magnetic lineations. Furthermore, according to the interpretation of the reflection data by Boeuf and Doust [4], the area of the Magnetic Quiet Zone is underlain by continental basement covered by a thick sequence of Mesozoic and Cenozoic sedimentary rocks, whereas the oceanic area further south is covered by Cenozoic sediments only.

## 3. SEISMIC REFRACTION MEASUREMENTS

During the 1976 VEMA cruise, thirty-six new refraction stations were established on the margin south of Australia and in the adjacent deep ocean basin. Their positions are plotted in Fig. 3. Recording were made with short and long-range sonobuoys shot with explosives and/or airgun. Measurements of sediment thickness, and basement and sub-basement velocities were obtained at nearly all stations, and mantle arrivals were recorded at nine locations.

Most of the seismic refraction stations were shot within the Magnetic Quiet Zone. A few were located south of magnetic lineation 22 in oceanic crust. Although two of the stations shot south of magnetic lineation 22 show anomalously high crustal velocities of 7.1 and 7.3 km/sec, these velocities are poorly determined and therefore there is no reason to suggest that the crustal structure of the oceanic areas is unusual. A sedimentary layer of variable thickness lies over what is presumably layer 2 with a velocity close to 5 km/sec which in turn overlies the main crustal layer with a velocity ranging from 6.6 to 7.0 km/sec. Mantle velocities are determined below the main crustal layer.

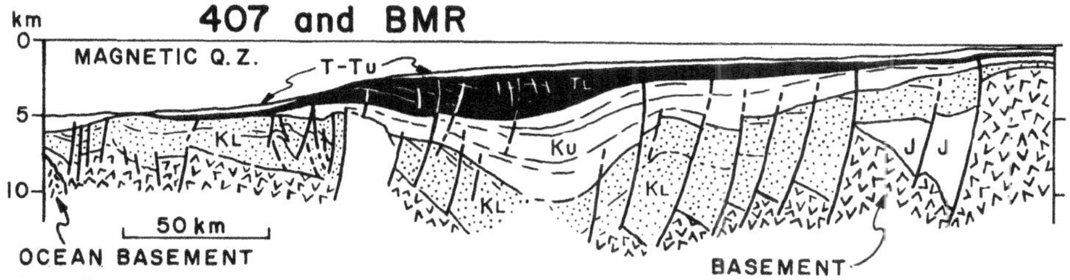


Fig. 5. Interpretation of Shell's PETREL line 407 (and an adjacent BMR line) Doust (1975). For location see Fig. 1. This section extends from the Ceduna Plateau to the continental rise.

A principal purpose of the experiment was to determine crustal velocities and layer thicknesses in the area of the Magnetic Quiet Zone. What we were looking for was some indication as to whether the crust in that area had evolved as an oceanic crust or had originated as a continental crust. Because of the large water depth and an absence of any very large gravity anomalies over the Magnetic Quiet Zone, approximate isostatic equilibrium must exist, which suggests that even if the crust had originated as a continental crust it would have to be substantially different at the present time. Such differences could arise from massive intrusions, thus making the crustal column heavier, or, an attenuation of the crust could have taken place, with most of the continental crust having been eroded from below and replaced by the mantle. (Beloussov [5], Bott [6], and Falvey [7], have considered various mechanisms of modification of the original continental crust.)

Our seismic results in the Magnetic Quiet Zone show a very large amount of variability and do not easily fall into any category. For the purpose of examination of the data we divided those refraction stations in the Magnetic Quiet Zone at which crustal velocities had been determined into three groups: Group I, in which the velocity of the main crustal layer ranges from 6.5 to 7.1 km/sec; Group II, in which the velocity of the main crustal layer ranges from 7.2 to 7.5 km/sec; and Group III, where the velocity ranges from 5.8 to 6.2 km/sec. When the stations are divided into groups in this fashion, a geographical distribution of groups does become apparent. Group II stations are mainly concentrated in the western part of the Magnetic Quiet Zone and Group III in the central part. Group I stations are found in both the eastern and the central part. We notice, however, that there are overlaps in all cases.

The velocities of Group I stations are not a typical of oceanic crust. A sedimentary sequence of variable thickness overlies a layer of velocity close to 5 km/sec which in turn overlies the main crustal layer with typical oceanic crustal velocities. More or less normal mantle velocities are found underneath. The velocities of Group I stations are therefore not too different from velocities found in the oceanic crust. However, the important difference is that the crustal thickness of Group I stations is appreciably larger than is found in the normal oceanic crust. The total crustal thickness ranges from about 13 to about 18 km which is 2 to 7 km larger than the generally observed value of ocean crustal thickness of about 11 km [8]. Nor is this increase in thickness simply attributable to thicker sediments. The crust below the sediments is appreciably thicker than the corresponding part of the normal oceanic crust.

The sections in Group II represent a fairly unusual crustal structure. The main crustal layer has a velocity greater than 7.2 km/sec. Such a layer has been elsewhere particularly in the Pacific [9]. However, in the Pacific and in other places where it has been determined earlier, this 7+ km/sec layer underlies a more or less normal oceanic crustal layer with velocity of about 6.7 km/sec. This is not true for the Group II sections where this layer underlies an upper crustal layer of velocity close to 5 km/sec.

Group III stations are the only ones that possess a thich main crustal layer with a velocity range 5.8 to 6.2 km/sec that is generally associated with continental crust. Owing to the thickness of this layer, mantle velocities were not obtained. Group III stations are on the average slightly shallower than the stations in Group I and Group II. However, the differences in crustal structure are so large and the differences in water depth generally so small that it cannot be claimed that the differences in the crustal structure in the three groups are simply related to changes in depth. We do note here that the water depth over the oceanic stations (south of lineation 22) is larger than the stations within the quiet zone which of course reflects the difference in depth between the abyssal plain and the continental rise.

To summarize the seismic refraction results we note firstly that the refraction stations in the area where the magnetic lineations point to sea-floor spreading show structure which is normal for oceanic crust. Lineation 22 which lies at the boundary between the sea-floor spreading-type anomalies and the Magnetic Quiet Zone thus also separates regions of different crustal structure. In other words, a major east-west trending structural boundary exists near magnetic lineation 22.

A more unexpected result is the east-west variation of crustal structure within the Magnetic Quiet Zone. We notice that the Group II type of stations are generally concentrated in the western part of the Quiet Zone, whereas Group I and III are located in the central and eastern part of the Magnetic Quiet Zone. This east-west difference in the Magnetic Quiet Zone is also reflected in the Magnetic profiles. The field is really smooth only in the eastern part, whereas in the western part, north of the Diamantina Zone, the magnetic field is quite disturbed.

In spite of these large east-west differences within the Magnetic Quiet Zone, we are reluctant to invoke a completely different origin for the western part and the eastern part. That is, we do not want to suggest, for example, that the western part is oceanic and the eastern part is continental. We believe that the two main structural discontinuities in the area are the E-W trending discontinuities, one of them at the Magnetic Trough and the other at anomaly 22 since these represent the most consistent geophysical lineations in the area. These discontinuities are expected to separate areas which have had different histories of evolution.

Representative stations from Group I, II and III and from the oceanic crust are presented for comparison in Fig. 6 together with a station obtained by Hinz [10] on the Inner Vøring Plateau. The Inner Vøring Plateau lies off Norway within a Magnetic Quiet Zone in an area which may have similarities to the Magnetic Quiet Zone south of Australia.

## 4. INTERPRETATION OF THE SEISMIC REFRACTION RESULTS IN TERMS OF THE NATURE AND EVOLUTION OF THE MAGNETIC QUIET ZONE

The interpretation of the Magnetic Quiet Zone crust as continental crust, which appears to be the prevailing view in the literature [7,4,11] is directly supported by sonobuoy refraction measurements only in the eastern part of the quiet zone. Here we recorded several crustal sections with a thick basement layer having an average velocity of 6.0 km/sec typical of continental crustal structure. To apply this solution to the quiet zone as a whole is difficult. Clearly the crustal structure in the central and western regions bears little resemblance to a normal continental crust. Although the crust is at present atypical of a continental crust, it might be argued that it was formerly continental, and has been altered to its present structure by some process of attenuation or "oceanization".

Deighton and others [11], among many others, have proposed a method of crustal thinning. It is based on an earlier model of Falvey [7], and they suggest it has application to the southern margin of Australia. However, it appears difficult to apply these schemes to the seismic results which yield sections of the Group II and Group I types. Group I sections show a crustal thickness of about 6 to 13 kms (excluding the water layer) and the proposed

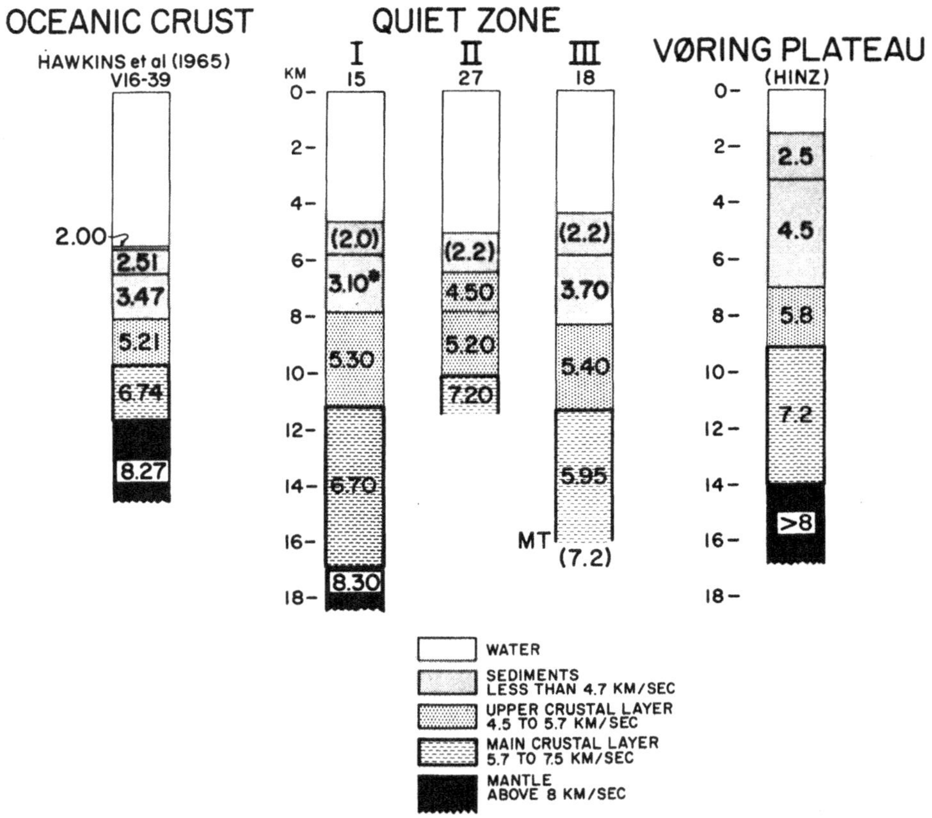


Fig. 6. Representative sections from Groups I, II and III in the Magnetic Quiet Zone and from the area of oceanic crust. Also included for comparison is a section obtained by Hinz [10] in the Magnetic Quiet Zone on the Inner Vøring Plateau off Norway.

schemes of crustal attenuation would be inadequate by a factor of two or three to achieve the required thinning. Group II sections show a velocity structure so different from typical continental sections that, again, the proposed attenuation scheme would be difficult to apply. Another objection to the thinning scheme is the absence of large amounts of Mesozoic sediments produced by erosion of the supposidly up-arched continental crust. No vast Mesozoic basins are located onshore in Australia in the region, and none have been discovered in Antarctica. The notion that crustal thinning has been achieved solely by erosion is considered wholly untenable.

It is possible to postulate a Mesozoic sea-floor spreading type of origin for the crust in the Magnetic Quiet Zone. The lack of magnetic anomalies can be attributed to one of the following hypothesis: 1) Generation of the Mesozoic ocean crust during an extended uniform polarity interval. 2) Thermal blanketing and/or metamorphism of the crust with attendant reduction of magnetic properties. 3) Very slow spreading. Boeuf and Doust's [4] assessment of the continental basement edge is based on a change in basement character from highly faulted to one defined by an envelope of diffractions. It could be argued that instead this reflects a boundary between Mesozoic and Tertiary sea-floor. The former could well have been faulted and subsided to great depth following cessation of Mesozoic spreading. The thick sedimentary sequence which exists beneath the Ceduna Plateau could then have developed in a narrow ocean basin rather than a wide rift-valley. The depositional environment might be very similar to a rift-valley, especially if sea-floor was generated near sea level and sediment infilling was rapid.

There are however two major problems with this thesis. The seismic station in Group II and Group III are so different from normal oceanic crustal sections that it seems difficult to accept their conversion from normal oceanic crust. Secondly, there does not appear to be any ready explanation for E to W variation of the nature of the Magnetic Quiet Zone crust if it is oceanic.

We have examined above two possible alternative structural interpretations of the quiet zone crust. In considering the implications of either of these choices we came upon obstacles which present serious difficulties with the interpretations.

We suggest that these difficulties arose because we restricted ourselves only to solutions which fell into the framework of conventional interpretations of the earth's crust in terms of "oceanic" or "continental" crustal types. We now propose that the quiet zone crust can best be treated as something completely unique, being neither continental nor oceanic.

The inner boundary of the quiet zone, as defined by the Magnetic Trough anomaly, is far from being a superficial or shallow crustal boundary. It appears to us to be associated with a primary discontinuity in the deep structure of the earth's crust. The outer boundary of the quiet zone is created by the junction of the quiet zone crust with normal oceanic crust of Tertiary age accreted at the SE-Indian ridge. It too forms a primary structural discontinuity. These two crustal discontinuities isolate the quiet zone from continental crust which adjoins it on the landward side, and oceanic crust which adjoins it on the seaward side.

Within the quiet zone the velocity layering of the crust is characterized by variability. The velocity structure is typically continental in some places, in others it is oceanic although the layer thicknesses are much larger than those for typically oceanic crust, and in still others neither typically oceanic nor continental. The distributions of these crustal types is not rigorously systematic and in many cases sections with different crustal types are closely juxtaposed geographically.

The quiet zone is seismically passive, and lack of recent faulting suggests it is presently stable. Therefore, it cannot be transitional in nature, at least not in a temporal sense.

As far as the evolution of the area is concerned, we envisage a Mesozoic tectonic regime marked by subsidence of the basement within a region defined by major boundary faults. Within the major graben taphrogenic basement deformation resulted in the block-faulting described by Boeuf and Doust [4] and Willcox [12] (in press). The basement elevation probably fell below sea level fairly early in the rifting history but marine conditions did not develop due to high sedimentation rates in the east and a barrier to the west. The structural and morphologic feature which is now the continental slope west of about $128^0$E was present during the middle to late parts of the Mesozoic graben tectonics.

It seems reasonable to suggest that while subsidence and block-faulting took place, the continental crust evolved for a period of several tens of millions of years prior to break-up. Igneous activity which is commonly associated with rifting occurred during this period and this activity must have varied along the rift-valley giving rise to east-west, that is, along-strike variations in the crustal structure and also to the close juxtaposition of areas with different crustal structures. A general difference between the seismic results in the Magnetic Quiet Zone and those from areas with normal oceanic crust is the presence of significant and abrupt discontinuities in crustal refraction travel time lines in the former area which may be attributed to the broken up nature of the crust resulting from tectonic and igneous activity. Thus we can visualize a crust in this Mesozoic rift-valley which is nearly continental in some areas, but in others it has been subjected to a very large amount of igneous activity which has appreciably altered the average crustal velocity as measured seismically giving rise to the Group I or II crustal sections. Greater igneous activity in the western part of the magnetic Quiet Zone presumably gives rise not only to higher crustal velocities there but to a more disturbed and irregular magnetic field as well.

The Mesozoic period of rifting and alteration of the continental crust was followed by a Tertiary period of continental break-up and sea-floor spreading. Thus the Austeralian-Antarctic Margin can be conceived as having evolved in two stages: a Mesozoic stage during which the rift geometry remained constant but the composition of the crust continuously evolved through igneous activity, and a Tertiary stage during which the geometry continuously evolved (that is, drift took place) but the composition of the newly-created crust presumably remained constant.

Recent studies of the Rhine Graben and African Rift system emphasize the confinement of the rift zones' anomalous deep structure to those areas immediately beneath the rift valley proper (Prodehl et al. [13] and Edel et al. [14] in the Rhine Graben, Maguire and Long, [15], Long and Backhouse [16] in the African Rift). These studies showed earlier ideas that the rifts were underlain by a broad region of up-arched mantle were incorrect. In the Long and Backhouse [15] study it was shown that essentially unaltered crust of the African Shield continues virtually to the rim of the rift, and "this steep boundary probably extends to a depth of at least 55 km". In the Rhine Graben study the formerly-held idea of a "rift-cushion" was dismissed in the light of new data. The new model shows a zone of anomalous crustal structure formed by a sharp rise in the depth of high velocity layers, occupying the region immediately beneath the rift proper. Virtually no crustal modification occurred in regions outside the rift.

Prodehl et al. [13], also commented on the peculiarity of the crust in the Rhine Graben in the following way. In discussing the crustal refraction data they noted "in 25 record sections prepared from the data... it is difficult to find two that look alike. These differences suggest a fine structure that is beyond the resolving power of the methods which can be used and which require certain assumptions not being true in all cases". A very similar statement could well have been made with reference to the extremely variable crustal structure of the Magnetic Quiet Zone of Australia's southern continental margin.

Although previous models of rift evolution have noted the similarity of rift graben structure to the basement structure of rifted continental margins, the results from the Australian Margin taken with the newer results from the African and German rifts provide the rifts real link between the deep structure of both systems. It seems very reasonable to us to suppose that the anomalous Quiet Zone crust is the relic of the anomalous rift zone crust. The magnetic trough marks the location of the rift zones bounding escarpments and these are underlain by a sharp discontinuity in the deep crust across which normal continental crust abruptly changes to the anomalous rift zone/quiet zone crust. The gross crustal structure of the Australian margin shows that the continent-to-ocean transition is achieved by a step-wise change from normal continental to quiet zone to oceanic type crust and this is the clear outcome of rupturing a rift zone with its anomalous crust and prominent boundary discontinuities.

The restriction of the anomalous crust to a sharply defined zone in the normal crust suggests that the agent of crustal modification be much more concentrated in shape and more intense than previous models have shown. It is more suggestive of the type of modification that might occur, for example, at the head of a rising diapir in the asthenosphere than would result from a domal uplift of isotherms.

We also believe that it would be appropriate to say that the Magnetic Quiet Zone is underlain neither by oceanic crust nor continental crust, but by "Rift Crust" and it would be useful to recognize Rift crust as a unique entity in discussing passive continental margins. A general model for evolution of rift crust would involve an initial stage when the rift geometry remains constant but the composition of the crust continuously evolves and is followed by a second stage during which drifting occurs by sea-floor spreading, but the composition of the newly created oceanic crust remains essentially constant.

ACKNOWLEDGMENT

This study was supported by Grant OCE 75-20438 from the National Science Foundation.

REFERENCES

1. M. Talwani, J. Mutter, R. Houtz and M. König, The crustal structure and evolution of the area underlying the Magnetic Quiet Zone on the margin south of Australia, Am. Assoc. Pet. Geol. (in press).
2. J.K. Weissel and D.E. Hayes, Magnetic anomalies in the southeast Indian Ocean, in: Antarctic oceanology II: The Australian-New Zealnad Sector, ed. by D.E. Hayes, Am. Geophys. Un. Antarctic Res. Soc., **19**, Washington D.C., 1972.

3. M. König and M. Talwani, Geol. Soc. Am., Bull., **88**, 1000, 1977.
4. M.G. Boeuf and H. Doust, Aust. Pet. Expl. Assoc. J., **15**, 33, 1975.
5. V.V. Beloussov, Some problems of development of the earth's crust and upper mantle of oceans, in: The crust and upper mantle of the Pacific area, ed. by L. Knopoff, G.L. Drake & P.J. Hart, Am. Geophys. Un. J., Monogr. **12**, Washington D.C., 1972.
6. M.H.P. Bott, Tectonophysics, **11**, 319, 1971.
7. D.A. Falvey, Aust. Pet. Expl. Assoc. J., **14**, 95, 1974.
8. R. Raitt, The crustal rocks, in: The Sea, ed. by M.N. Hill, Wiley, 1963.
9. G.H. Sutton, G.L. Maynard and D.M. Hussong, Widespread occurrence of a high-velocity basal layer in the Pacific crust found with repetitive sources and sonobuoys, in: The structure and physical properties of the earth's crust, ed. by J. Heacock, AGU, Man. **14**, Washington D.C., 1971.
10. K. Hintz, The seismic crustal structure of the Norwegian margin in the Vøring Plateau, in the Norwegian deep sea, and on the eastern flank of the Jan Mayen Ridge between 66° and 68°N, 24th Int. Geol. Congr., Sect. 8, 1972.
11. I. Deighton, D.A. Falvey and D.J. Taylor, Aust. Pet. Expl. Assoc. J., **16**, 25, 1976.
12. J.G. Willcox, The Great Australian Bight: A regional interpretation of gravity, magnetic, and seismic data. (In press).
13. C. Prodehl, J. Ansorge, J.B. Edel, D. Emter, K. Fuchs, S. Mueller and E. Petterschmitt, Explosion seismology research in the central and southern Rhine Graben - A case history, in: Explosion seismology in central Europe: Data and results, ed. by P. Giese, C. Prodehl and A. Stein, Eu. Seis. Comm., Monogr. **1**, Springer, New York, 1976.
14. D.A. Edel, K. Fuchs, C. Gelbke and C. Prodehl, J. Geophys., **41**, 333, 1975.
15. P.H.K. Maguire and R.E. Long, Geophys. J. Roy. Astron. Soc., **44**, 661, 1976.
16. R.E. Long and R.W. Backhouse, Geophys. J. Roy. Astr. Soc., **44**, 677, 1976.

# PASSIVE CONTINENTAL MARGINS AND INTRA-CRATONIC RIFTS, A COMPARISON

Willem J.M. van der Linden

Vening Meinesz Laboratorium
Rijksuniversiteit Utrecht
The Netherlands

ABSTRACT. Within a complete Plate Tectonic cycle, which leads to the formation of an ocean basin and of (passive) continental margins, there are two phases; first, a rifting phase with the attenuation of continental lithosphere; later, a drifting phase with sea-floor spreading. Attenuation and spreading both take place under tension, whereby the lithosphere stretches and breaks apart over the underlying expanding asthenosphere. The rifting or attenuation phase of margin and basin formation conforms to the evolution of continental rifts. There are a number of lithological and structural similarities, combined in the form of a rift cushion and its surrounding aureole of magmatic intrusives, that appear permanently frozen in passive margins and continental rifts. They produce characteristic density and velocity contrasts and sometimes also a typical magnetic anomaly pattern. Detection of these characteristics is often difficult due to their shape and to the limitations of geophysical prospecting technology.

## 1. INTRODUCTION

Geophysical measurements, specifically seismic refraction studies, over the Labrador continental margin have led to the suggestion that under Atlantic type margins the transition from continent to ocean occurs over a distance of several hundred rather than over a few tens of kilometers. Also that the formation of such margins to a large extent preceded sea-floor spreading and that it progressed conformal to the evolution of intra-continental rifts [1].

The spreading model developed for the Labrador Sea (Fig. 1) holds that, during an initial period of rifting, attenuation of the

*I. B. Ramberg and E.-R. Neumann (eds.), Tectonics and Geophysics of Continental Rifts, 221-227.*

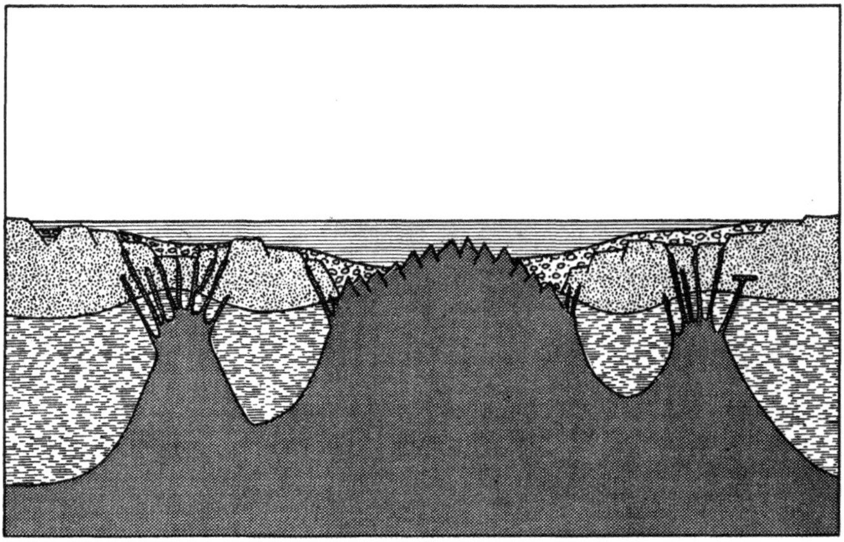

Fig. 1. Schematic model of passive continental margins and ocean basins, the result of attenuation of continental lithosphere and sea-floor spreading. Stretching and fracturing of the lithosphere during an initial rifting phase, are interpreted to have been accompanied by the intrusion of asthenosphere derivatives. Where the asthenosphere broke through in bulk, during a subsequent drifting phase, sea-floor spreading completed the separation of lithosphere elements. The striped zone represents the hydrosphere; the lithosphere is divided into a sedimentary layer, a metamorphic and crystalline crust, and the upper mantle; dense shading marks the asthenosphere and its derivatives.

lithosphere occurred and was accompanied by wide spread magmatic intrusion. Thinning and stretching of the lithosphere during this period is thought to have produced a number of sub-parallel relatively thick singular or composite elongate ridges, separated by singular or composite elongate basins that are underlain by relatively thin crust. Ultimately, during a subsequent drifting phase, separation of the lithosphere occurred centred on what developed to become the Mid Labrador Sea Ridge (sea-floor spreading). This stage was accompanied by further margin subsidence under increased loads of land-derived sediments. The structure of the SW Pacific region, the African Rift system and the Arctic Archipelago add substance to the suggestion that during rifting the pattern of ribbon-like fragmentation depicted for the Labrador Sea might be fairly common for Atlantic type margins [2].

Ziegler's [3] reconstruction of the continental fit of the North Atlantic ocean (this volume) shows a complex predominantly northerly-trending system of larger and smaller troughs and graben. In this Mesozoic system one recognizes the configuration of the future mid-ocean ridges or spreading axes as well as what are now intra-continental rifts that never passed through a subsequent drifting or sea-floor spreading stage. In other words his reconstruction supports the concept that the rudimentary forms of Atlantic type ocean basins and intra-continental rifts are probably identical.

## 2. EXPANSION OF THE ASTHENOSPHERE AND MAGMATIC STOPING

The similarity between passive margins and rift zones is not restricted to the upper parts of the crust but extends deep into the lithosphere, incorporating at least the upper mantle. In contrast to the 'normal' seismic stratigraphy of continental areas refraction observations show that divergent margins and continental rifts are both underlain by anomalous crust. First of all the crust in these regions is relatively thin; secondly typical mantle velocities are often not observed and furthermore anomalous velocities somewhere between those of crystalline crust (6.5 to 7.2 km/s) and mantle (7.8 to 8.2 km/s) are encountered frequently, if not always. Intermediate velocities of about 7.5 km/s measured under active continental rifts seem to belong to pillow-shaped bodies a few tens of kilometers thick that laterally extend for more than a hundred kilometers [4-7]. High heat flow associated with active rifts suggests, that these low velocity (low density) rift cushions may be result of thermal expansion. This would imply according to Osmaston [5] and Ziegler [7], that after the causative heat anomaly is exhausted the process should reverse itself. The low density material would thereby be resorbed by the mantle and thus after cooling, normal densities and velocities should again be encountered. There are, however, strong arguments that suggest that rift cushions are not purely thermal.

If it is true that incipient ocean-basin formation and intra-cratonic rifting are indeed the same processes, then it is logical to expect that the deep crustal or upper mantle anomalous velocities encountered under Atlantic type margins represent the same substance as the cushions that are found under intra-cratonic rifts. Since anomalous velocities are still measured under now inactive (i.e. dead and cold) margins such as the Labrador Sea, it follows that at least those marginal rift cushions can not be thermal anomalies. They must therefore represent anomalous lithologies that are permanently frozen at the crust-mantle interface.

In the attenuation model developed for Atlantic type margins [1,2], which is strongly influenced by evolutionary models of the

Red Sea and Thyrrhenian Sea, rifting processes are thought to be induced by thermally and chemically expanding tumors or mantle diapirs. These 'asthenoliths' interact with the overlying lithosphere through physical and chemical processes, for which Daly's [8] term 'magmatic stoping' is very appropriate. The physical part is the earlier mentioned attenuation or stretching and thinning of the lithosphere over the expanding asthenosphere, through brittle fracturing in the upper cold region, and through plastic or ductile flow deeper down. Partial melting and fracturing of the lithosphere and the generation and intrusion of basaltic magmas (differentiates presumably of gabbros and periodotites) help blend and fuse the crust and mantle into an 'irreversible' low velocity - low density rift cushion. These cushions are surrounded by an aureole of dykes and sills of low viscous volcanics, some of which completely penetrate the crust and flow out at the surface.

## 3. DETECTION OF RIFT CUSHIONS AND THE SEISMIC REFRACTION METHOD

Why are there passive margins where rift cushions (or their appropriate velocities) have not been encountered? There are reasons that suggest that their absence in those cases is only 'apparent' because of the poor resolution of most refraction experiments. Or, although they have not been detected, it does not mean that the rift cushions are not present after all.

Igneous petrologists and ore geologists who have mapped the outline of magmatic intrusives and ore-bodies can testify that usually such an outline is a very irregular surface, which reflects the high mobility or low viscosity of magmatic melts and is also a consequence of stoping. In this respect it is important to realize that, although there are scale differences, the rheology of a granitic intrusive for instance is comparable to that of a salt diapir [9]. The shape of salt bodies as is well known can be fantastic, sometimes with tear-shaped blobs completely detached from the mother-salt. By inference one can expect that magmatic intrusives and thus rift-cushions may have similar morphologies and it is probably realistic to compare the outline of such bodies to thunderclouds. This supposition is backed up by the results of a number of intersecting refraction lines, shot in the Gulf of St. Lawrence. The Gulf is a Palaeozoic, sediment filled, intra-cratonic basin with many, if not all, of the characteristics of rift zones. These include the accumulation of evaporites with associated volcanic activity and in places the presence of deep crustal or subcrustal rocks with anomalous velocities of about 7.4 km/s under a thin crystalline crust [10-13]. These velocities are found rather consistently on long refraction lines but only occasionally on the short ones. This, however, agrees rather well with the expectation that long i.e. 'deep' lines have a better chance to reach the crust-mantle mix or rift-cushion, than the short ones. The 'shallow' lines will only sound the tops of the 'thundercloud'. Of course, the seismic refraction method presupposes that velocity contrasts occur along more or less planar surfaces, and it is therefore not particularly suited to try and define the presence and outline of magmatic intrusions.

## 4. THE MAGNETIC SIGNATURE OF ATTENUATED CONTINENTAL CRUST

Another characteristic of a rift-cushion or more precisely of its surrounding aureole of volcanic intrusions should be the magnetic signature. This can be expected to be a pattern of elongate positive and negative anomalies, parallel to the continental margin or to the rift axis, since that is the dominant direction of fissures or magma conduits that opened by normal faulting. Especially when the volcanics are not in the form of plateau lavas of large horizontal extent, (are only intrusives or subaqueous extrusives), the resulting magnetic pattern should be predominantly linear. Support for this proposition is found in the remarkable coincidence or at least proximity of the positions of rifted margins and quiet magnetic zones (QMZs). This coincidence suggested to Poehls et al. [14] that the marginal QMZs owe their origin to the initial break-up of the ancient craton (Pangea). They further suggested that marginal QMZs are areas in which oceanic anomalies are attenuated at sea level due to the depth of the magnetic rocks under a thick wedge of sediments. This explanation is also implied in the work of Barrett and Keen [15], but it is in contrast with the thesis in which magnetic anomalies within marginal QMZs are thought to be caused by magmatic intrusions in fracturing and thinning continental crust [1, 2]. Subsidence of the margin under increasing sediment loads would of course have a further attenuating influence on 'continental' magnetic anomalies, in the same way as subsidence of the ocean floor would have on oceanic anomalies. The anomaly pattern of QMZs is still ambiguous and their origin is difficult to explain. There are most likely different causes or a combination of causes that can produce similar patterns. Marine geophysicists, as much as other scientists, are influenced by territorial imperatives and they are thus inclined to find evidence for oceanic crust in the subtle lineations of QMZs. The match of anomalies from line to line is often questionable and a reasonable correlation with established magneto-stratigraphy can only be achieved by assuming extensive variability in spreading rates. For continental lineations there are no such problems and that consideration favours marginal QMZs to be zones of attenuated and intruded continent in the manner explained here and elsewhere [1, 2, 16].

For continental rifts, however, the pattern seems more chaotic and the vague trends apparent in the Rhinegraben magnetics for instance run obliquely to the graben direction [17]. The Red

Sea shelf is a QMZ with anomalies aligned parallel to the central rift, but that is convincing evidence according to Girdler and Styles [18] for the presence of oceanic crust and sea-floor spreading. Apparently attenuation has to be far enough advanced for magmatic intrusions to be abundant and to reach sufficiently shallow levels in the crust to accentuate the tectonic grain.

## 5. CONCLUSION

It is probably fair to conclude that the agreement in structure between intra-continental rifts and passive continental margins is such that it would be very surprising to find that their formation is not controlled by basically the same processes. Formation of passive continental margins is rigidly coupled to the genesis of an ocean basin; they are mutually inclusive processes. Formation of continental rifts could be characterized as an aborted or not yet completed attempt to margin and basin formation. It is by no means clear, however, what the formative processes are, nor, to state a more fundamental problem, what causes them. Preference for one or another model of continental margin formation, and by implication of continental rifting, is founded on ambiguous evidence only. Unequivocal support for one interpretation or model over another is conspicuously absent. On the problem of ocean basin formation, earth scientists have for a long time been divided into two opposing camps; the fixists or oceanizers and the mobilists or drifters. Arguments for or against either position have been raging for most of this century and even the 'new global tectonics' have not yet settled the controversy.

For the continental margins and to a lesser extent for continental rifts, however, the suggestion is that oceanisation and rifting both take place and that in those areas they are complimentary processes.

With this statement I share the growing ranks of those that advocate a more flexible interpretation of plate tectonics. As stated absolute proof for such an interpretation unfortunately is conspicuously lacking, but this is true also for possible alternatives. This is likely to remain the case until there is geological information from deep wells through the attenuated crust. The increasing interest in passive margins and rift zones, because of their high hydro-carbon potential, should encourage the acquisition of such evidence.

## ACKNOWLEDGEMENT

I thank Dr. Dick Haworth, Atlantic Geoscience Centre, Bedford Institute of Oceanography, Canada, for his encouragement and criticism.

REFERENCES


1. W.J.M. van der Linden, Earth Planet.Sci.Lett., **27**, 409, 1975.
2. W.J.M. van der Linden, Mar.Geophys.Res., **3**, 1977.
3. P.A. Ziegler, Pro. NATO Adv. Study Inst., 1977.
4. M.E. Artemjev and E.V. Artyushkov, Jour. Geophys. Res., **76**, 1197, 1971.
5. M.F. Osmaston, Tectonophysics, **11**, 387, 1971.
6. I.B. Ramberg, Norg. Geol.Unders., **325**, 1, 1976.
7. P.A. Ziegler, Am.Assoc.Pet.Geol.Bull., **59**, 1073, 1975.
8. R.A. Daly, Am.J.Sci., **15**, 269, 1903.
9. G. Richter-Bernburg, Geol.Jahrb., **88**, 1, 1969.
10. A.M. Dainty, C.E. Keen, M.J. Keen and J.E. Blanchard, Am. Geophys. Union Monogr., **10**, 349, 1966.
11. G.N. Ewing, A.M. Dainty, J.E. Blanchard and M.J. Keen, Can. J. Earth Sci., **3**, 89, 1966.
12. R.D. Howie and M.S. Barss, Geol. Surv. Canada Paper 74-30, **II**, 35, 1975.
13. R.E. Sheridan and C.L. Drake, Can. J. Earth Sci., **5**, 337, 1968.
14. K.A. Poehls, B.P. Luyendyk and J.R. Heirtzler, Jour. Geophys. Res., **78**, 6985, 1973.
15. D.L. Barrett and C.E. Keen, Jour.Geophys.Res., **81**, 4875, 1976.
16. P.F. Burollet and R.S. Byramjee, C.F.P. Notes et Mémoires, **11**, 71, 1974.
17. W. Bosum and A. Hahn, Interpretation der Flugmagnetometervermessung des Oberrheingrabens, in: Graben Problems, ed. by J.H. Illies and St. Mueller, Schweizervart, Stuttgart, 1970.
18. R.W. Girdler and P. Styles, Nature, **247**, 7, 1974.

# OBSERVATIONS ON THE MARGIN OFF NORWAY (66-70°N) AND THE HISTORY OF EARLY CENOZOIC RIFTING

Olav Eldholm

Department of Geology, University of Oslo,
Oslo, Norway.

ABSTRACT. The early history of rifting and subsequent formation of an ocean are reflected in the structural and depositional history of the passive continental margin. The margin off Norway between the Jan Mayen and Greenland-Senja fracture zones exhibits features that relate to the process of rifting. Among these are: marginal basement highs, escarpments, depth anomalies, prominent block faults and areas of opaque smooth basaltic basement. Recent investigations have also revealed areas of typical sub-basement reflectors. The margin is separated into two regions representing different styles of development. This is probably caused by the pre-rifting geological framework and the very early phase of sea floor spreading in the Norwegian Sea.

## 1. INTRODUCTION

The concept of plate tectonics implies that the normal sea floor spreading stage is preceded by a sequence of events associated with the break-up of the continental crust. Hence, evidence of the early development of non-failed rifts will have to be found beneath the passive continental margins. Important questions relate to the extent of the continental crust and the structural and compositional changes across the margin. As a passive margin matures, the original rifting structures become modified and subdued caused in part by deposition of sediments, subsidence and lithospheric modification. Detailed geophysical information, in particular deep seismic penetration, is needed to study problems of this kind.

Because its young age, the Red Sea has become a type area

*I. B. Ramberg and E.-R. Neumann (eds.), Tectonics and Geophysics of Continental Rifts, 229-236.*

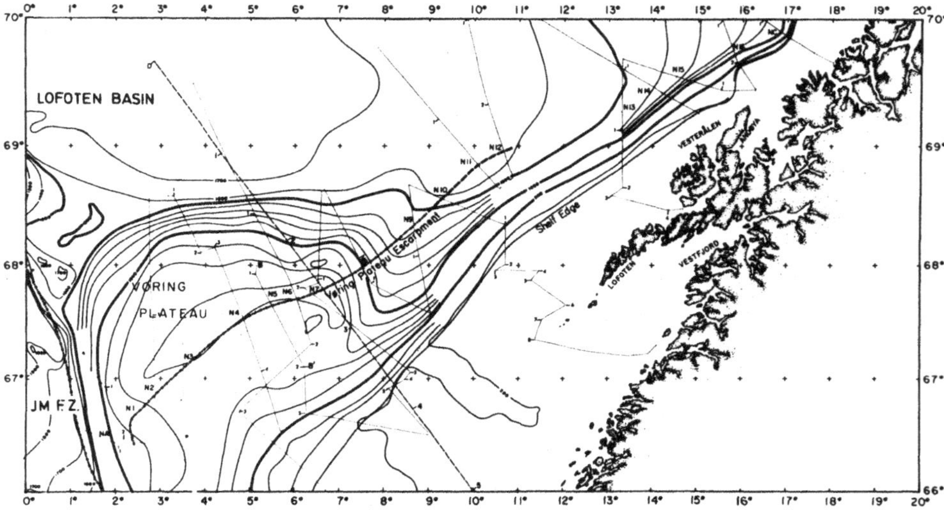


Fig. 1. Bathymetric map from Talwani and Eldholm [1]. Contours in uncorrected fathoms.

exhibiting many of the features that characterize rifting and early sea floor spreading. On the other hand, recent studies of older Cenozoic passive margins have yielded data which will have to be interpreted in the context of the initial development of a non-failed rift. Here, we focus on the segment of the Norwegian continental margin located between the Jan Mayen and Greenland-Senja fracture zones (66-70$^{\circ}$N). A tentative history of the early Cenozoic development is suggested based on available data. The tentative nature of the ideas is stressed because the virtual absence of information about the lower crust and upper mantle.

## 2. RESULTS FROM MARINE GEOPHYSICAL INVESTIGATIONS

The physiography of the margin changes significantly between 66 and 70$^{\circ}$N. In the south there is a marginal plateau, the Vøring Plateau, at a depth of about 1500 m; but north of 68$^{\circ}$N a typical continental slope exists between the shelf and the Lofoten Basin (Fig. 1). Both the shelf and slope narrow progressively northwards and the slope attains a width of only 25 km off the Vesterålen islands. The Lofoten-Vesterålen shelf forms an unique sedimentary province on the shelf off Norway. Apparently, the Cenozoic and part of the Mesozoic section are missing and the total sediment thickness is smaller than elsewhere on the shelf [1].

Among the main structural elements of the margin is a major

buried structural high underneath the outer part of the Vøring Plateau [1]. The high is bounded landward by a steeply dipping interface, the Vøring Plateau Escarpment (Fig. 1). The escarpment has been associated with prominent changes in the geophysical parameters. There are significant contrasts in the sediment thickness and the velocity-depth function across the escarpment. The thickness of sediments increases from a few hundred meters on top of the high to 6-8 km in the landward basin. Furthermore, both a gradient in the isostatic gravity anomaly and a magnetic edge anomaly bounding a landward quiet zone are related to the seismic expression of the escarpment. The surface of the acoustic basement appears to be continuous with the oceanic basement in the Lofoten Basin, but its opaque surface is clearly much smoother than what is typical for the oceanic basement [1].

Except for a linear narrow basement ridge along the base of the slope off Vesterålen, little is known about the structural fundament of the continental margin north of 69°N [2]. On the juxtaposed Greenland margin a basement high similar to that on the Vøring Plateau, is buried underneath the lower continental slope just south of the Greenland Fracture Zone. These marginal highs appear to be continuous in a reconstructed pre-anomaly 23 configuration [3].

Talwani and Eldholm [1] interpreted the escarpments as representing the location of Cenozoic rifting in the Norwegian-Greenland Sea, hence separating areas of subsided continental crust from young oceanic crust created by the process of sea floor spreading. DSDP drilling results have confirmed a basaltic nature of the marginal smooth acoustic basement reflector [4].

Multi-channel seismic reflection measurements have revealed sub-basement reflectors beneath the inner part of the Vøring Plateau structural high [5]. Although the seismic signature of layer 2 has not been established and the fact that drilling results from layer 2 may be consistent with reflectors of this kind, the observations cannot easily be overlooked because these reflectors have only been defined in a zone adjacent to the continental margin. A multi-channel profile recorded by the Seismological Observatory off Lofoten (line I, Fig. 2) also reveals sub-basement reflectors in a zone seaward of the base of the slope. A recent study of all the available data [6] shows that the margin consists of two regions having different structural styles. A southern region extending to about 69°N is characterized by the escarpment at the Vøring Plateau. The basement high plunges north of the plateau proper, but a 55-90 km wide zone of smooth acoustic basement in part underlain by typical sub-basement reflectors lies on the seaward side of the escarpment which becomes progressively less distinct. Linear pre-23 magnetic anomalies lie in this zone. The northern region is characterized by a narrow margin underlain by major

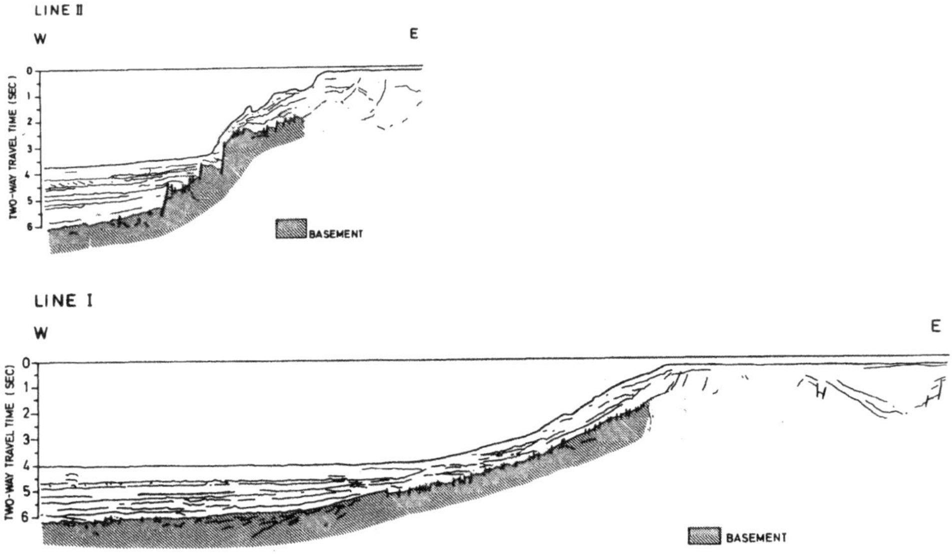


Fig. 2. Line drawings of multi-channel profiles across the margin off Lofoten (I) and Vesterålen (II) [6].

block faults trending along the bathymetric contours (line II, Fig. 2). In the north the smooth acoustic basement appears to continue all the way from the Lofoten Basin to the shelf edge. South of Lofoten the edge of the basement turns seaward and loses distinction approaching the northern flank of the Vøring Plateau (Fig. 3).

Poles of rotation describing the Cenozoic relative plate motion in the Norwegian-Greenland Sea have been calculated by Talwani and Eldholm [3]. The earliest pole of rotation is the finite difference pole between anomalies 21 and 23. The pre-anomaly 23 opening has been determined by assuming that the escarpments mark the line of rifting continuing the 21-23 rotation an amount which make the Greenland and the Vøring Plateau escarpments line up. The time of onset of normal sea floor spreading was set to the negative polarity interval between anomalies 24 and 25 (57-58.5 mybp [7]). Both in the Norway and Lofoten basins a pre-anomaly 23 shift in the spreading axis is indicated by the consecutive plate rotations [3].

## 3. DISCUSSION

These observations appear to have a close relationship to the

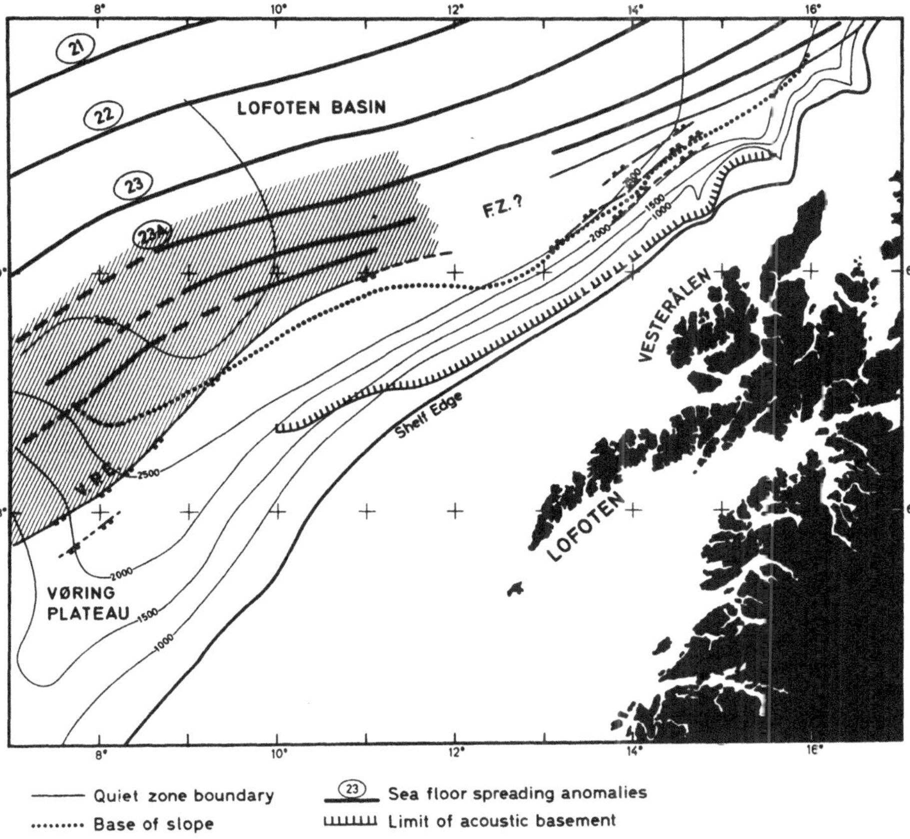


Fig. 3. Major features of the Lofoten-Vesterålen margin. Shading indicates area of smooth acoustic basement in part underlain by sub-basement reflectors. Simplified from Eldholm, Sundvor and Myhre [6].

rift formation and the initial accretion of oceanic crust by sea floor spreading. Among the outstanding questions are the nature of the marginal basement high and the zone of smooth basement including the sub-basement reflectors, how has the Vøring Plateau obtained its shallow elevation and what is the significance of the shallow acoustic basement underlying the slope in the north.

The overall geophysical data, the fact that well-defined linear magnetic anomalies overlie the zone of sub-basement reflectors and the geometrical constraints indicate that the marginal escarpments reflect a major geological boundary associated with the transition from ocean to continent. This does not rule out a complex uppermost part of the oldest oceanic crust most probably caused by basalt flowage. Pre-Cenozoic spreading implying oceanic crust landward of the escarpment has been favored by Bott [8] and Russell [9] suggesting Mesozoic and lower Permian episodes, respectively. The geophysical data and exploration results from the Norwegian shelf [10] and the North Sea [11] do not support the existence of oceanic crust landward of the Vøring Plateau Escarpment. In our opinion, there has been no earlier period of normal sea floor spreading prior to the Cenozoic. However, Mesozoic crustal extension by down-faulting and graben formation possibly related to the North Sea Mesozoic phase of crustal tension [11], is consistent with the data. Thus, it appears that the stage of oceanic crust formation was never achieved.

The onset of normal sea floor spreading took place between anomalies 24 and 25, with a subsequent shift in the spreading axis. Detailed analysis leads to the conclusion that the shift occurred just prior to anomaly 23A time, the new axis being the incipient Mohns Ridge. This evolutionary model is consistent with the lineation pattern and the plate geometry. It is not known whether the original spreading axis was entirely continuous or offset by a transform fault (Fig. 3). It appears that the initial phase of spreading was accompanied by an exceptionally high rate of basaltic extrusion forming the smooth acoustic basement surface. This may have been a continuous process or connected to the shift in spreading axis in such a way that the smooth surface represents flows and pyroclastic material overlying slightly older oceanic crust. The period of smooth basement formation ended prior to anomaly 23 time and the basalts may have been extruded subaerially. The volcanic material also overflowed the neighbouring continental crust in the young rift zone. The smooth reflector appears to be bounded landward by the flanks of the rift depression. No similar reflector has yet been identified landward of the Vøring Plateau Escarpment, but may exist below the thick section of Cenozoic sediments. The smooth acoustic basement does not exhibit any appreciable magnetic or gravity signature landward of the escarpment.

The rifting took place within a sedimentary basin of epicontinental nature, the rift zone crossing the basin diagonally in such a way that the central basin lies on the Norwegian side in the south and on the Greenland side in the north. Thus the initial rift was located on the basin margin off Vesterålen. This possibly together with the early shift in the spreading axis, may explain the different structural styles of the margin. In the north the subsidence by thermal contraction has occurred in a narrow region between the rift axis and the emerged continent causing the displacement to take place along major fault zones. Flexuring and minor fault displacement dominate in the south where the distance to the basin margin becomes considerably larger (Fig. 2). An early Tertiary age for maximum subsidence is indicated by the absence of fault structures in the overlying sediments [6].

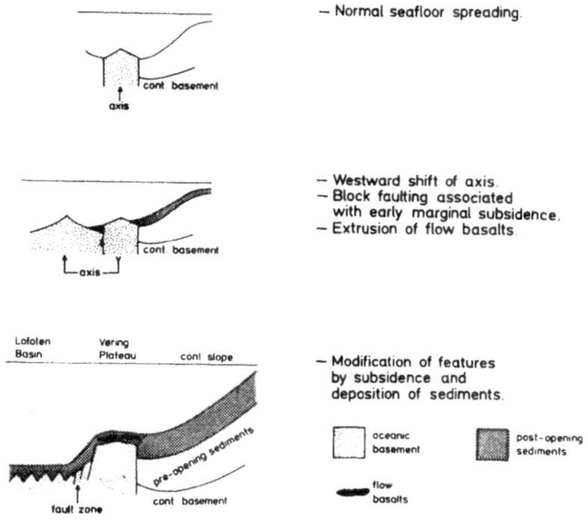


Fig. 4. Development of the Vøring Plateau. Schematic only.

We finally turn to the question of the high elevation of the marginal structures and note that the Vøring Plateau is a prominent depth anomaly both with regard to the acoustic basement and the deepest underlying reflector. This may be ascribed to one or more of the following causes:

- initial high rate of basaltic extrusion in conjunction with onset of sea floor spreading.
- "drag" effect by reversal of the oceanic subsidence curve adjacent the passive margin due to coupling between oceanic and continental crust.
- extinct spreading ridge high.
- spreading axis within a hotspot region [3].
- faulting in the newly created oceanic crust.

The latter mechanism reflects the existence of several scarps and troughs on the outer flank of the Vøring Plateau and the fact that block faulting contributes significantly to the marginal subsidence during the early period of rifting, whereas flexuring becomes more dominant as the margin matures. Block faulting has generally been depicted in the continental crust, but there is no obvious reason why it should not occur also in the virginal oceanic crust. This mode of development is shown schematically in Fig. 4.

## 4. CONCLUSIONS

The process of rifting in the continental crust is reflected in the structures and composition of the passive margin. Geophysical data from a segment of the Norwegian margin between 66 and 70°N have revealed that there is a narrow transition between the sea floor spreading type crust and the continental crust. The adjacent continental crust is thinned and stretched, but contrasts clearly with the oceanic crust. Migration of the spreading axis occurred about 2 m.y. after the onset of sea floor spreading and was possibly accompanied by extrusion of flow basalts and deposition of volcanic material defining a smooth opaque acoustic basement surface which in places is underlain by sub-basement reflecting interfaces. This event may be associated with the activity in the Brito-Arctic volcanic province reflecting the existence of a regional hotspot. The structural style of the subsidence in the marginal rift zone reflects the interaction between rift location and the existing geological framework.

## REFERENCES

1. M. Talwani and O. Eldholm, Geol. Soc. Am. Bull., **83**, 3572, 1972.
2. O. Eldholm and C.C. Windisch, Geol. Soc. Am. Bull., **85**, 1661, 1974.
3. M. Talwani and O. Eldholm, Geol. Soc. Am. Bull., **88**, 969, 1977.
4. M. Talwani and G. Udintsev et al., Initial Reports of the Deep Sea Drilling Project, **38**, 1256, 1976.
5. K. Hinz and J. Weber, Zeitschrift Erdöl und Kohle, Erdgas, Petrochemie, 3, 1976.
6. O. Eldholm, E. Sundvor and A. Myhre, Inst. Geol. Oslo, Intern Skr. Ser., **13**, 1977.
7. J.L. LaBrecque, D.V. Kent and S.C. Cande, Geology, **5**, 330, 1975.
8. M.P.H. Bott, Norges Geol. Unders., **316**, 195, 1975.
9. M.J. Russell, Scott. J. Geol., **12**, 315, 1976.
10. H. Rønnevik and T. Navrestad, GeoJournal, **1**, 33, 1977.
11. P.A. Ziegler, GeoJournal, **1**, 7, 1977.

# PELAGIC SEDIMENTATION IN IMMATURE OCEAN BASINS [x]

J. Thiede

Department of Geology, University of Oslo,
Blindern, Oslo 3, Norway.

ABSTRACT. Young ocean basins or their remains are relatively uncommon features of the earth's surface. Whereas the world's old ocean, the Pacific, is surrounded almost entirely by active continental margins which have 'eaten' the early, probably pre-Mesozoic rifted margins of its primordial predecessors, passive margins characterize almost all fringes of young ocean basins which have originated since mid-Mesozoic times (Southern Ocean, Arctic Ocean, Indian Ocean, Atlantic Ocean). Such basins have evolved from rifts cutting through relatively stable continental crust. Evidence of the early history of passive continental margins, and of the invasion of the sea into the young, growing ocean basins has to be sought today along the continental rise and slope of these margins. Generally the early history of the Mesozoic basins lies deeply buried. However, Cenozoic analogues can be studied in the Gulf of Aden/Red Sea system as well as in the Gulf of California, and in the Arctic Ocean. Although a truly oceanic or pelagic environment is established only after these rifts have attained considerable depth and width, and after the initial volcanism has died out, some of the early lithofacies typify restricted depositional environments. The original distribution of such facies is controlled by the physical shape of the evolving ocean basin, which can be approximated using principles of plate rotation and subsidence of oceanic crust. The nature of these restricted lithofacies is probably entirely climate-controlled. End-members of such restricted lithofacies are halites, which have been sampled in the Red Sea and which probably occur along major segments of

x This study has been supported by a grant from the Norwegian Petroleum Directorate.

*I. B. Ramberg and E.-R. Neumann (eds.), Tectonics and Geophysics of Continental Rifts, 237-248.*

the Atlantic margins, or fine-grained black shales, which have been deposited in anaerobic paleoenvironments and which therefore have preserved their high contents of organic carbon (examples Red Sea, Indian and Atlantic Oceans). Such restricted depositional environments exist as long as the exchange of bottom water masses between the youthful basins and the open ocean is hampered by narrow, and/or shallow sills. The tectonic evolution of such circulation barriers is therefore one of the keys to the understanding of the depositional paleoenvironment of the "restricted" pelagic or hemipelagic sediment facies. It is these sediments which have documented the history of the early oceanic stages of old and new rifts.

## 1. INTRODUCTION

This symposium discusses ubiquitous and important tectonic features which mark elongate regions of general crustal extension [1,2]. This paper will restrict itself to rifts which are cutting through or have cut through continental crust to create a marine basin with an underlying oceanic crust (back-arc basins are therefore excluded); it will discuss the depositional environment in such rapidly deepening, but usually narrow and elongate immature ocean basins. Since the evidence from Paleozoic or older passive continental margins is ambiguous and disputable, this study will focus upon the available modern, Cenozoic and Mesozoic cases (Fig. 1). These are only a very few examples of spreading centers which presently intersect continental margins, or where this has happened in the past and where the early history of the young ocean has been preserved [3]. Evidence for the latter cases are generally found - often deeply buried under thick sedimentary sequences - along the continental rise of passive margins.

## 2. MORPHOLOGY AND WATER DEPTH OF YOUNG OCEANS

### 2.1 Gulf of Aden, Red Sea, Gulf of California and Arctic Ocean

In the modern ocean, active spreading centers intersect continental crust in three regions (Fig. 2): a) the Gulf of Aden/Red Sea complex with its branches into the East African rift system [4] and into the Arabian peninsula [5], b) the Gulf of California with the complex plate-tectonic situation at its northern end [6] and c) the eastern end of the Eurasian Basin (Arctic Ocean) where the Nansen Cordillera intersects the continental margin of the Laptev Sea [7].

Much attention has been paid recently to the pattern of rifting and sea floor spreading around the triple junction/hot spot area of Afar [8] and to the evolution of the oceanic basins

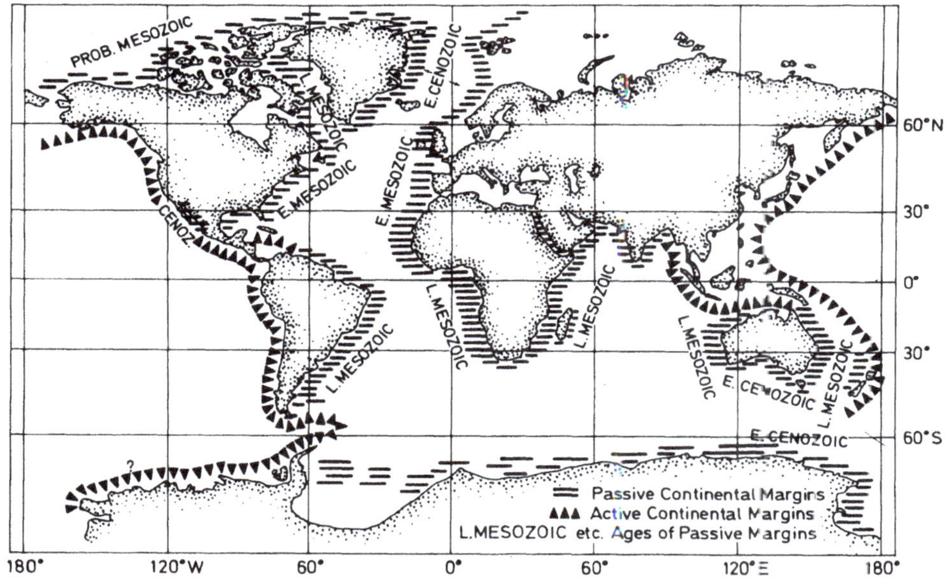


Fig. 1. Distribution of modern active and passive continental margins and their approximate ages (after various sources).

of the Red Sea and of the Gulf of Aden [9]. The Gulf of Aden and the Gulf of California are morphologically similar since both can exchange their bottom water mass with the meso- and bathypelagic water masses of the adjacent open ocean (Fig. 2a and d), because they are not bounded by shallow sills. In the inner Gulf of Aden the spreading axis can be traced into the Tadjoura Trench which shallows landwards, breaking up into several basins. The Red Sea is today believed to be nearly all oceanic in crustal structure because large amplitude, long wavelength magnetic lineations of the sea-floor spreading type occur along its margins [10]; the magnetic anomalies mapped in the Red Sea seem to suggest two phases of spreading, an early Late Eocene - Early Oligocene phase and a Late Plio-Pleistocene one. The inner Red Sea with the Gulf of Aqaba (Fig. 2b) is similar to the inner Gulf of Aden because the central rift zone also breaks up into a series of isolated and landward shallowing troughs. However, the long and narrow basin of the entire Red Sea is separated from the Gulf of Aden by the shallow (137 m) Straits of Bab el Mandeb (Fig. 2a). The dominant rock formations on both sides of Bab el Mandeb are composed of Late Cenozoic volcanics [11] which are covered only by a thin veneer of Recent calcareous sands in the straits [12]. The shallow depth of this young ocean crust suggests that it may in the future develop into a feature similar to aseismic ridges in older ocean basins.

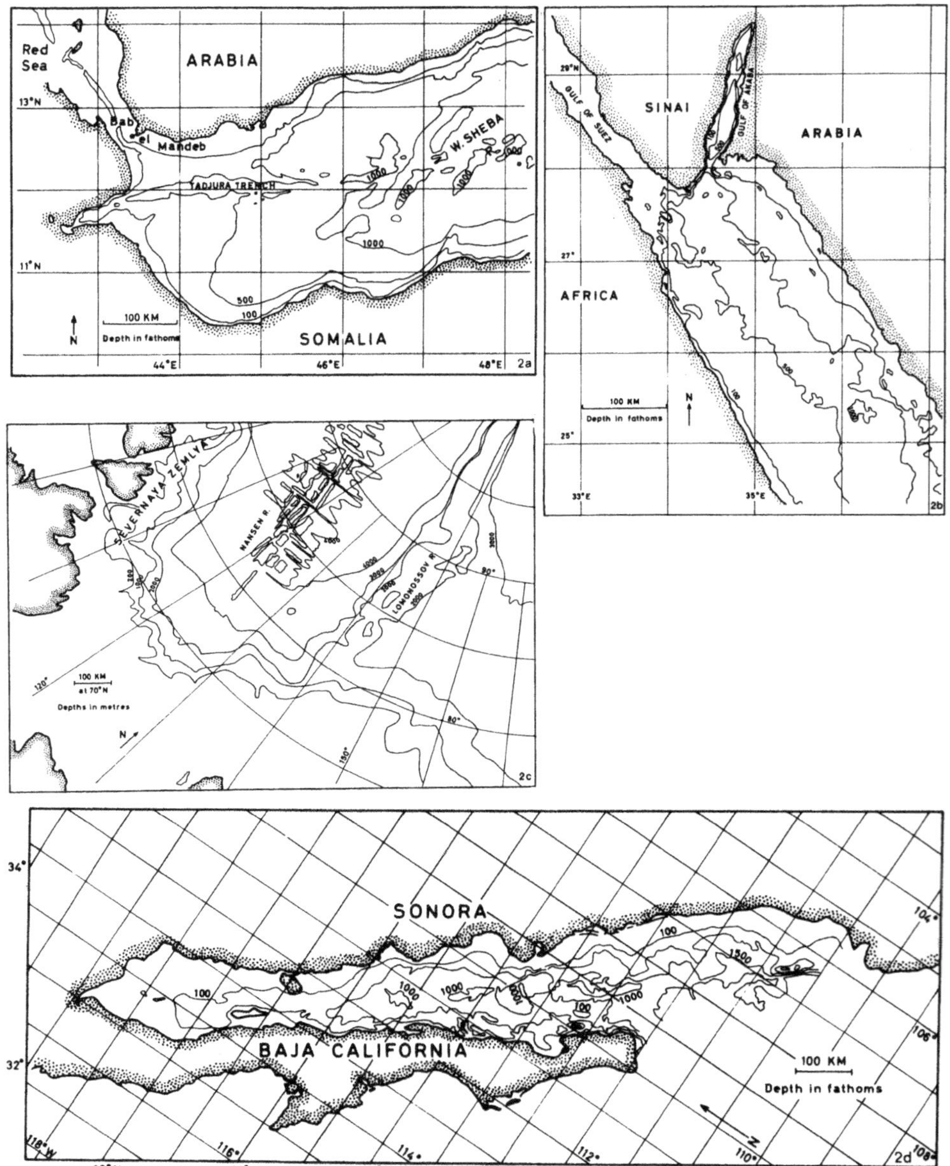


Fig. 2. Bathymetry of four youthful ocean basins, a) inner Gulf of Aden, simplified after Laughton in [9], b) the inner part of the Red Sea, simplified after [13], c) the intersection of the Eurasia Basin spreading center with the Laptev Sea continental margin [14], d) the Gulf of California [15,16].

The history of the opening of the Gulf of California is different from that of the inner Gulf of Aden and Red Sea since its basin shape is largely due to movements along a system of transform faults which offset short segments of axial accretion (spreading centers) [6]. The basin itself (Fig. 2d) is open to the epi-, meso- and bathypelagic water masses of the tropical Northeast Pacific Ocean, but it contains a series of closed troughs along its deep axis. These troughs deepen with decreasing distance to the mouth of the approx. 1000 km long Gulf of California. The gulf itself it characterized by extreme changes in relief, with steep boundary slopes and shallow ridges and banks subdividing it into the above mentioned troughs [15].

The Arctic ocean spreading center which intersects the Laptev Sea continental margin (Fig. 2c) possibly resembles the situation in the Red Sea and Gulf of Aden. However, the nature of spreading along the Nansen Cordillera is not well understood [17]. The seismicity of this spreading center is very low when approaching the Laptev Shelf, and it is not known how Siberia has absorbed $\sim 0.4 \times 10^6$ km$^2$ of crustal extension required by spreading in the Eurasia Basin since Early Tertiary time [7]. The Laptev Sea is floored by the extensive terrigenous input from the Lena and other large Siberian rivers [18] which has apparently obscured any clear morphological or structural evidence how this rift intersects with the continental margin. It is possible that the Sadko Trough [7] is in part a graben - like structure developed in Cenozoic shelf sediments.

## 2.2 Mesozoic and Early Cenozoic Margins

The nature of the depositional paleoenvironments in fossil young ocean basins has to be studied along the foot regions of all passive continental margins, where the history of the early stages of newly created ocean basins is buried (Fig. 1). The age of the examples depend upon the time of rifting and initial spreading. Cenozoic cases can be studied in the Norwegian Sea and Arctic Basin, between Antarctica and Australia, and in the Gulf of Aden and in the Red Sea. Mesozoic margins are observed in the North and South Atlantic and Indian Oceans. In the case of the Indian Ocean [19], early stages of basin development are virtually unknown because magnetic anomalies of the oldest crust over large parts of this ocean basin are poorly mapped and difficult to interpret. It is not easy to reconstruct the paleomorphology of the fossil young basins, mainly because the crustal subsidence along passive continental margin is poorly understood [20]. However, deposits of the early oceanic stages, which represent small spatial and temporal targets, have been sampled at a few sites by the Deep Sea Drilling Project.

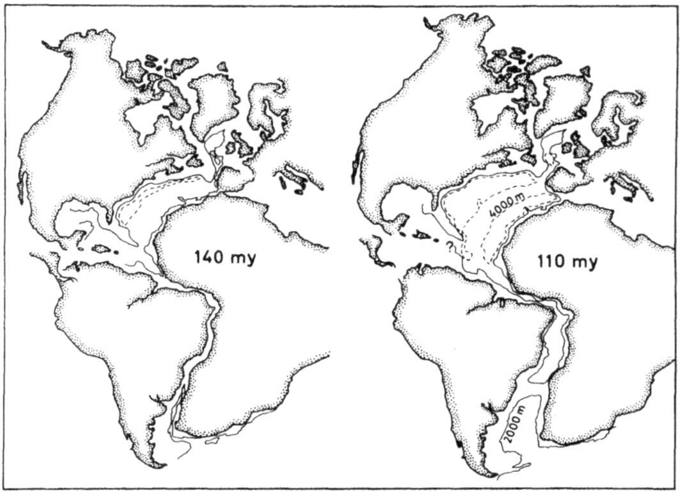


Fig. 3. Paleobathymetry of the young North and South Atlantic Oceans, simplified after [21].

The morphologic development of the known fossil cases seems to confirm to what has been learned from the modern ones (compare Figs. 2 and 3). The early Atlantic [21] and Indian [19] ocean basins consisted of a series of troughs, which developed along the trace of the widening rift and which then deepened rapidly due to the subsidence of the floor of the ocean basin [22] and of the adjacent continental margins [20]. As in the modern Red Sea, for example, the water exchange between parts of the young ocean or with the open world ocean can be inhibited by structural highs (often volcanic features) which in some case were initially emergent [23] and which seem to subside in a fashion similar to oceanic crust [24]. The development of large transform faults, which allow strike slip movements between opposite continental margins - as in the early South Atlantic between NE South America and Central Africa (see Fig. 3) - will also lead to relatively shallow areas separating deeper parts of the developing ocean basins.

## 3. THE DEPOSITIONAL ENVIRONMENT IN YOUNG OCEAN BASINS

Once the shape of a young ocean basin has been established, its depositional environment is probably dominantly controlled by the mode of water exchange with the open ocean [25]. The water masses filling the Gulf of California, the inner Gulf of Aden and the Red Sea are known to have special physical and chemical characteristics. Sediments deposited in these basins differ from regular pelagic sediment facies because young oceans are much shallower than old ocean basins, whose bottom water masses are prevented from entering. Depending upon the respective sill depth only epi-, meso- and bathypelagic waters can feed the youthful marginal basin. Prominent depth indicators such as the calcite compensation depth or the lysoclines of various calcareous fossil groups [26] of the open ocean can therefore not be traced into the marginal basin, even if isolated depressions drop well below the regional water depth of these interfaces. This is illustrated by the existence of calcareous sediments covering the floor of the over 5 km deep Alula-Fartak-Trench [27] in the outer Gulf of Aden. By comparison the calcite compensation depth occurs at 4-4.5 km in the adjacent basins of the Indian Ocean [26]. The terrigenous input into these young oceans is relatively modest despite the steep topographic slopes of their passive margins and despite the elevation of the flanking continental rims which reflects the positive temperature anomaly of the lithosphere and asthenosphere in such a region.

The origin of young ocean basins due to crustal spreading which shapes the basin, as well as the climate-controlled exchange of its water masses with the open ocean, give rise to the development of a number of typical sediment facies. Such sediments are probably not unique in a global sense, but they seem to characterize the special depositional environment in an immature ocean basin. Since we are far from having examples of all possible modern and historic climatic end members at out disposal, the topic of pelagic sedimentation in such basins can only be discussed conceptually at the present time. The sedimentation of those basins responds to a complicated network of boundary conditions which are controlled by the tectonics and geology of the basin surroundings, by the oceanography of the adjacent ocean basin and by the meteorology of its climatic zone.

### 3.1 Restricted biota

In many areas the pelagic/hemipelagic sediments in young oceans are dominantly biogenous. Their components are produced by calcareous and siliceous planktic faunas and floras which are clearly different from those of the adjacent open ocean. It usually requires tedious and sophisticated micropaleontological studies of such faunas and floras to show that they have been deposited in a restricted basin. The ranges of surface water temperature and salinity fluctuations of the marginal basin exceed those of the open ocean, resulting in lower diversities and higher dominances in the restricted area as has been illustrated by the faunal/floral compositions and the occurrence of specific phenotypes for planktic foraminifers in the modern Gulf of California [28] and in the Red Sea [29,30,31,32]; in the Red Sea this also holds for calcareous nannoplankton, dinoflagellates, radiolaria and pelagic gastropods [31,33]. The restriction of basins is also reflected in particular oxygen isotope ratios of shell material from organisms which have grown in such environments [34, 35].

### 3.2 The development of anoxic depositional environments

Sediments which apparently were deposited in an anoxic environment have been found in almost all young oceans. Until now they have been sampled in the Quaternary [33] as well as Miocene/ Pliocene [36] Red Sea and Gulf of Aden [37], in the Cretaceous North and South Atlantic and Indian Oceans [38]. The question arises whether such anoxic sediments have been deposited in a stagnant basin, as advocated for the mid-Cretaceous North Atlantic Ocean [39], or under the influence of a mid-water oxygen minimum, as could be observed in the Cretaceous Pacific [40] and South Atlantic Oceans [41] and as can be studied in the modern Gulf of California [42]. A stagnant basin would require a very shallow sill depth, whereas oceanic mid-water oxygen minima develop even without a barrier between the marginal basin and the open ocean. The reason for the sudden and apparently basinwide coeval deposition of anoxic sediments with high concentrations of organic carbon several times during Mesozoic and Cenozoic time in the oceanic realm has yet to be explained satisfactorily [38,43], mainly because the composition and origin of the organic matter in these sediments is not well known. It now appears that the mechanisms of organic carbon preservation [44] might be important for the genesis of anoxic sediment facies.

### 3.3 Deposition of evaporites

The occurrence of evaporites in an ocean basin without doubt requires the presence of a shallow barrier regulating the inflow of saltwater from the open ocean into the youthful basin, and an excess of evaporation over precipitation. However, such barriers did not block the water exchange in all young oceans. Thus, although the climate was sufficient, evaporites are missing in the Gulf of Aden, the Indian Ocean and in parts of the Atlantic Ocean [45]. Evaporites believed to overly oceanic crust have been found in the Red Sea, where they have been sampled by the Deep Sea Drilling Project, and in the South and North Atlantic, where they have been inferred from seismic data [46,47,48]. However, in many areas it has been problematic to establish if the evaporites are autochthonous or if they have been tectonically implaced; evidence from the Red Sea [49] seems to suggest this latter possibility. Even if it can be established that the salt is still in its original depositional location over the oceanic basement one has to decide whether salt deposition was possible in a water-filled relatively deep young ocean basin [50], or if the basin had to dry up to allow salt deposition, as has been proposed for the Miocene salts in the Mediterranean [51].

### 3.4 Metalliferous sediments

Metalliferous sediments have been found in many oceanic drill sites, where they overly directly the basaltic crustal rocks. However, they have only rarely been observed in drill sites penetrating the sedimentary record of young ocean basins. Probably the only region where their formation in such a geologic situation can be observed at the present time, is the brine deeps of the Red Sea [52,53]. Due to the physical properties of such metalrich sediments, they are easily eroded; a favorable depositional environment, in isolated troughs of young ocean basins with a very sluggish bottomwater circulation, creates a stratified water column above the soft metalliferous sediments and shields them from erosion [54].

## 4. CONCLUSIONS

The history of pelagic/hemipelagic sedimentation in young ocean basins begins considerably after the initial rifting and spreading, after the initial volcanism has died away, and after the evolving basin has deepened to several hundred if not thousand meters water depth.

Such young ocean basins are narrow troughs with steep walls and with a rugged terrestrial topography surrounding them. Their connection to major basins of the world ocean is narrow; sometimes they are separated from it by shallow sills. Since the oceanic crust of such basins is very young, the basins are not as deep as adjacent parts of the world ocean. These marginal appendices of the adjacent old basins are therefore filled by oceanic intermediate or surface water masses.

The long and narrow ocean basins generated by the rifting process exchange their surface water masses with the open ocean through narrow entrances. The circulation pattern in the young ocean is climate controlled and its surface water masses rapidly attain their own physical and chemical characteristics, resulting in biota with faunal and floral assemblages which are different from those of the corresponding pelagic environment.

The basins themselves are often subdivided into smaller troughs by structural highs. The sills between these troughs can be considerably shallower than the basins themselves, and thereby inhibit the renewal of bottom waters. Despite this restriction of surface and bottom water renewal, these basins are not entirely anoxic but can allow benthic life at the basin floor (Red Sea, Gulf of Aden, Gulf of California). The fact that anoxic sediments can be deposited under euxinic conditions or in a mid-water oxygen minimum layer is an important clue for the understanding of the history of structural barriers separating young ocean basins from the open world ocean.

Evaporites overlying oceanic crust require the presence of a shallow sill depth during their deposition and a climate where evaporation exceeds precipitation. The extent and thickness of the evaporites is probably intimately linked to the nature and stability of the sill depth that regulates the inflow of saltwater over geologic time.

## REFERENCES

1. J.H. Illies, Taphrogenesis and Plate Tectonics, in: Approaches to Taphrogenesis, ed. by J.H. Illies and K. Fuchs, Schweizerbart, Stuttgard, 1974.
2. E.E. Milanovsky, Tectonophysics, 15, 65, 1972.
3. E.D. Schneider, Geol. Soc. Am. Mem., 132, 109, 1972.
4. B.H. Baker, P.A. Mohr and L.A.J. Williams, Geol. Soc. Am.Spec. Pap., 136, 1972.
5. L. Dybertret, Phil. Trans. Roy. Soc. Lond., A 267, 9, 1970.
6. R.L. Larson, H.W. Menard and S.M. Smith, Science, 161,781, 1968.
7. P.R. Vogt and O.E. Avery, Tectonic history of the Arctic Basins, in: Marine Geology and Oceanography of the Arctic Seas, ed. by Y. Herman, Springer, Berlin, 1974.
8. A. Pilger and A. Rösler, Temporal relationship in the tectonic evolution of the Afar Depression (Ethiopia) and the adjacent Afro-Arabian rift system, in: Afar between Continental and Oceanic Rifting, ed. by A. Pilger and A. Rösler, Schweizerbart, Stuttgart, 1976.
9. A.S. Laughton, R.B. Whitmarsh and M.T. Jones, Phil. Trans. Roy. Soc. Lond., A 267, 227, 1970.
10. R.W. Girdler and P. Styles, The relevance of magnetic anomalies over the Southern Red Sea and Gulf of Aden, in: Afar between Continental and Oceanic Rifting, ed. by A. Pilger and A. Rösler, Schweizerbart, Stuttgart, 1976.
11. Z.R. Beydoun, Phil. Trans. Roy. Soc. Lond., A 267, 267, 1970.
12. G. Einsele and F. Werner, "Meteor"Forsch.-Ergbn., C10, 39,1972.
13. A.S. Laughton, Phil. Trans. Roy. Soc. Lond., A 267, 21, 1970.
14. Map of the Arctic Region, Sheet 14 World 1:5,000,000, ed. by American Geographical Society, New York, 1975.
15. T.H. van Andel, Am. Assoc. Pet. Geol. Mem., 3, 216, 1964.
16. G.A. Rusnak, R.L. Fisher and F.P. Shepard, Am. Assoc. Pet. Geol. Mem., 3, 59, 1964.

17. R.M. Demenitskaya and K.L. Hunkins, Shape and Structure of the Arctic Ocean, in: The Sea, 5, 223, ed. by A.E. Maxwell, Wiley - Interscience, New York, 1970.
18. M.L. Holmes and J.S. Creager, Holocene History of the Laptev Sea Continental Shelf, in: Marine Geology and Oceanography of the Arctic Seas, ed. by Y. Herman, Springer, Berlin, 1974.
19. J.G. Sclater, D. Abbott and J. Thiede, Am. Geoph. U. Spec. Publ. (in press, 1977).
20. A.B. Watts and W.B.F. Ryan, Tectonophysics, 36, 25, 1976.
21. J.G. Sclater, S. Hellinger and C. Tapscott, J. Geol., 85, 509, 1977.
22. J.G. Sclater, R.N. Anderson and M.L. Bell, J. Geophys. Res., 76, 7888, 1976.
23. J. Thiede, Am. Assoc. Pet. Geol. Bull., 61, 929, 1977.
24. R.S. Detrick, J.G. Sclater and J. Thiede, Earth Planet. Sci. Lett., 34, 185, 1977.
25. E. Seibold, Geol. Rundsch., 60, 73, 1970.
26. W.H. Berger, Biogenous Deep Sea Sediments: Production, Preservation and Interpretation, in: Chemical Oceanography, 5, 265, ed. by J. P. Riley and R. Chester, Academic Press, London, 1976.
27. J. Thiede, Geol. Jb., D17, 93, 1976.
28. F.L. Parker, J. Foram. Res., 3, 70, 1973.
29. H. Risch, Geol. Jb., D17, 3, 1976.
30. Y. Herman, Evidence of Climatic Changes in Red Sea Cores, in: Means of Correlation of Quaternary Successions, ed. by R.B. Morrison and H.E. Wright, University of Utah Press, Salt Lake City, 1968.
31. W.A. Berggren, Micropaleontologic Investigations of Red Sea Cores - Summation and Synthesis of Results, in: 'Hot Brines and Recent Heavy Metal Deposits in the Red Sea', ed. by E.T. Degens and D.A. Ross, Springer, New York, 1969.
32. Z. Reiss and E. Halicz, Isr. J. Earth Sci., 25, 27, 1976.
33. C. Müller, Geol. Jb., D17, 33, 1976.
34. W.G. Deuser and E.T. Degens, $O^{18}/O^{16}$ and $C^{13}/C^{12}$ Ratios of Fossils from the Hot-Brine Deep Area of the Central Red Sea, in: Hot Brines and Recent Heavy Metal Deposits in the Red Sea, ed. by E.T. Degens and D.A. Ross, Springer, New York, 1969.
35. M. Schoell and H. Risch, Geol. Jb., D17, 15, 1976.
36. P. Stoffers and D.A. Ross, Init. Rep. Deep Sea Drill. Proj., 23, 849, 1974.
37. E. Olausson and I.U. Olsson, Palaeogeogr. Palaeoclimat., Palaeoecol., 6, 87, 1969.
38. W.B.F. Ryan and M.B. Cita, Marine Geol., 23, 197, 1977.
39. I.N. McCave, Init. Rep. Deep-Sea Drill. Proj., 43, (in press, 1977).
40. S.O. Schlanger and H.C. Jenkyns, Geol. Mijnb., 55, 179, 1976.
41. J. Thiede and T.H. van Andel, Earth Planet. Sci. Lett., 33, 301, 1977.

42. S.E. Calvert, Am. Assoc. Petrol. Geol., Mem., 3, 311, 1964.
43. A.G. Fischer and M.A. Arthur, Soc. Econ. Paleont. Mineral. Spec. Publ. (in press, 1977).
44. B. Tissot, Phil. Trans. Roy. Soc. Lond. (in press, 1977).
45. K. Burke, Geology, 3. 613, 1975.
46. P. Stoffers and R. Kühn, Init. Rep. Deep Sea Drill. Proj., 23, 821, 1974.
47. P. Lehner and P.A.C. de Ruiter, Am. Assoc. Petrol. Geol., Bull., 61, 961, 1977.
48. G. Pautot, V. Renard, J. Daniel and J. Dupont, Am. Assoc. Petrol. Geol., Bull., 57, 1658, 1973.
49. R.W. Girdler and R.B. Whitmarsh, Init. Rep. Deep Sea Drill. Proj., 23, 913, 1974.
50. R.F. Schmalz, Am. Assoc. Petrol. Geol., Bull., 53, 798, 1969.
51. W.B.F. Ryan, K.J. Hsü et al., Init. Rep. Deep Sea Drill. Proj., 13, 514, 1973.
52. R.D. Bignell, D.S. Cronan and J.S. Tooms, Geol. Assoc. Can. Spec. Pap., 14, 147, 1976.
53. E.T. Degens and D.A. Ross, Hot Brines and Recent Heavy Metal Deposits in the Red Sea, Springer, New York, 1969.
54. H. Bäcker and M. Schoell, Nature Phys. Sci., 240, 153, 1972.

# NORTH SEA RIFT AND BASIN DEVELOPMENT

P.A. Ziegler

Shell Internationale Petroleum Mij.
Den Haag, The Netherlands

ABSTRACT. The intracratonic Mesozoic North Sea rift formed part of the Arctic-North Atlantic rift system, which was emplaced during the Upper Paleozoic. Carboniferous and Permian rifts in the North Sea became inactive at the onset of the Mesozoic. During the Triassic a new complex graben and trough system formed in NW Europe. In the North Sea the Viking and Central Grabens remained active throughout the Mesozoic. With the onset of sea-floor spreading in the Norwegian-Greenland Sea the North Sea rift became inactive and the Cainozoic North Sea basin began to subside.

## 1. INTRODUCTION

The North Sea occupies a large part of the intracratonic Northwest European basin. Sediments underlying the North Sea reach thicknesses of up to 9 km and range in age from Paleozoic to Quaternary.

Consolidation of the Northwest European craton became effective with the Caledonian diastrophism. The post Caledonian evolution of the North Sea area was intricately linked with the development of the North Atlantic-Norwegian-Greenland Sea. The Variscan and Alpine diastrophism influenced the North Sea only around the margins. During the Mesozoic a major rift system developed in the North Sea. This rift formed part of the Arctic-North Atlantic rift system. With the onset of sea-floor spreading in the Norwegian-Greenland Sea during the early Tertiary the North Sea rift became inactive and the saucer-shaped North Sea basin started to subside; in this area Cainozoic series attain thicknesses of about 3.5 km. The geometry of the North Sea basin

*I. B. Ramberg and E.-R. Neumann (eds.), Tectonics and Geophysics of Continental Rifts, 249-277.*

is strongly influenced by tectonic patterns established during the late Paleozoic and Mesozoic.

The stratigraphical record of the North Sea offers a unique opportunity to trace the evolution of an intracratonic rift system from its emplacement through its main rifting stages to its subsidence after becoming inactive.

## 2. PALEOZOIC TECTONIC FRAMEWORK

The North Sea is underlain by a continental crust consisting of Pre-Cambrian and Caledonian basement elements [1]. The front of the Variscan fold-belt skirts the North Sea to the south (Fig. 1).

The late Silurian-early Devonian Caledonian orogeny resulted in the fusion of the North American-Greenland plate with the Fennoscandian-Baltic plate along the suture of the Arctic-North Atlantic Caledonides [2]. In the Appalachian-Mauretanides-Variscan geosyncline the Caledonian orogenic pulse was, however, less intense and did not lead to the consolidation of a major fold-belt. The Innuitian fold-belt in the Canadian Arctic also remained active after the Caledonian diastrophism until the early Carboniferous. Consolidation of the Variscan fold-belt was only achieved after a number of further orogenic phases, during the late Carboniferous and early Permian. By this time the North American-European craton and the South American-African plate were welded together to form the Pangaean megacontinent.

Viewed against this megatectonic framework the North Sea occupies a keystone position at the junction of the Fennoscandian-Baltic Pre-Cambrian shield, the Caledonides and the Variscides. The Pre-Cambrian London-Brabant massif assumed the role of a stable intramontane platform.

## 3. EARLY HISTORY OF THE ARCTIC-NORTH ATLANTIC RIFT

In both the Appalachian-Variscan and the Innuitian geosynclines orogenic movements continued after consolidation of the Arctic-North Atlantic and Mid-European Caledonides.

During the Devonian and Carboniferous the southern North Sea and its eastward adjacent onshore areas formed part of the Variscan foredeep in which a thick sedimentary wedge was accumulated on the apparently rapidly degraded Mid-European Caledonides. In contrast the Arctic-North Atlantic Caledonides were affected during the Devonian and Carboniferous by strong post-orogenic uplifting, accompanied by transcurrent movements along a complex fault system, subparallel to their axes [3-8]. This resulted

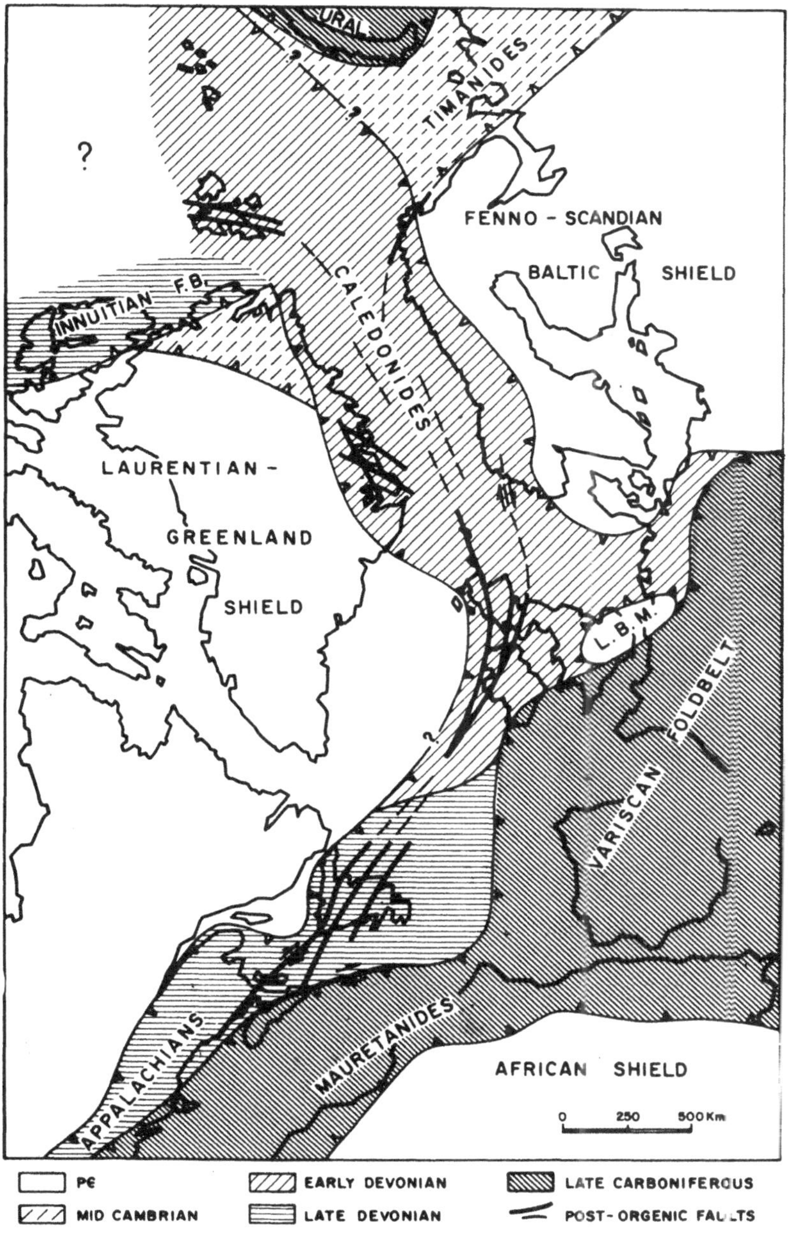


Fig. 1. Spacial relationship of Paleozoic orogenic belts in the North Atlantic realm (Legend gives age of consolidation).

in the rapid subsidence of the intramontane Old Red "Molasse" basins of Ireland, Scotland, West Norway, East Greenland and Spitzbergen. In these basins continental clastics and in part lacustrine series attain thicknesses of 6 to 8 km. Subsidence of these partly faultbounded basins was in many areas repeatedly interrupted by folding related to wrench movements. In East Greenland these deformations were accompanied by the intrusion of Devonian granites [5]. In the northern British Isles deposition of the Old Red was parallelled in the Midland Valley by the extrusion of calc-alkaline volcanics and the intrusion of post-orogenic granites [9].

Folding of the Old Red series was roughly contemporaneous with periods of intensified orogenic activities in the Appalachien-Variscan and in the Innuitian geosyncline [4]. The magnitude and in part also the direction of the Devonian wrench movements between the North American-Greenland plate and the Fennoscandian Baltic plate remain a subject of controversy.

During the Carboniferous transcurrent movements in the area of the Norwegian-Greenland Sea gave way to extension tectonics [5]. In the British Isles the largely faultbounded Midland Valley and Northumberland-Solway basins continued to subside rapidly during the Carboniferous. This was parallelled by a predominantly alkaline rift volcanism that contrasts with the Devonian calc-alkaline volcanism [9,10]. Thus an extensional setting has to be assumed for the Carboniferous development of these graben-shaped basins. However, strike-slip movements appear to have continued on a minor scale through parts of the Carboniferous.

Contemporaneous extension tectonics are documented from the Canadian Maritime Provinces as well, where a number of Post-Acadian successor basins started to subside during the Late Devonian [11,12]. However, also in these basins Carboniferous transcurrent movements appear to have played a significant role [13].

This illustrates that the regional stress systems in the Arctic-North Atlantic realm gradually changed during the Carboniferous from the Devonian strike slip/compressive setting to an extensional one that persisted throughout the Upper Paleozoic and Mesozoic, leading during early Tertiary to crustal separation between the North American-Greenland plate and the European plate.

## 4. PERMIAN EXTENSION TECTONICS

With the late Carboniferous final phases of the Variscan Orogeny the long-standing compressional setting of Central and Western Europe came to a close. The Permian development of north-western Europe was dominated by extension tectonics.

The Permian post-orogenic uplift of the Variscan fold-belt was accompanied by its partial collapse and the deposition of thick continental clastics in intramontane "molasse" basins. At the same time a NW-SE striking fault system, sub-parallel to the Tornquist-Teisseyre lineament (Fennoscandian border zone), came into evidence [14]. Emplacement of this fault system, which transects the Variscan deformation front and extends into the northern foreland of the Variscides, (Fig. 2) is thought to be related to regional tensional stresses affecting NW Europe as a whole. Such stresses may have originated both in the early opening phases of the Proto-Tethys [15] and in extension movements in the Arctic-North Atlantic rift zone.

The Lower Perminan uplift and faulting of the Variscides and their foreland was accompanied by widespread volcanic extrusions. Complex differentiation sequences ranging from basalts to porphyries and quartzporphyries are common [16]. Maximum volcanic thicknesses of about 130 have been recorded in northern Germany [17,18]. Lower Permian volcanics are, however, largely restricted to the Variscan fold-belt and its foredeep, the margins of the Ringkøbing-Fyn-Møns High, the Fennoscandian border zone and the Oslo Graben. Their absence below the northern North Sea, indicates that the Carboniferous rifts had become inactive.

In the Oslo Graben volcanism started during the Upper Carboniferous to Lower Permian and terminated in the Lower Triassic [19,20]. Towards the southwest the Oslo rift may find its extension in the Bamble Trough underlying the Skagerrak and possibly in the Horn Graben located in the Danish offshore. This is supported by the occurrence of Permo-Carboniferous volcanics in the coastal areas of southern Norway [21] and the presence of thick Permian volcanics in the northern Horn Graben.

The Oslo Graben and the Bamble Trough are located at the north-Western termination of the Tornquist-Teisseyre lineament, which crosses Scania; from there Permo-Carboniferous dykes subparallel to the strike of this lineament have been reported [22,23]. It is envisaged that minor wrench movements along the Tornquist-Teisseyre lineament triggered the emplacement of the Oslo-Bamble Rift and its underlying crustal hotspot. Volcanics in the Oslo Graben display the distinctly alkaline composition of intracontinental rifts [24-26]. In contrast, Permian volcanics in the Variscan foreland are only mildly alkaline [27]. Volcanics in the Variscan domain are characterised by a calc-alkaline chemistry, that is typical for a post-orogenic volcanism.

## 5. PERMIAN VARISCAN FORELAND COLLAPSE

With the onset of the Saxonian (Upper Lower Permian), volcanism

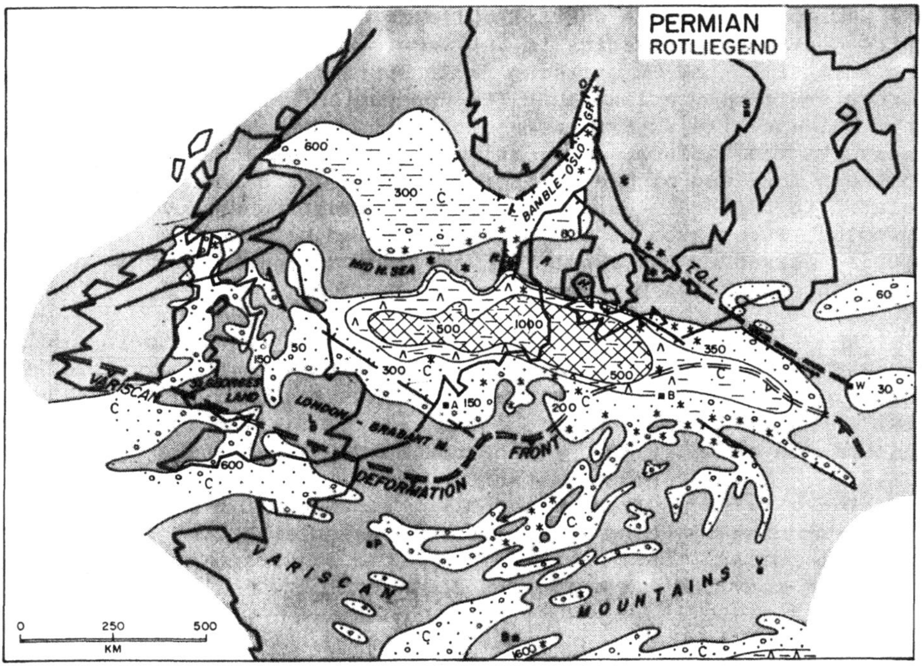


Fig. 2. Rotliegendes paleogeography of NW Europe (for legend to paleogeographic maps see Fig. 14)

in the Variscan foreland came to an end and two large essentially east-west oriented basins started to subside in northwestern Europe (Fig. 2). These basins were separated by the Mid-North Sea-Ringkøbing-Fyn-Møns trend of highs.

The southern Permian basin, which overlies the Variscan foredeep and encroaches onto the Variscids themselves in the DDR and Poland, showed a greater tendency of subsidence than the northern Permian basin. It remains an open question whether in the case of the southern Permian basin a causal relationship exists between the termination of the Lower Permian volcanism and its subsidence.

In the southern Permian basin several en echelon trending subsidence centres are recognised, reflecting continued activity along faults emplaced during the early Permian [28]. Clastics derived from the rapidly degraded Variscan mountains were deposited in the southern Permian basin under eolian and Sabka conditions as the Rotliegendes sequence of red beds and halites, [29-31]. Maximum sedimentary thicknesses are of the order of 1000 m to 1500 m.

In the northern Permian basin red beds ranging from conglomerates to shales are dominant as well. Evaporites appear to be missing. Maximum Rotliegendes thicknesses are of the order of 600 m.

In the southern and northern Permian basins subsidence exceeding the rate of deposition lead to the development of large topographical depressions, located well below the sea level of the Permian oceans.

At the onset of the Upper Permian rifting movements in the area of the future Norwegian-Greenland Sea resulted in the establishment of a seaway between the Arctic Permian basins (Sverdrup, Barents Sea, Pechora) and the two Permian basins of NW Europe [32]. The transgression of the Zechstein seas into the latter may have taken place under catastrophic conditions. This transgression entered the NW European Permian basins through the Northern North Sea; however, there is no evidence that a Permian gràben had started to subside in this area.

The rapid Zechstein transgression was followed by the accumulation of thin deep-water carbonates and laminated anhydrites in basinal areas and thick shallow-water carbonate and evaporite banks on the basin margins and on local highs. An arid climate, a restricted influx of seawater and periodic eustatic sealevel changes led to the infilling of the two NW European Permian basins with thick Zechstein salts and polyhalites. These reach thicknesses of about 1500 m in the southern Permian basin and about 1000 m in the northern one.

The present day Zechstein salt edge (Fig. 12) illustrates that differential subsidence of the Horn Graben had started during the Zechstein. Differential subsidence of the Viking Graben during the Upper Permian remains questionable since the zero edge of the Zechstein salts in this area may have been considerably modified by Mesozoic erosion.

Minor Zechstein volcanic activity has been recorded from the eastern parts of the Mid-North Sea High, an area that subsided rapidly during the Triassic. Neither the Bamble Trough nor the Oslo Graben show signs of differential subsidence during the Upper Permian.

At the transition from the Permian to the Triassic the marine connection between the Permian basins of the Arctic and those of NW Europe was again interrupted, possibly due to the emplacement of a rift dome in the Norwegian-Greenland Sea area.

This led to the reestablishment of a continental regime in the North Sea.

## 6. TRIASSIC RIFT SYSTEMS

During the Triassic, the Permian basin framework of NW Europe was gradually modified by the emplacement of a number of new grabens and flexure-bound troughs (Fig. 3). These transect in part the Permian structural elements. Emplacement of these grabens and troughs was contemporaneous with intensified rifting movements in the North Atlantic and the Norwegian-Greenland Sea area [33, 34]. In the Tethys Triassic extension movements caused the subsidence of a complex graben system and widespread marine transgressions [35, 36].

Northwestern Europe appeared during the early Mesozoic as a triangular meta-stable platform bounded on two sides by the active rift zones of the Tethys and the Arctic-North Atlantic. The third side of this triangle was formed by the Tornquist-Teisseyre line which separates the metastable European craton from the stable Russian platform.

In the North Sea area Triassic extension movements caused the rapid subsidence of a complex graben and trough system. Major elements of this system were the Viking-Central Graben rift in the North Sea proper and the Danish-Polish Trough parallel to the Tornquist-Teisseyre lineament.

Bordering the North Atlantic rift zone, the West Shetland-Minch Graben subsided rapidly during the Triassic. Branching off from the Bay of Biscay Fracture Zone are the Porcupine Trough, the Western Approaches and the Celtic Sea Graben, which all started to subside during the Triassic. The Severn Graben and the partly fault-bounded Manx-Furness Basin provided an offshoot of this graben system into the Irish Sea. Along the southern margin of the southern Permian basin a number of smaller, essentially NNE-SSW striking troughs came into evidence during the early Triassic; these are from east to west the Saale, Weser and Hessian depressions, and the Emsland and the West Netherlands troughs. The latter may have its northward extension in the Horn Graben and in the southern parts of the Central Graben, both of which breached the Permian barrier of the mid North Sea-Ringkøbing-Fyn High during the Triassic. In the southern North Sea the NW-SE striking Sole Pit and NW Netherlands basins, which were already established during the Permian, continued to subside during the Triassic. Other Triassic subsidence centres were the Glückstadt Graben in Schleswig-Holstein (northern Germany), the Rønne Graben located adjacent to the west coast of Bornholm [37] and the Bamble Trough in the Skagerrak.

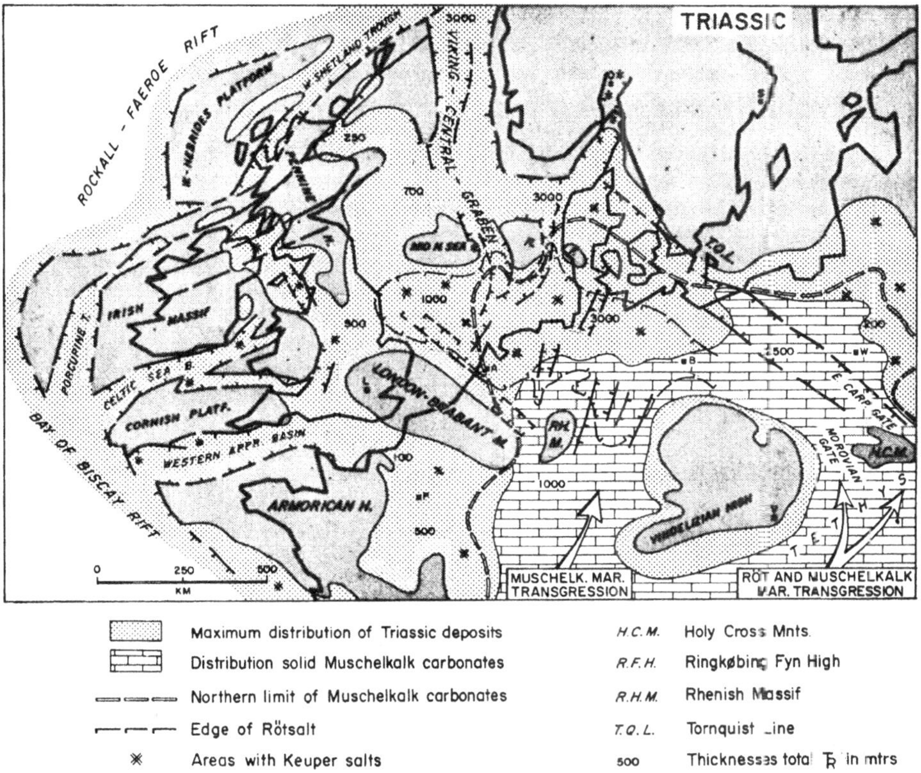


Fig. 3. Triassic paleogeography of NW Europe.

Although Triassic volcanism was widespread in the Tethys [36] and is of importance along the Atlantic rift zone [38, 39], only very minor amounts of Triassic volcanics have been recorded in NW Europe. Volcanism in the Oslo Graben terminated during the Lower Triassic. Minor Triassic volcanics have been reported from the Sunnhordland region on the west coast of Norway [20], off-setting the Horda fault zone and the Triassic Horda-Egersund basin. Traces of Triassic volcanism have recently been encountered in the Central Graben, at its intersection with the Mid North Sea-Ringkøbing-Fyn High. Yet the general lack of an intense Triassic rift volcanism in the North Sea is all the more striking as the dimensions of the Viking-Central Graben rift and the Danish-Polish Trough are comparable with those of the volcanic Rhine-Rhône rift or segments of the African rift system.

In the Viking Graben, the Danish Embayment and in northern Germany Triassic sediment thicknesses of more than 3000 m have been recorded. Up to 2500 m thick Triassic series accumulated in the down-faulted Polish Trough. In the Permian basins diapirism of the Zechstein and Rotliegendes salts was triggered by these thick series during the Upper Triassic.

In the southern Permian basin where the Triassic is present in its classical Germanic facies, the onset of rifting movements can be dated as Upper Scythian, whereby the Hardegsen unconformity gives evidence for a first regionally significant tectonic pulse [40] which resulted in the uplifting of positive elements and the downwarping of the intervening lows. Although the effects of this early rifting pulse can be traced from Poland to the southern North Sea, its significance in the northern North Sea can not yet be documented for want of sufficient well penetrations and a lithostratigraphic zonation of the thick Triassic red bed series.

Triassic sedimentation significantly overstepped the Permian basin edges. By the end of the Triassic the Variscan fold-belt was degraded to such a point that only disjointed low relief highs emerged from the immense, monotonous tidal flat/flood plain that covered much of northwestern Europe.

## 7. TAPHROGENIC STAGE OF THE NORTH SEA RIFT

In the North Sea the northern 1000 km long Viking and Central Grabens rift developed during the Jurassic into the dominant structural element. This graben system continued to subside during the Jurassic and Cretaceous in several distinct and basinwide phases; these are correlative with major rifting phases in the northern North Atlantic and the Norwegian-Greenland sea [41, 42]. The North Sea rift can be regarded as an integral part of the complex northern Atlantic-Norwegian Greenland Sea-European Arctic rift system (Fig. 4) which gradually opened during the Mesozoic, leading ultimately in the early Tertiary to crustal separation between the North American, Greenland and the European plates and the onset of sea-floor spreading [43, 44].

In the Tethys the Jurassic-Cretaceous extension movements resulted in the opening of oceanic basins. This was parallelled by renewed marine transgressions over wide areas of the NW-European craton.

At the Keuper-Rhaetian boundary the "Early Kimmerian" tectonic pulse affected the Arctic-North Atlantic rift system as well as the NW European and Tethian regions. This regional extension phase preceded the Rhaeto-Liassic marine transgression, which

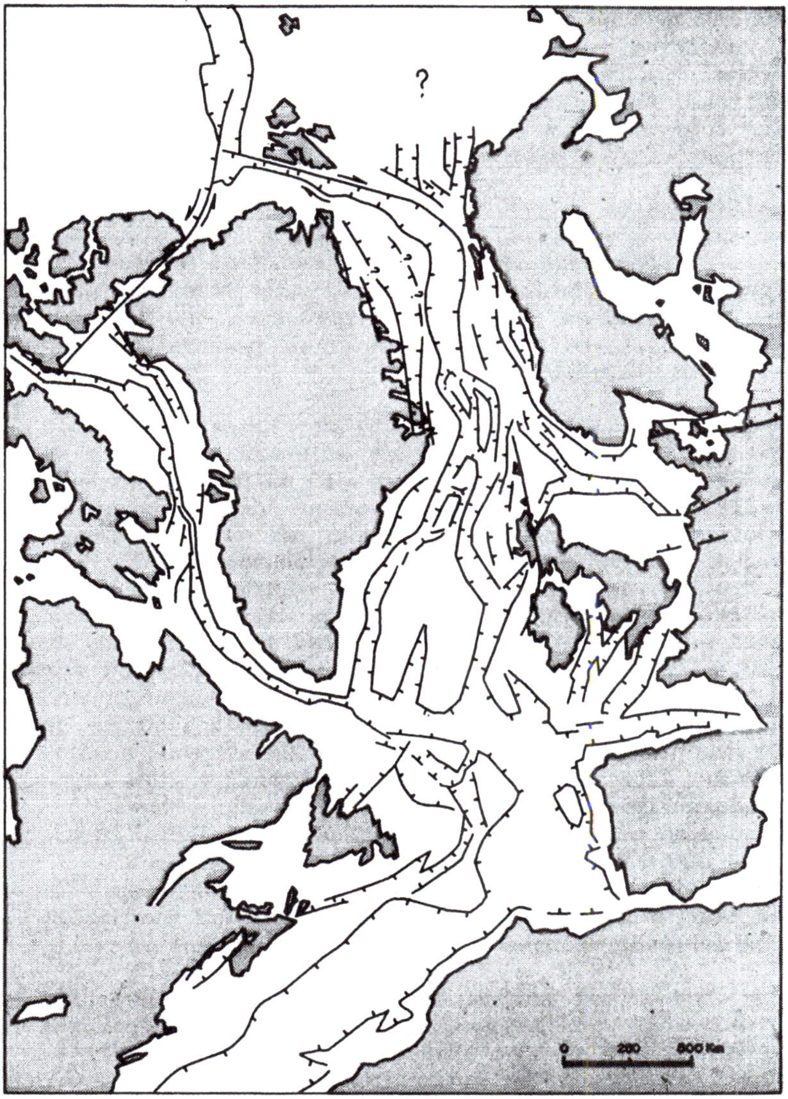


Fig. 4. Schematic pattern of the Mesozoic Arctic-North Atlantic rift system.

inundated much of NW Europe. In the North Sea area the Early Kimmerian tectonism caused a mild accentuation of the major positive elements from which clastics were shed into such continuously subsiding areas as the Viking Graben and the Danish-Polish Trough. During the Pliensbachian fully marine conditions were established throughout the North Sea with transgressions originating from both the Tethys and the by now firmly established North Atlantic-Norwegian Greenland-Arctic seaway.

Liassic deposits in the North Sea area are generally characterised by open marine shales (Fig. 5). Only minor amounts of sand were derived from sources in Poland, East Germany, southern Sweden, the Shetland platform and the Pennine High. Subsurface data indicate that the Viking Graben, the Moray Firth Basin and the Danish-Polish Trough continued to subside differentially during the Liassic.

During the Dogger a number of rifting pulses, summarised under the term "Mid-Kimmerian phases", affected the entire metastable northwest European platform as well as the Arctic-North Atlantic rift system and the Tethian areas. At the Aalenian-Bajocian boundary a major tectonic event, the main Mid-Kimmerian pulse, caused fundamental paleogeographic changes in the North Sea area. In the central North Sea a large high encompassing the Mid North Sea and the Ringkøbing-Fyn Highs was uplifted, resulting in a renewed separation of the northern and southern North Sea basins (Fig. 6). This uplift, which can be considered as a rift dome, was transected by the Central Graben, the southern parts of the Viking Graben and the Moray Firth Basin fault system. At the triple junction of these grabens, a major volcanic centre was emplaced [46]. Subsidiary mid-Jurassic volcanic centres appeared in the southern Viking Graben, the Egersund Basin, the Sunnhordland area of SW Norway and in Scania [20, 23]. This attests to a reactivation of the Horda fault zone and the Fennoscandian border zone by the Mid-Kimmerian tectonism. In contrast, the Horn Graben, the Glückstadt Graben, and the Emsland and West Netherlands troughs apparently became inactive.

In the area of the Central North Sea rift dome Lower Jurassic and older sediments were subjected to extensive truncation. Erosion products were shed northward into the continuously subsiding Viking Graben and the Danish Embayment, where they were deposited under largely paralic conditions. Clastics shed southward from this dome, were deposited in the incipient Sole Pit-West Netherlands and Lower Saxony basins in which marine conditions prevailed.

During the late Middle and the Upper Jurassic the Central North Sea rift dome gradually foundered. However, several, in part minor yet distinct, rifting phases (subsequent Mid-Kimmerian phases) interrupted this process. Continued differential

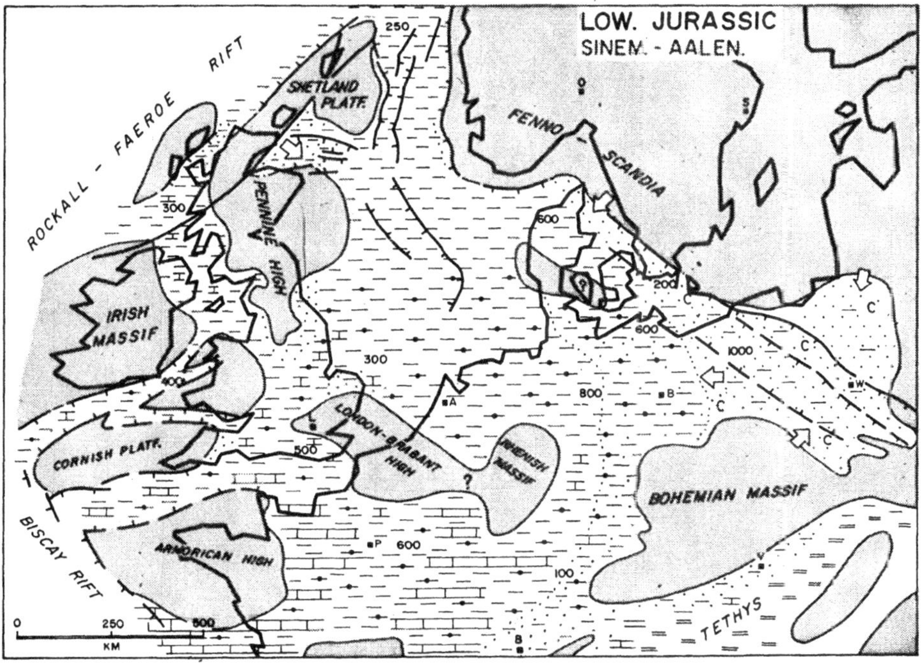


Fig. 5. Lower Jurassic paleogeography of NW Europe.

subsidence of the Viking and Central Grabens was accompanied by the progressive transgression of the Jurassic seas which reached their maximum extent in the Kimmeridgian. By this time deep-water conditions were established in the Viking and the Central Grabens. In large parts of the central and northern North Sea the Kimmeridgian is developed as shaly facies rich in organic material (Fig. 7). During the Upper Jurassic transgression shallow-water sands derived from the Shetland Platform and the Pennine High accumulated on the western shelves flanking the North Sea Graben system. Deep-water sands, derived from these shallow-water sand accumulations and transported by density currents, were locally deposited in the Central and Viking Grabens.

Progressive subsidence of the Central North Sea rift dome during the Upper Jurassic gave rise to a renewed marine connection between the northern and southern North Sea basins. In the latter, Upper Jurassic series display a wide variety of facies ranging from carbonates and evaporites in the Mecklenburg-Brandenburg, Subhercynian and Lower Saxony basins to paralic series in the West Netherlands Basin and, ultimately, to shales containing

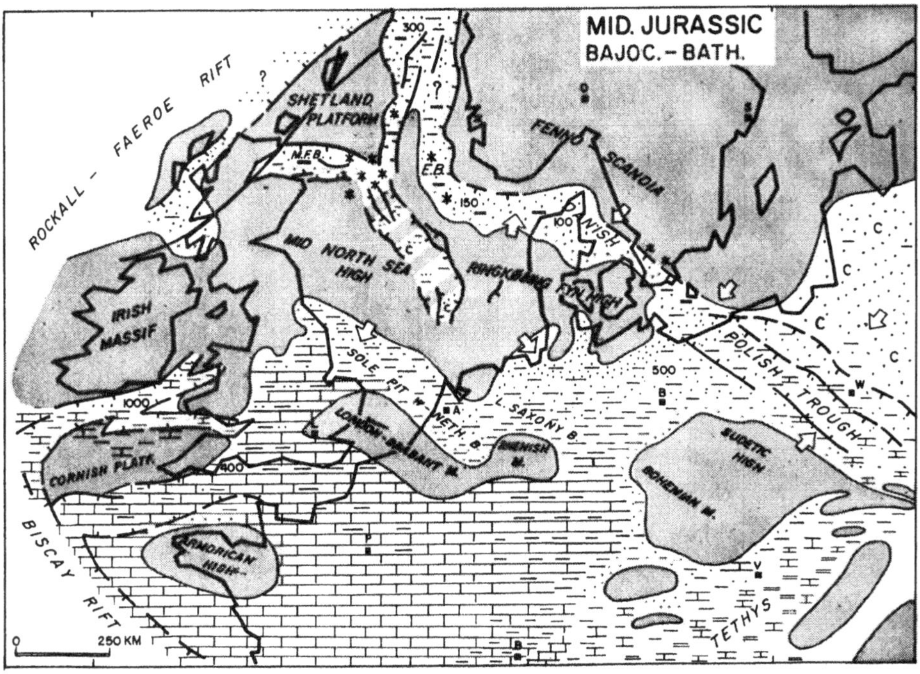


Fig. 6. Middle Jurassic paleogeography of NW Europe.

only minor carbonates and clastics in the Sole Pit Basin.

These basins, which to the north separate the London-Brabant and the Rhenish-Bohemian massifs, are referred to as "Marginal Troughs" ("Randtröge", [47]). Their subsidence was probably caused by minor right-lateral wrench movements between the Danish-North German block to the north and the Variscan massifs to the south in response to crustal distension in the North Sea Rift and in the Polish Trough [41]. Fault patterns already established during the Lower Permian may have played a role in the localization of the Marginal Troughs.

Rift volcanism in the central North Sea ceased with the onset of subsidence of the Central North Sea rift dome. However, minor volcanic centres developed during the Upper Jurassic in the Dutch Waddensea [48], at the edge of the West Netherlands Basin, and in Scania [49]. Furthermore, Upper Jurassic dates have been obtained from dykes in the northernmost part of the North Sea and Upper Jurassic ash beds have been reported from the Isle of Skye in Western Scotland [50].

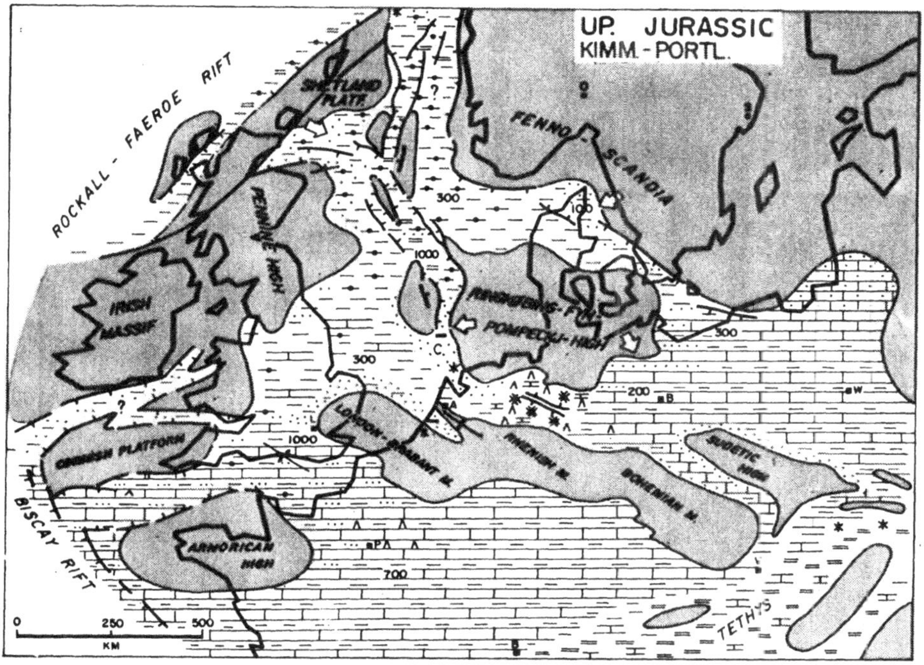


Fig. 7. Upper Jurassic paleogeography of NW Europe.

At the Jurassic-Cretaceous boundary a third major rifting phase, the "Late Kimmerian phase" affected the entire North Sea area. In the Viking Graben this phase resulted in the accentuation of the already existing sea-floor topography, giving rise to a strong block-faulted relief of 2000 to 3000 m (Fig. 8). In the Central Graben rift tectonics are, at least at shallow levels, less obvious due to interference of the intense Zechstein salts diapirism.

The Late Kimmerian downfaulting of the Viking and the Central Grabens was accompanied by a temporary uplifting of their rift flanks. In the southern North Sea area the Late Kimmerian tectonism caused the emergence of the large Ringkøbing-Fyn-Pompeckj Platform, the sharp accentuation of the Marginal Troughs and the uplifting of the Brabant-Rhenish-Bohemian Massif (Fig. 9).

In large parts of the North Sea the base of the Cretaceous

Fig. 8. Structural cross-sections through Viking and Central Grabens (for location of sections see Fig. 14).

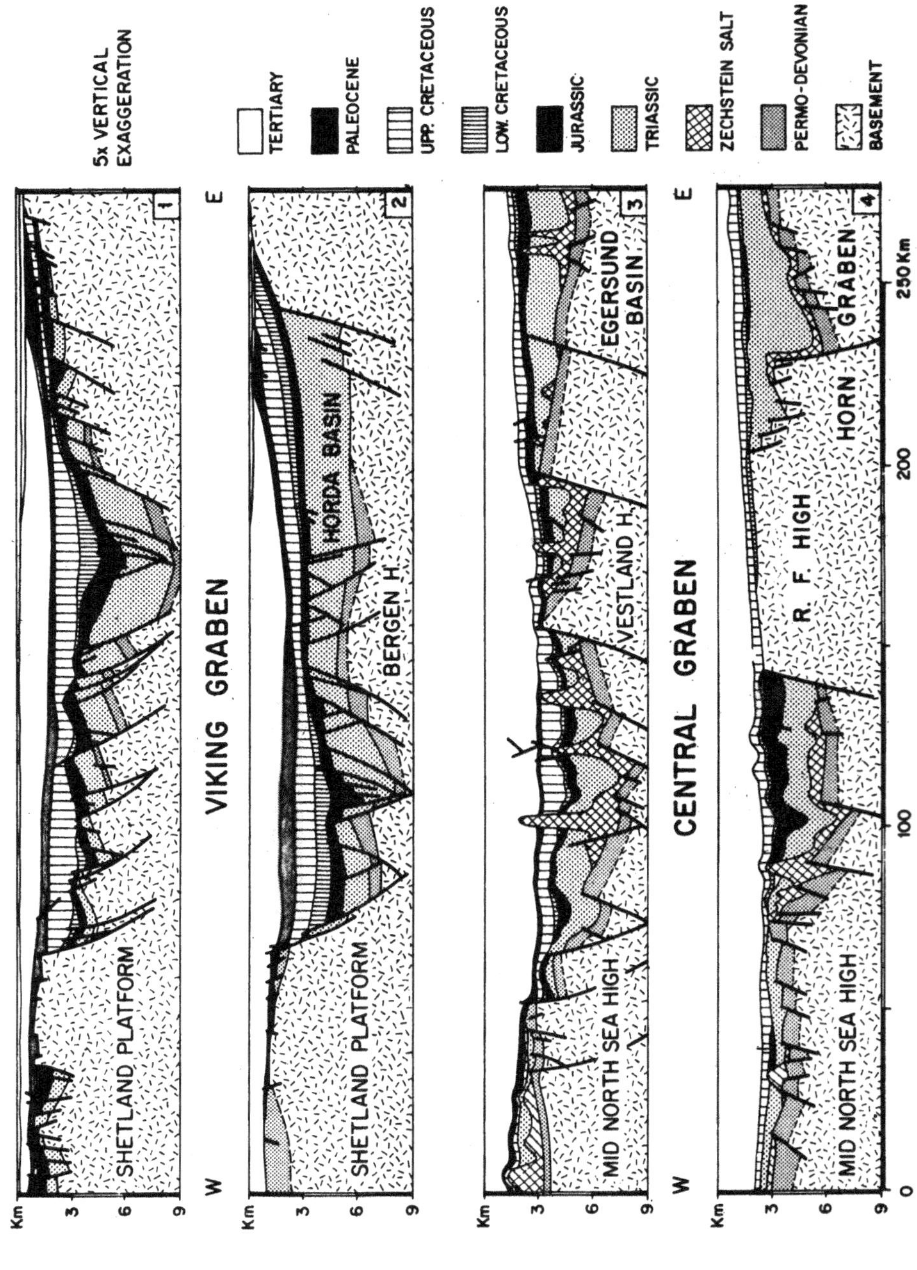

5x VERTICAL EXAGGERATION
TERTIARY
PALEOCENE
UPP. CRETACEOUS
LOW. CRETACEOUS
JURASSIC
TRIASSIC
ZECHSTEIN SALT
PERMO-DEVONIAN
BASEMENT
VIKING GRABEN
CENTRAL GRABEN
SHETLAND PLATFORM
HORDA BASIN
BERGEN H.
MID NORTH SEA HIGH
VESTLAND H.
EGERSUND BASIN
R. F. HIGH
HORN GRABEN
W
E
Km
3
6
9
1
2
3
4
0
100
200
250 Km

is marked by an unconformity referred to as the "Late Kimmerian Unconformity". In the Viking and Central Grabens this unconformity is largely of a sub-marine nature. In uplifted areas the Late Kimmerian break in sedimentation was followed by a rapid transgression, which inundated many of the Late Jurassic and Early Cretaceous highs in the North Sea. With this the sand supply to the Central and Viking Grabens was essentially shut off. During the Lower Cretaceous the topography of these rifts was gradually infilled by up to 1200 m thick Lower Cretaceous shales. In the continuously subsiding Horda-Egersund Basin Lower Cretaceous shales reach thicknesses of 500 to 1000 m. In the North Sea area the occurrence of Lower Cretaceous sands is restricted to the Moray Firth Basin, the Danish Embayment, the Dutch part of the Central Graben and the Marginal Troughs. The occurrence of Lower Cretaceous clastics in the Moray Firth Basin and in the Danish Embayment testifies to the Late Kimmerian reactivation of both the Great Glen fault and the Fennoscandian border zone.

Although crustal distension apparently continued in the North Sea rift during the Lower Cretaceous, there is no evidence of contemporaneous volcanism. Early Cretaceous extrusives are, however, known from Scania [23, 49] and from the Cornish Platform [51].

During the Aptian a minor rifting phase (Austrian phase) affected the North Sea rift system, causing a renewed short-lived uplifting of its rift flanks. Austrian tensional movements were accompanied by a sharp accentuation of the Marginal Troughs. In the Lower Saxony Basin the Austrian tectonism led to the emplacement of the deep-seated, presumably basic Bramscher Massif Laccolith, which intruded along faults already active during the Upper Jurassic and Lower Cretaceous [52]. Aptian volcanic activity has also been reported from Scania, indicating that the Fennoscandian Border Zone was again activated during the Austrian phase [23, 49]. Throughout the North Sea area the Austrian phase was followed by the Aptian-Albian transgression.

## 8. LATE TAPHROGENIC STAGE OF THE NORTH SEA RIFT

In the Lower Cretaceous, crustal separation was achieved in the North Atlantic between the Azores-Gibraltar and Charlie-Gibbs fracture zones. However, in the Norwegian-Greenland Sea crustal separation was only established during the Early Tertiary [43].

In the North Sea, Upper Cretaceous rifting movements were considerably less intense than during the Jurassic and Lower Cretaceous. At last distinct rifting pulse affected the central and northern North Sea during the Paleocene (Laramide phase). This phase was followed by the onset of sea-floor spreading in the northern North Atlantic and in the Norwegian-Greenland Sea. With

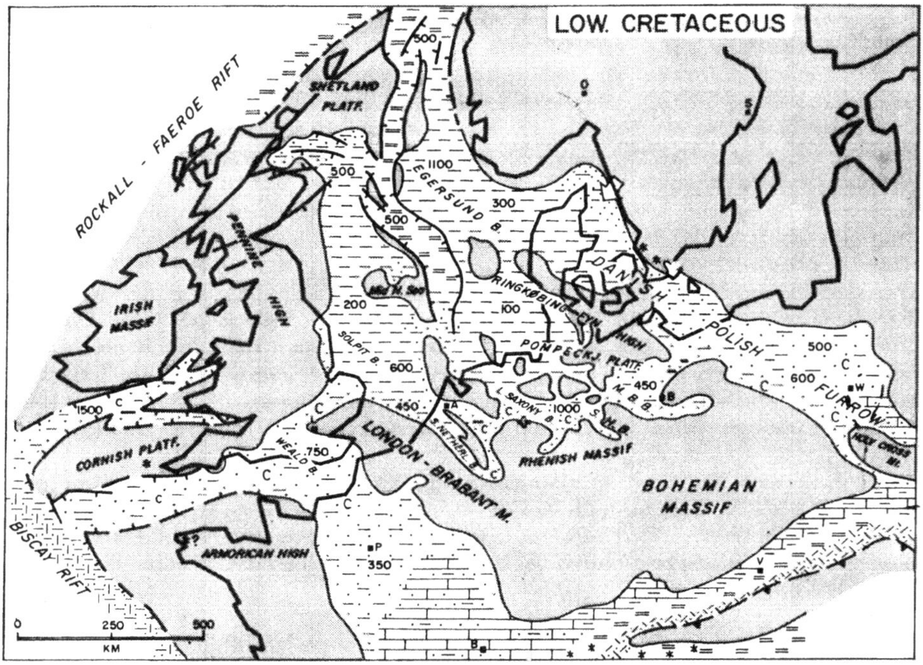


Fig. 9. Lower Cretaceous paleogeography of NW Europe.

this the active extensional setting that governed the Mesozoic evolution of the North Sea area came to a close.

Following the Aptian-Albian transgression, sedimentation in the North Sea area changed from shales to carbonates. The purity of the Upper Cretaceous limestones and chalks indicates the extent to which the surrounding highlands were inundated by the seas, which reached their maximum extension during the Senonian (Fig.10).

In the central and northern North Sea, Upper Cretaceous times correspond to a period of tectonic quiescence that was only interrupted by the Mid-Senonian Subhercynian rifting phase which gave rise to a sub-regional disconformity. Both the Viking and the Central Grabens continued to subside differentially during the Upper Cretaceous. Up to 1200 m thick Chalk series, consisting mainly of coccolith oozes, accumulated in the Central Graben and in the southern Viking Graben. Northward these carbonates grade into equally thick marls and calcareous shales. On the highs flanking the North Sea rift Upper Cretaceous series reached thicknesses of only 200 to 300 m. By the close of the Cretaceous most of the North Sea rift topography had apparently been infilled.

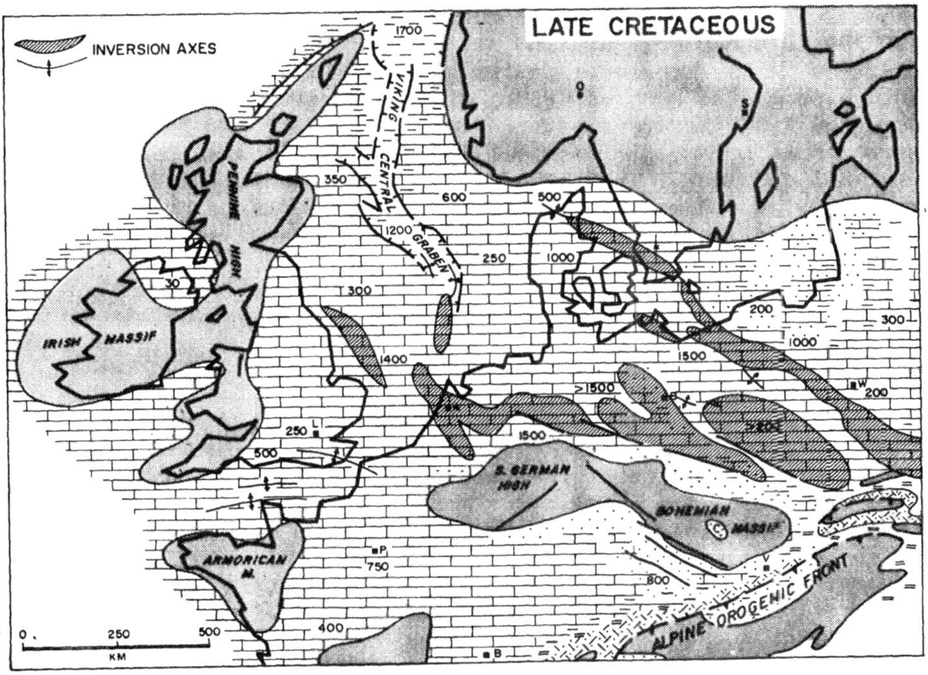


Fig. 10. Upper Cretaceous paleogeography of NW Europe.

During the Early Paleocene Laramide rifting phase the Viking and Central Grabens were once more down-faulted. This was accompanied by the temporary uplifting and erosion of their rift flanks. Regional paleogeographical changes induced by the Laramide tectonisms caused a return to a clastic depositional regime in the entire North Sea area at the end of the Danian.

## 9. SUBHERCYNIAN AND LARAMIDE INVERSION

The Subhercynian and Laramide tectonic movements were purely extensional in the northern and central North Sea but were of a tangential compressional nature in the Marginal Troughs and the Danish-Polish Trough as well as in the southern parts of the North Sea rift system (Fig. 10). This resulted in the inversion of these troughs whereby their basin fill was folded and uplifted above the erosional base level. In such strongly inverted basins as the West Netherlands [53] and the Lower Saxony basin [54] steep reverse faults and small-scale overthrusts play a significant role in the structural style. In less intensely inverted basins such as the

Sole Pit basin and the Dutch part of the North Sea Central Graben, inversion tectonics are characterised by regional warping at shallow levels and block-faulting at deeper levels, whereby movements on pre-existing tensional faults were in part reversed, [55, 56]. Inversion of the Central Graben dies out northward in the Danish offshore. Similarly, inversion tectonics along the Fennoscandian Border Zone become progressively less intense toward the northwest and fade out in northern Jutland. In Scania, where the Pre-Cambrian basement was involved in major upthrusted fault blocks, Upper Cretaceous volcanism has been reported [49].

Parallel with the inversion of the Marginal Troughs the Rhenish-Bohemian massif was accentuated, whereby major structural elements were uplifted and tilted in response to transcurrent and upthrust movements along a set of essentially NW-SE and SW-NE striking faults.

The Late Cretaceous to Early Tertiary change of the regional stress pattern in the Alpine foreland from an extensional setting to a tangential compressive one coincided with major orogenic phases in the Alpine geosyncline [41]. These are considered to be the expression of continent to continent collision between the Eurasian and African plates. Resulting compressional stresses apparently also affected the fragmented Alpine foreland, causing the tilting and uplifting of the Variscan massifs and the inversion of the main Mesozoic troughs located up to 800 km to the north of the present Alpine deformation front. Once these basins were inverted, they became largely inactive.

## 10. CAINOZOIC NORTH SEA BASIN

With the onset of sea-floor spreading in the Norwegian-Greenland Sea during the Early Eocene the North Sea rift became inactive.

The Cainozoic development of the North Sea was dominated by regional subsidence, which apparently is still continuing. The structural configuration of the Cainozoic North Sea basin is summarised in Fig. 11, a structural map drawn at the base of the Tertiary clastics [57].

As referred to above, the Laramide tectonism caused profound paleogeographic changes in the North Sea area. Apart from renewed down-faulting of the North Sea rift, Laramide extension tectonics reactivated the Faeroe-Rockall trough; this was accompanied by the uplifting of the Scottish Highland and the West Shetland platform, possibly in response to the emplacement of a rift dome. Uplifting of these areas gave rise to an eastward directed drainage pattern, which deposited an outbuilding clastic foreset wedge on the shelves east of the Shetland and Orkney Islands

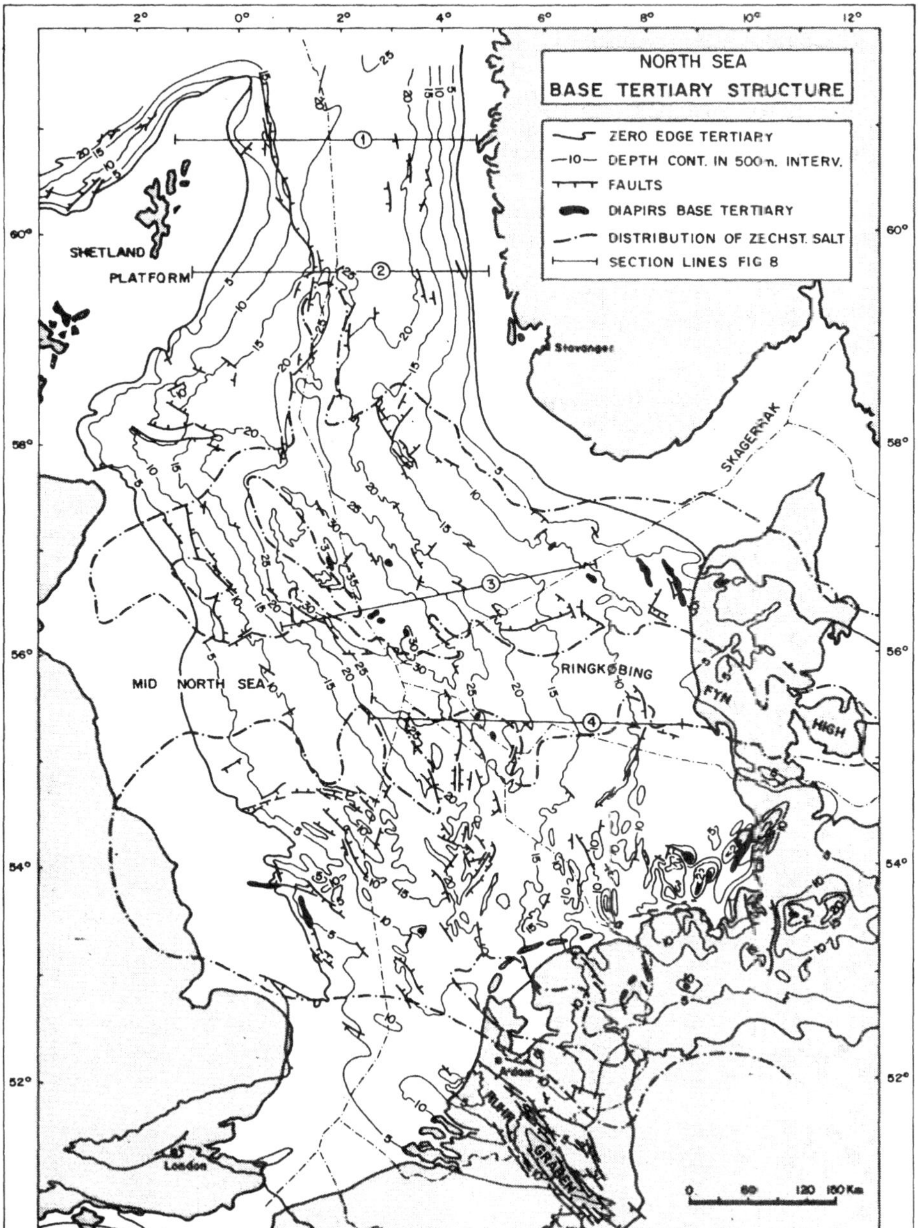


. 11. Base Tertiary structure map.

during the Paleocene and Eocene. Slope failure of this foreset sequence triggered density currents, which supplied sands to the downfaulted rift valleys of the Viking and Central Grabens in which water depths were in the order of several hundred metres [58]. In the inverted areas of the North Sea as well as on the uplifted rift flanks a major erosional phase preceded the transgression of the Upper Paleocene and the Eocene.

At the Paleocene-Eocene transition, volcanism along the Faeroe-Rockall rift reached its peak, scattering tuffaceous material over much of the North Sea area. This gave rise to the deposition of the nearly basinwide Lower Tertiary tuffmarker. Time equivalent volcanism is suspected also to have occurred in the Skagerrak [59].

During the Eocene the supply of clastics from the Scottish Highlands and the West Shetland platform gradually decreased, possibly due to the gradual subsidence of the Faeroe-Rockall rift dome. In the central and southern North Sea Paleocene and Eocene series are represented mainly by clays and silts. A regional hiatus marks both the base and the top of the Oligocene. These breaks in sedimentation which are thought to have been caused by eustatic sea-level changes, rather than by local tectonic events, are mainly expressed in the marginal areas of the North Sea basin. Oligocene series consisting predominantly of clays and silts reach thicknesses of up to 1000 m in the Central North Sea.

Miocene and younger series are made up by shallow marine and paralic sands and clays as well as of Quaternary glacial material. Neogene and Quaternary sediments attain maximum thicknesses of 2000 m in the Central North Sea.

Overall Oligocene and younger time-stratigraphic units expand in thickness from the margins of the North Sea basin towards its centre; its basin axis coincides with the trace of the Mesozoic Viking and Central Grabens [57].

Pleistocene uplifting of the Fennoscandian shield resulted in the truncation of the northwestern edge of the Tertiary North Sea basin. Scouring out of the sea bottom relief of the Skagerrak and the Norwegian channel is attributed to glacial activity.

The Neogene Rhine Rift does not extend into the North Sea but peters out to the north in the Dutch onshore areas (Fig. 11).

## 11. CONCLUSIONS

As pointed out the Mesozoic North Sea rift formed part of the complex Arctic-North Atlantic mega-rift system. The driving mechanisms behind the latter are thought to be upwelling mantle convection cells, which were at least intermittently active since the late Paleozoic. However, as long as crustal separation between the North American-Greenland and European plates was not yet accomplished, extensional stresses originating from such a convection system apparently affected a wide area around the eventual ocean spreading centres. This resulted in the development of a complex graben system whereby fracture propagation, following in part pre-existing lineaments, may have played a significant role. Such lineaments were provided by the Variscan and Caledonian fold belts as well as by the Arctic-North Atlantic Devonian transcurrent fault system. With progressive extension certain grabens became inactive whereas others remained active until crustal separation became effective. In the North Sea area a complex graben system began to form during the Triassic and became accentuated during the intensified Jurassic rifting activities. The North Sea rift and the Polish-Danish trough emerged as the dominant fracture zones. During the Middle Jurassic a large rift dome developed in the central North Sea. This was accompanied by the emplacement of several volcanic centres. At the same time a number of subsidiary Triassic grabens became inactive. The volcanic stage of the North Sea rift was, however, relatively short. During the Cretaceous the central graben system of the North Sea rift suffered rapid differential subsidence, and the rift dome gradually foundered. By Late Cretaceous times the Mid-Jurassic rift dome had essentially disappeared. The Cainozoic isopach map (Fig. 11) illustrates that the North Sea rift system had ceased to subside differentially during the Tertiary. Instead, a symmetrical, saucer-shaped basin developed, the axis of which closely follows the trace of the Mesozoic North Sea rift system (Fig. 12).

A refraction profile (Fig. 13) from the Shetland Islands to a point on the Norwegian coast located some 120 km north of Bergen shows that the crust thins drastically under the North Sea rift [60]. Under this attenuated crust the mantle velocity is of the order of 8.3 km/sec. This refraction profile crosses the Viking Graben where it flares out to the north and grades into the West Norway Shelf Basin. Furthermore, in the eastern part of this profile the basin was severely truncated by the Pleistocene Fennoscandian uplift. Despite this, and although the Southern Viking and the Central Grabens are considerably narrower and show less extension at shallow levels, it it thought that this profile is essentially representative for the crustal configuration of the North Sea rift. However, it should be kept in mind that this is the only refraction profile available across the North Sea graben system.

The structural configuration of the Viking and Central Grabens indicates at the Late Kimmerian unconformity level that thinning

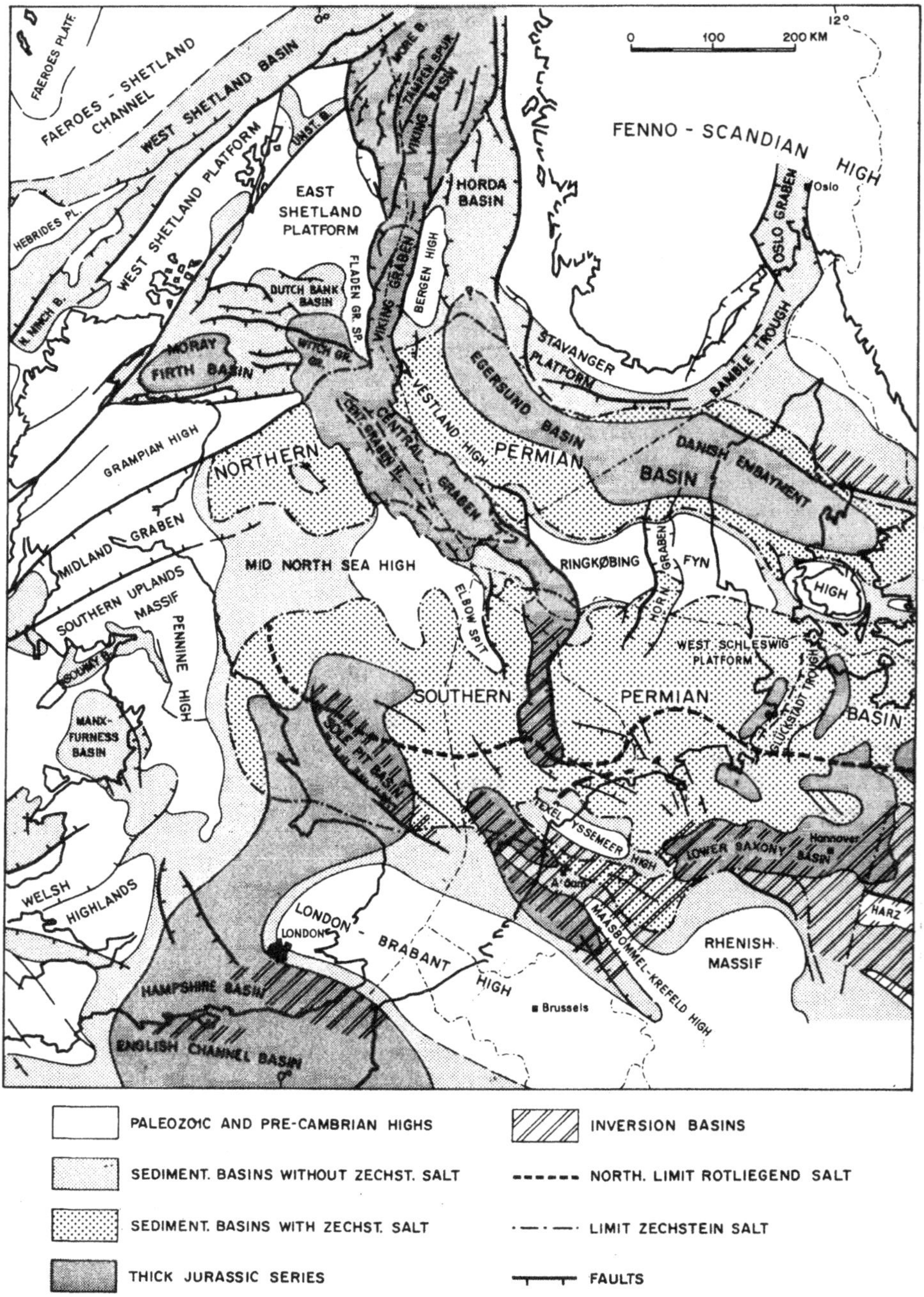


Fig. 12. Permian and Mesozoic tectonics of the North Sea.

of the brittle upper crust was achieved by extensional faulting. Thinning of the lower crust was probably achieved by concommittant ductile deformation [61] whereby subcrustal erosion may also have played a role.

It has been suggested that development of a Middle Jurassic rift dome in the Central North Sea was associated with the emplacement of a low-density, low-velocity rift cushion at the crust/mantle interface. In this respect the model of the currently active Rhine Graben rift is invoked where refraction data permit the mapping of such a rift cushion (Fig. 13). The width of the Rhine Graben rift dome corresponds to the width of the underlying rift cushion [62]. The physical properties of such a rift cushion can be explained by melting processes. Magmas intruding the crust from a rift cushion may eventually reach the surface where they display at first the chemical characteristics of a typical alkaline intial rift volcanism. If extension tectonics persist, this volcanism will gradually change to a tholeiitic basalt type. Triple junctions where faulting and thinning of the crust are most

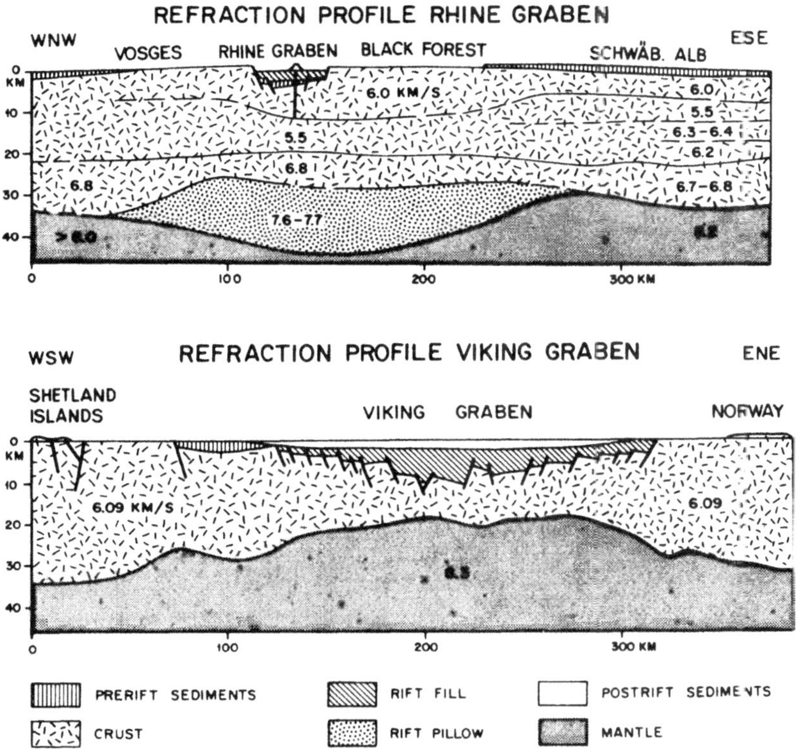


Fig. 13. Refraction profiles through Rhine Graben and Northern Viking Graben.

intense are likely places for an early manifestation of rift volcanism. However, as in the Rhine Graben, volcanic centres are not necessarily restricted to the actual rift zone (i.e. Hegau and Urach volcanic centres), yet lateral volcanic centres appear to remain within the areal confines of the rift cushion.

In the North Sea the main Mid-Jurassic volcanic centre is located at the triple junction of the Viking and Central Grabens and of the Moray Firth Fault system. Secondary volcanic centres occur in the Egersund basin. The Mid-Jurassic volcanics of the North Sea consist of undersaturated porphyritic olivine basalts that display an alkaline affinity [64]. The subcrop of Lower Jurassic and older series against the "Mid-Kimmerian" unconformity, as well as regional paleogeographical considerations give evidence that a rift dome centred over the Central North Sea existed during the Mid-Jurassic.

It is therefore hypothesised that the North Sea Graben system was underlain during its Jurassic rifting stage by a rift cushion related to a rift dome and an associated volcanism. However, the limited refraction data available indicate that at present the mantle displays a normal velocity under the North Sea rift. Thus it must be assumed that the rift cushion postulated above, disappeared during the late Mesozoic and Tertiary, presumably by cooling and solidification. This process was apparently paralleled by the gradual subsidence of the bulged-up crust/mantle interface, resulting in the development of the broad, saucer-shaped Cainozoic North Sea Basin. The width of this basin may have corresponded approximately to the width of the original rift cushion. However, subsidence of such a basin continues until isostatic equilibrium is reached. In this respect consideration has to be given to the intrusion of heavy mantle material into the crust during the rifting stage.

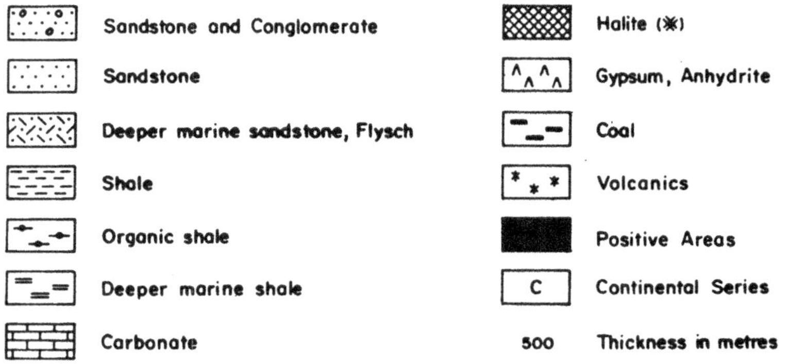


Fig. 14. Legend to paleogeographic maps.

REFERENCES


1. P.A. Ziegler, Geojournal, 1, 7, 1977.
2. J.F. Dewey, The Geology of the Southern Termination of the Caledonides, in: Ocean Basins and Margins, Vol. 2, The North Atlantic, ed. by A.E.M. Nairn and F.G. Stehli, 1974.
3. W.B. Harland, Am. Assoc. Pet. Geol., Mem., 19, 599, 1973.
4. W.B. Harland, Can. Soc. Pet. Geol., Mem., 4, 235, 1975.
5. J. Haller, Tectonic Map of East Greenland, Medd. Grønland, 11, 1970.
6. N. Rast and R. Grant, Am. J. Sci., 272, 542, 1973.
7. W.A. Morris, Can. J. Earth Sci., 13, 1236, 1976.
8. K.M. Storetvedt, Geol. Mag., 111, 23, 1974.
9. M.R. Leeder, Tectonophysics, 36, 167, 1976.
10. F.H. Francis, Geol. Rundsch., 57, 219, 1968.
11. R.D. Howie and M.H. Barss, Paleogeography and Sedimentation in the Upper Paleozoic, Eastern Canada, in: Canada's continental Margins and Offshore Petroleum Exploration, ed. by C.J. Yorath et al., Can. Soc. Pet. Geol. Mem., 4, 1975.
12. W.H. Pool, Can. Geol. Surv., Pap., 76-1B, 113, 1975.
13. K. Bell and J. Blenkinsop, Nature, 265, 616, 1977.
14. W.A. Busch, R.G. Garetskiy, Y.A. Ivanov and L.G. Kiryuchin, Geotectonics, 1, 8, 1973.
15. H. Termier and B. Termier, C.R. Acad. Sc. Paris, 283 Série D, 139, 1976.
16. F.J. Eckhardt, Geol. Jb., 85, 227, 1969.
17. W.A. Busch and L.G. Kiryuchin, Z. angew. Geol., 18, 323, 1972.
18. W.W. Gluschko et al., Z. angew. Geol., 21, 253, 1975.
19. K.S. Heier and W. Compston, Lithos, 2, 133, 1969.
20. R.B. Færseth, R.M. MacIntyre and J. Naterstad, Lithos, 9, 331, 1976.
21. E. Halvorsen, Nor. Geol. Tidsskr., 50, 157, 1970.
22. H.N.A. Priem et al., Phys. Earth Planet. Interiors, 1, 373, 1968.
23. I. Klingspor, Geol. Fören. Stockh., Förh., 98, 195, 1976.
24. C. Oftedahl, Permian Rocks and Structures of the Oslo Region, in: Geology of Norway, ed. by O. Holtedahl, Norg. Geol. Unders. 208, 1960.
25. C. Oftedahl, Geol. Rundsch., 57, 203, 1968.
26. I.B. Ramberg, Norg. Geol. Unders., 325, 1976.
27. W. Kramer, Z. Geol. Wiss., Berlin, 5, 7, 1977.
28. G. Katzung, Z. Geol. Wiss. Berlin, 3, 1453, 1975.
29. K.W. Glennie, Am. Assoc. Pet. Geol., Bull., 56, 1048, 1972.
30. J. Pokorski and R. Wagner, Warsau Geol. Inst., Bull., 252, 115, 1975.
31. M. Lutz, J.P.H. Kaasschieter and D.H. van Wijhe, Geological Factors Controlling Rotliegendes Gas Accumulations in the Mid-European Basin. Proceed. 9th World Petrol. Congr. Tokyo 5/12-16, 3, 131, 1975.
32. T. Birkelund and K. Perch-Nielsen, Late Paleozoic-Mesozoic

evolution of Central East-Greenland, in: Geology of Greenland, ed. by A. Escher and W.S. Watt, Geol. Surv. Greenland, 1976.
33. A. Hallam, J. Geol., 79, 129, 1971.
34. L.F. Jansa and J.A. Wade, Can. Geol. Surv. Pap., 74-30, 51, 1974.
35. W.S. Olson and R.J. Leyden, North Atlantic Rifting in Relation to Permian-Triassic Salt Deposition, in: The Permian and Triassic Systems and their Mutual Boundaries, ed. by A. Logan and L.V. Hills, Can. Soc. Petrol. Geol. Mem., 2, 1973.
36. H. Laubscher and D. Bernoulli, Mediterranean and Tethys, in: The Ocean Margins, Mediterranean, ed. by A.E.M. Nairn, 1977 (in press).
37. O.B. Andersen, B. Larsen and S.W. Platou, Geol. Soc. Den., Bull., 24, 45, 1975.
38. P.R. Vogt, Early Events in the Opening of the North Atlantic, in: Implications of Continental Drift to the Earth Science, 2, ed. by D.H. Tarling and S.K. Runcorn, 1973.
39. R.D. Ballard and E. Uchupi, Am. Assoc. Pet. Geol., Bull., 59, 1041, 1975.
40. W.H. Ziegler, Outline of the Geological History of the North Sea, in: Petroleum and the Continental Shelf of NW Europe, Vol. 1, Geology, ed. by A.W. Woodland, 1975.
41. P.A. Ziegler, Am. Assoc. Pet. Geol., Bull., 59, 1073, 1975.
42. T. Birkelund, Geology of East Greenland: A Review, Offshore North Sea Technology Conference and Exhibition, Stavanger 1976, Handout ONS-76-G-W-3, 1976.
43. A.S. Laughton, Norg. Geol. Unders., 316, 169, 1975.
44. X. le Pichon, Tectonophysics, 38, 169, 1977.
45. E. Bölau, Geol. Fören. Stockh., Förh., 95, 165, 1973.
46. F. Howitt, E.R. Aston and M. Jacqué, The Occurrence of Jurassic Volcanics in the North Sea, in: Petroleum and the Continental Shelf of NW Europe, Vol. 1, Geology, ed. by A.W. Woodland, 1975.
47. E. Voight, Dtsch. Geol. Ges. Z., 114, 378, 1962.
48. A. Cottencon, B. Parant and G. Flaceliére, Lower Cretaceous Gas Fields in Holland, in: Petroleum and the Continental Shelf of N-W Europe, Vol. 1, Geology, ed. by A.W. Woodland, 1975.
49. I. Printzlay and O. Larsen, Geol. Fören. Stockh., Förh., 94, 259, 1972.
50. R.W. Knox, Nature, 265, 323, 1977.
51. J.G. Mitchell, R.M. MacIntyre and I.R. Pringle, Geol. Mag., 112, 55, 1975.
52. G. Stadler and R. Teichmüller, Fortschr. Geol. Rheinld. u. Westf., 18, 547, 1971.
53. P. Heybroek, Geol. Mijnbouw, 53, 43, 1974.
54. I. Nodop, Fortsch. Geol. Rheinld. u. Westf., 18, 411, 1971.
55. D.G. Blair, Structural Styles in North Sea Oil and Gas Fields, in: Petroleum and the Continental Shelf of NW Europe, Vol. 1, Geology, ed. by A.W. Woodland, 1975.

56. P. Heybroek, On the Structure of the Dutch Part of the Central North Sea Graben, in: Petroleum and the Continental Shelf of NW Europe, Vol. 1, Geology, ed. by A.W. Woodland, 1975.
57. P.A. Ziegler and C.J. Louwerens, Tectonics of the North Sea, in: The Quaternary History of the North Sea, ed. by E. Oele et al., Acta Univ. Ups., Symp. Univ. Ups. Annum Quingentesiumum Celebrantis 2, Uppsala 1977 (in press).
58. J.R. Parker, Lower Tertiary Sand Development in the Central North Sea, in: Petroleum and the Continental Shelf of NW Europe, Vol. 1, Geology, ed. by A.W. Woodland, 1975.
59. K. Åm, Norg. Geol. Unders., 287, 1, 1973.
60. M. Solli, En seismisk skorpeundersøkelse Norge-Shetland, Thesis University of Bergen, Seismological Observatory, 1976.
61. M.P.H. Bott, Tectonophysics, 36, 77, 1976.
62. S. Müller, E. Peterschmitt, K. Fuchs, D. Emter and J. Ansorge, Tectonophysics, 20, 381, 1973.
63. M.J. Le Bas, Nature Phys. Sci., 230, 85, 1971.
64. F.G.F. Gibb and R. Kanaris-Sotirious, Nature, 260, 23, 1976.

# A COMPARISON OF MAGNETIC ANOMALIES IN THE RED SEA AND THE GULF OF ADEN

D.J. Noy

Department of Geophysics & Planetary Physics,
School of Physics, University of Newcastle upon Tyne,
NE1 7RU, England

ABSTRACT. As there is no evidence for any shear along, or major extension across, the Ethiopian Rift, the spreading histories of the Red Sea and the Gulf of Aden should be very similar. To test this, magnetic anomaly profiles for the two regions are converted to the pole for direct comparison. There appear to be some considerable differences and possible reasons for these are discussed.

## 1. INTRODUCTION

Evidence for a two stage formation of the Red Sea came from observations of shear along the Dead Sea [1]. These show a total of 105 km of post-Cretaceous, left lateral shear, of which 45 km is Plio-Pleistocene. Support for this has come from studies of Red Sea magnetic anomalies [2]. The large axial anomalies have been interpreted as due to sea-floor spreading over the last 4.5 m.y. [2,3]. The smaller amplitude anomalies on either side have proved much more difficult to date.

## 2. INTERPRETATION OF A GULF OF ADEN PROFILE

A magnetic anomaly profile from the recent cruise of the R.R.S. Shackleton to the Gulf of Aden has been converted to the pole (Fig. 1). It extends across the full width of the Gulf and has satellite and radar navigation. The synthetic profile shown in Fig. 1 is generated from the model proposed by Girdler and Styles [4]. As can be seen, the correlations seem to be very good across the whole profile.

*I. B. Ramberg and E.-R. Neumann (eds.), Tectonics and Geophysics of Continental Rifts, 279-287.*

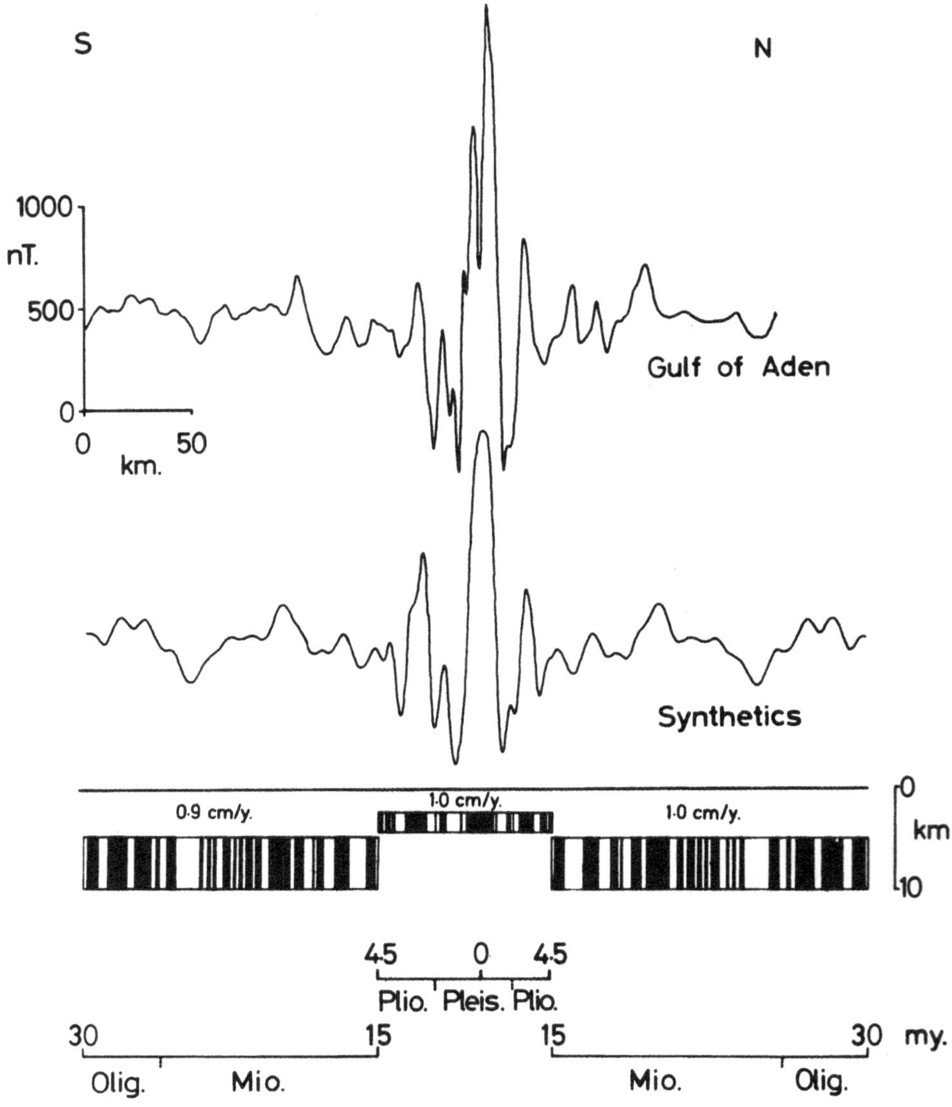


Fig. 1. Sea floor spreading history of the western Gulf of Aden after Girdler and Styles [4]. The upper profile is the observed one (profile F, SHACKLETON 2/75) converted to the pole. The lower profile is the synthetic profile generated using the reversal timescale of Tarling and Mitchell [5] which is used throughout.

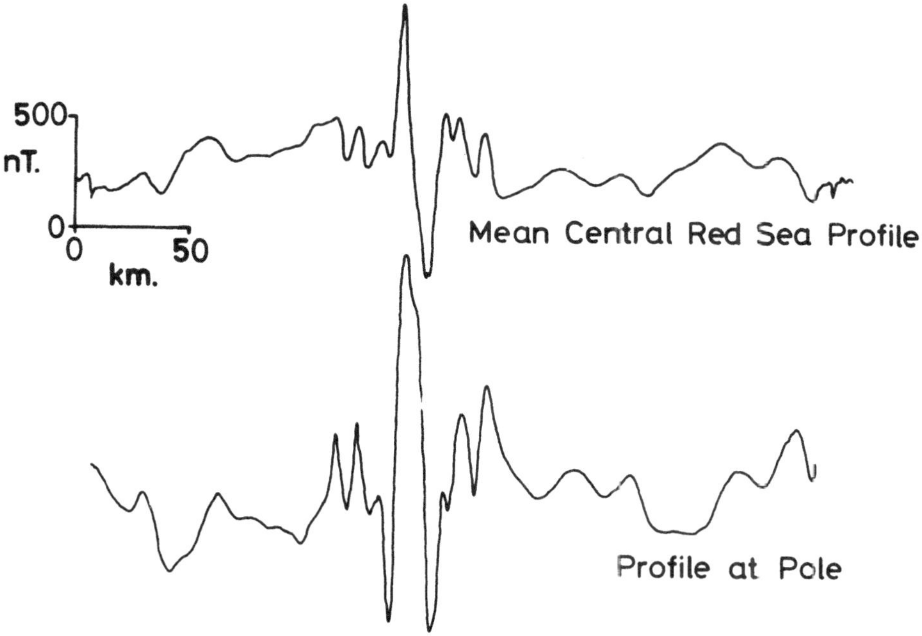


Fig. 2. A composite total intensity magnetic anomaly profile across the full width of the Red Sea, with its "reduced to the pole" analogue.

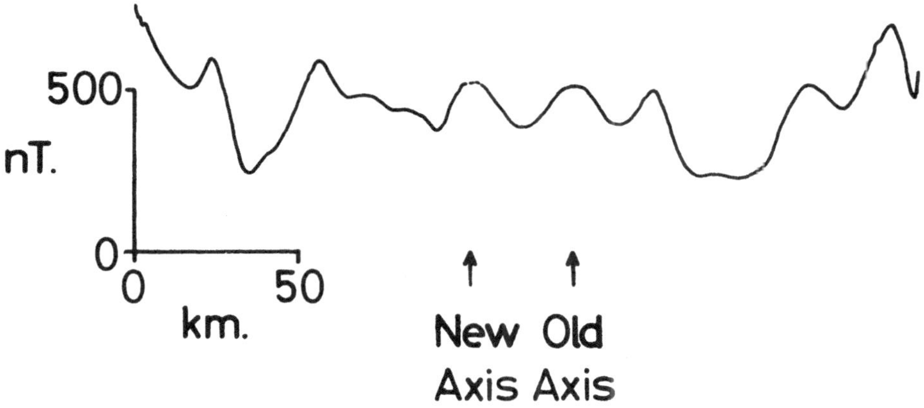


Fig. 3. The Red Sea profile (lower part of Fig. 2) with the magnetic anomalies due to the recent spreading phase (0 to 4½ m.y.) removed. The remaining anomalies are found not to be symmetrical about the present spreading axis.

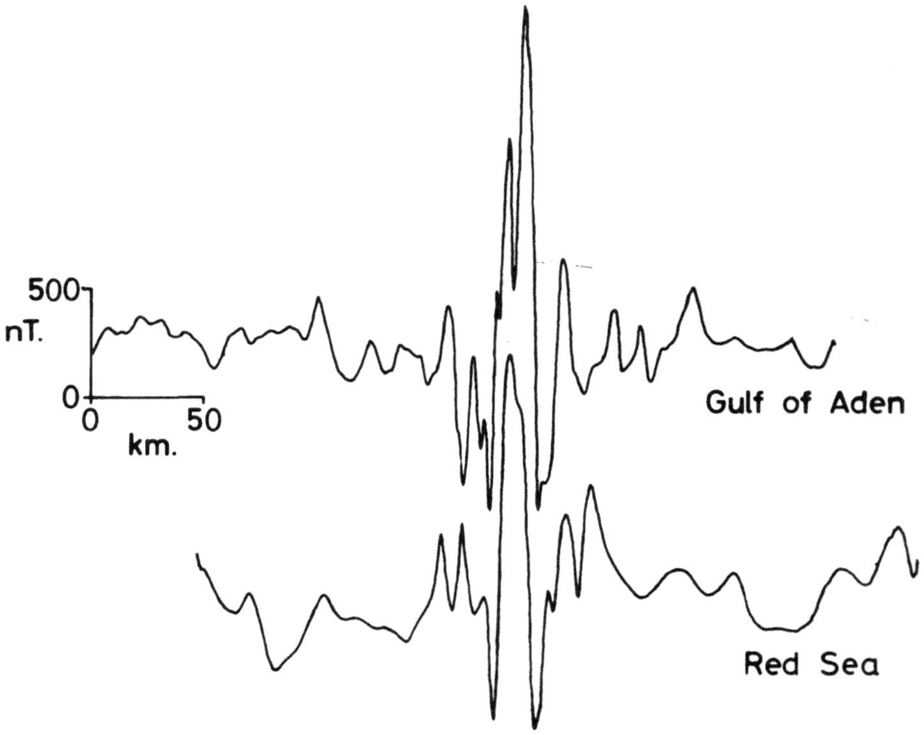


Fig. 4. The Red Sea and the western Gulf of Aden magnetic profiles converted to the pole to facilitate direct comparison.

## 3. CONSTRUCTION OF A RED SEA PROFILE AND ITS COMPARISON WITH THE GULF OF ADEN

There is a shortage of well navigated magnetic data for the Red Sea, especially from shore to shore. The best that can be done is to compile a full width profile for the Central Red Sea where two aeromagnetic surveys overlap. Six profiles were stacked and the resulting mean profile is shown in Fig. 2.

The observed profiles over the Red Sea and the Gulf of Aden may not be directly compared because the spreading axes are almost at right-angles, and it is necessary to convert them to the pole [5]. When this transformation is applied to a profile taken across a continuously spreading ocean ridge the result should become symmetrical about the ridge. Fig. 2 shows how the large anomalies in the Red Sea transform in this way.

This figure also shows that the marginal anomalies are smaller

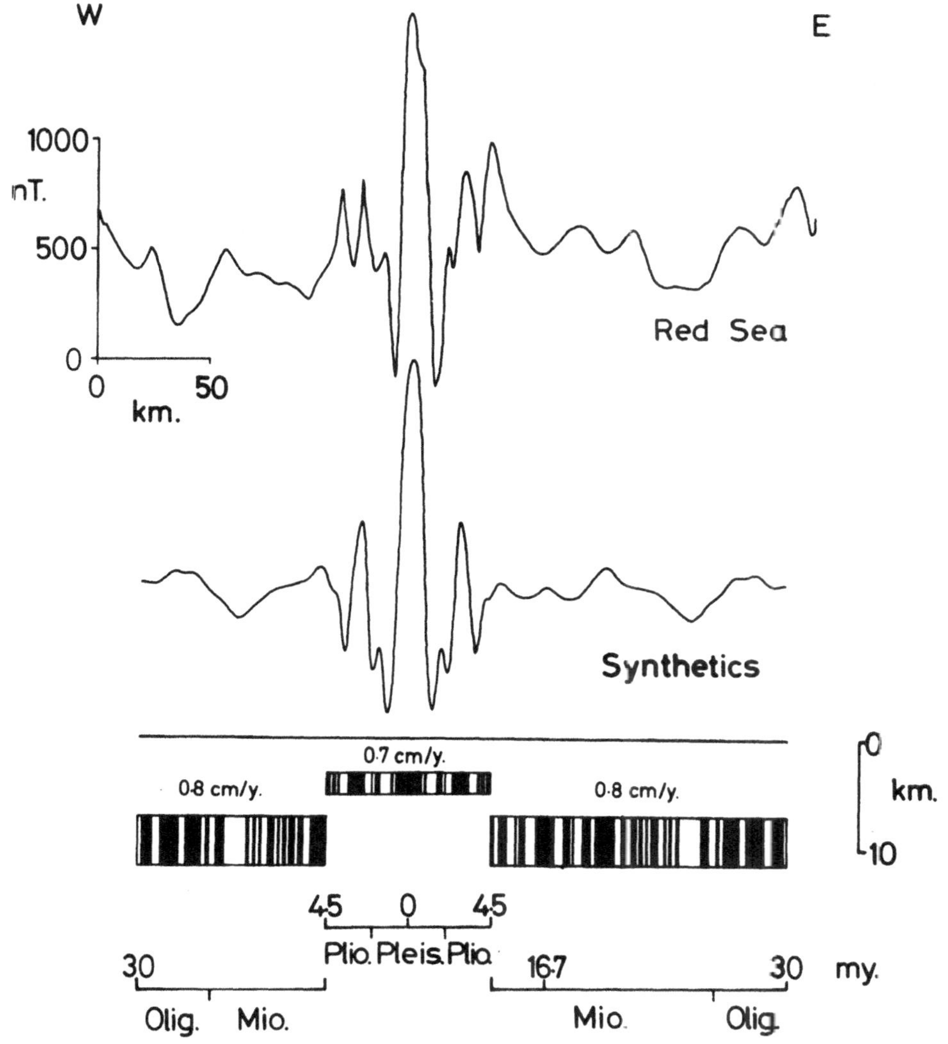


Fig. 5. A sea floor spreading model for the Red Sea using the ages proposed for the Gulf of Aden [4]. A westward shift of the spreading centre of 24 km is required at the start of the second phase.

in amplitude and longer in wavelength than the axial anomalies. This is probably because their source is much deeper, lying beneath at least 5 km of evaporite [6]. The symmetry of these smaller anomalies becomes clear when the large anomalies are removed (Fig. 3). This suggests that when spreading resumed, it began about 30 km west of the old spreading axis.

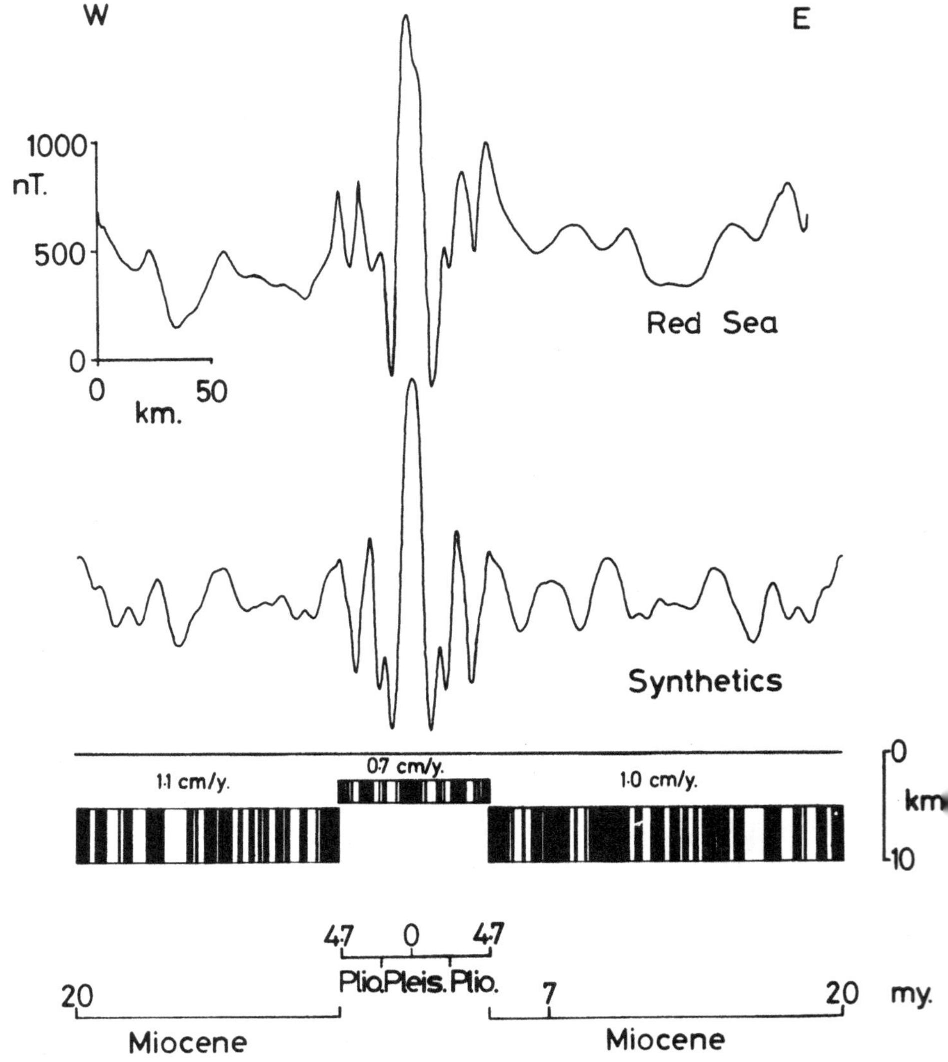


Fig. 6. An alternative model for the Red Sea. The same dates are used for the recent phase of spreading but younger dates for the earlier phase. In this model, a 27.5 km westward shift of the spreading centre is required at the re-initiation of spreading.

Comparing this Red Sea profile with the Gulf of Aden profile (Fig. 4) it is seen that correlations between the marginal anomalies are difficult. Even allowing for the shift of spreading centre in the Red Sea, several large discrepancies can be seen.

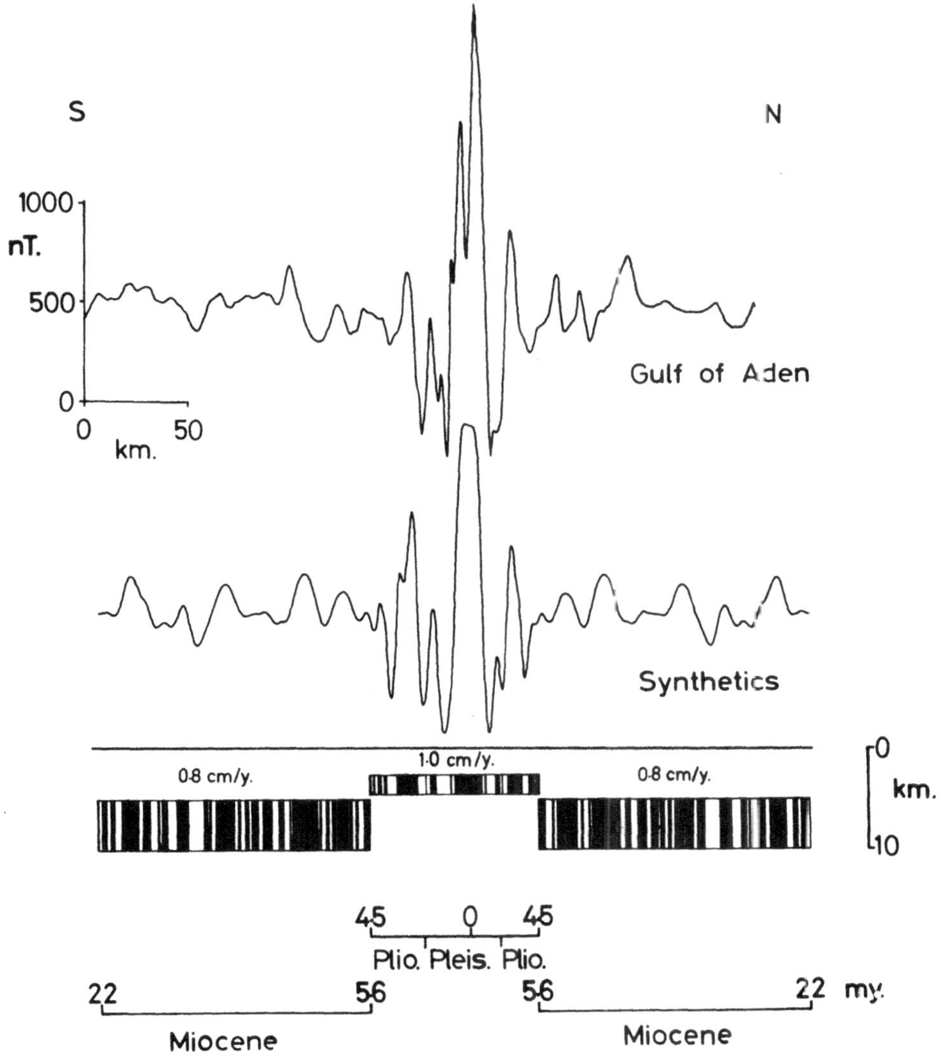


Fig. 7. Re-interpretation of the Gulf of Aden profile using the ages assumed for the Red Sea in Fig. 6.

## 4. INTERPRETATION OF THE RED SEA PROFILE

The question must now be asked whether the observed differences between the Red Sea and the Gulf of Aden can be accommodated within the sea-floor spreading history for the Gulf of Aden, allowing different depths to bodies and slower spreading rates in the Red

Sea. Fig. 5 shows the model obtained in this way for the Red Sea. It can be seen that the correlations are less convincing than was the case for the Gulf of Aden, and there are several differences between the large scale features.

## 4.1 Alternative Models

Many synthetic profiles were generated and Fig. 6 shows the model that seems to give the best fit to the Red Sea data. Most of the large scale features are reproduced, though the finer details are less convincing. When this history is applied to the Gulf of Aden (Fig. 7) the positions of the major anomalies are found to correlate quite well, but their shapes are less satisfactory.

To distinguish between these two models further information is required. For the Gulf of Aden, the bathymetric record shows that the new crust forms a rugged central region within a flat plain [4] that is covered with 0.5 to 1 km of sediments [7]. This indicates that there was sufficient time for the old material to cool before spreading restarted. It seems unlikely that this would be possible in the gap of 1.1 m.y. allowed by the model of Fig. 7. A similar argument can be made with regard to the great thickness of evaporite in the Red Sea, though uncertainty over the rate and history of deposition would seem to make this less conclusive. The evidence for 2-stage shear along the Dead Sea is impressive, but more accurate dating of the early phase is required. K-Ar dates for volcanic rocks form Red Sea boreholes have been given [8], but may be unreliable [9]. Dates from the Afar margins [10] show considerable scatter between 27 m.y. and 9 m.y. which again could agree with either model.

## 5. CONCLUSIONS

This study shows

i) The magnetic anomalies over the Red Sea and the Gulf of Aden display significant differences.

ii) It is possible to obtain similar interpretations (i.e. an early phase of spreading from 30 to 15 m.y. ago) for both regions, but not entirely satisfactorily.

iii) If different ages are chosen for each region (i.e. 30 to 15 m.y. in the Gulf of Aden and 20 to 7 m.y. in the Red Sea) a slightly more convincing fit can be obtained, but this is less satisfactory from the point of view of other geophysical and geological data.

ACKNOWLEDGEMENTS

I am indebted to Dr. S.A. Hall and P. Styles for the magnetic anomaly profiles. I also wish to thank Dr. R.W. Girdler and P. Styles for many helpful discussions, and for critically reading this manuscript. This research is funded by the Natural Environment Research Council.

REFERENCES

1. R. Freund, Geol. Mag. 102, 189, 1965.
2. R.W. Girdler and P. Styles, Nature, 247, 7, 1974.
3. H.A. Roeser, Geol. Jb., D13, 131, 1975.
4. R.W. Girdler and P. Styles, Nature, In preparation.
5. R. Blakely and A. Cox, J.G.R., 77, 4339, 1972.
6. S.B. Frazier, Phil. Trans. Roy. Soc., 267, 131, 1970.
7. A.S. Laughton, R.B. Whitmarsh and M.T. Jones, Phil.Trans. Roy. Soc., 267, 1970.
8. J.D. Lowell and G.J. Genik, Bull. Am. Assoc. Petrol. Geol., 56, 247, 1972.
9. R.F. Nelson, Personal communication.
10. P. Jones, Nature, 261, 567, 1976.

# PALAEOMAGNETISM IN THE OSLO RIFT ZONE

K.M. Storetvedt, S. Pedersen, R. Løvlie & E. Halvorsen

Department of Geophysics, University of Bergen,
N-5014 Bergen, Norway

ABSTRACT. Recent palaeomagnetic results (both published and unpublished) from the Oslo region, including data from plutonics, rhomb porphyry lavas, interbasaltic sedimentary horizons and dike rocks, are reviewed. These results suggest that in the southern and western areas of the rift zone there has been substantial magmatism of pre-Kiaman age (i.e. pre ca. 300 m.y. B.P.). In pre-Mesozoic times the areas of main igneous activity appear to have propagated northward. Thus, volcanic activity seems to have swept the Oslo district in the late Carboniferous and/or early Permian while in the northern extremity of the rift (at Brumunddal) outburst of lava flows took place in the latest Permian. Tectonic deformation is seen to post-date the volcanism and is probably of Mesozoic age. At least two pulses of dike intrusions appear to have affected the Oslo region in post-Palaeozoic times, probably in the Jurassic.

## 1. INTRODUCTION

The original palaeomagnetic study of the Oslo volcanics [14] was for a long time considered as one of the most elaborated palaeomagnetic studies ever undertaken. However, by analysing the data more closely it has become clear that even after alternating field treatment up to several hundred oersteds, within-site magnetizations frequently exhibit meridional distributions. This may imply that one actually is dealing with unresolved multi-component remanences, possibly related to a long magmatic history. These aspects formed the impetus for further palaeomagnetic work in the Oslo region. Of newer published data there are those from 9 diabase dikes in the Kragerø Archipelago [9] and from 3 rhomb porphyry lavas in the

*I. B. Ramberg and E.-R. Neumann (eds.), Tectonics and Geophysics of Continental Rifts, 289-296.*

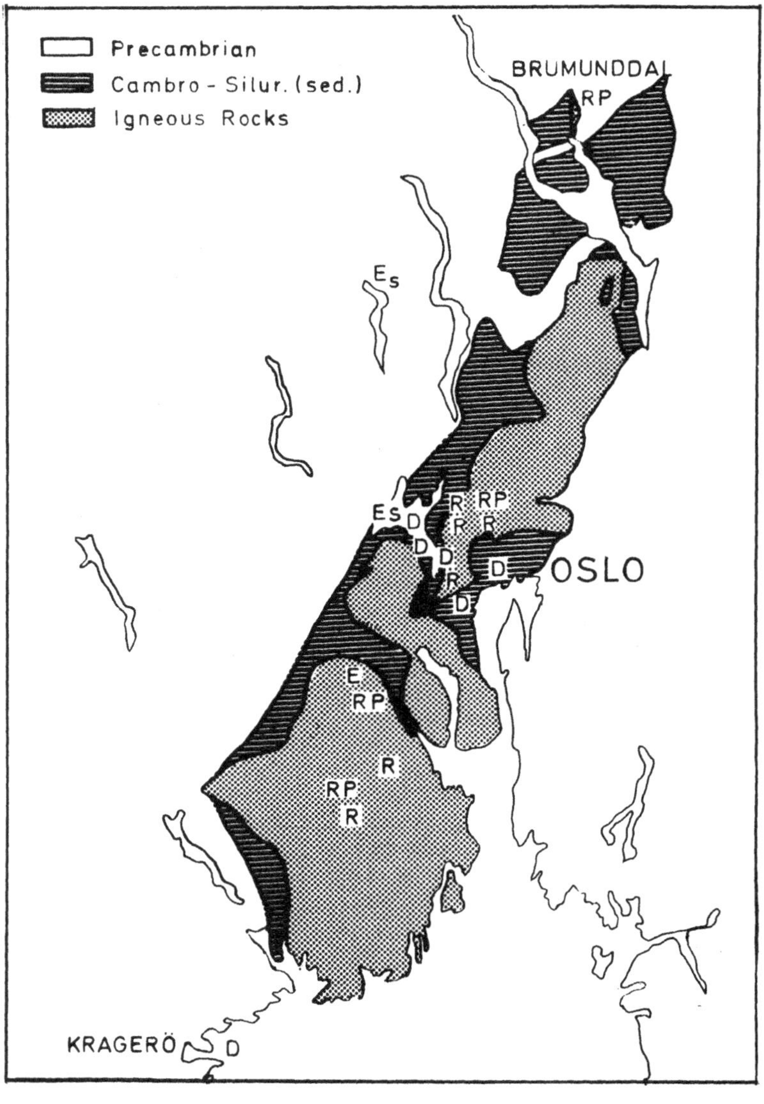


Fig. 1. Geological sketch map of the Oslo rift zone including areal distribution of investigated rocks. R: Red interbasaltic sediments; RP: Rhomb porphyry lavas; E: Ekerite; Es: Essexite; and D: Diabase dikes. In some local areas where more than one site of a particular rock type have been studied these are only shown by a single symbol.

northern extremity of the rift zone [11]. Further data are now available from 23 sites, embracing 3 rhomb porphyry (RP) lavas, 8 interbasaltic sedimentary horizons (baked by overlying RP flows), 1 ekerite (soda granite), 2 essexites and 9 diabase dikes. Distribution of sampling sites is shown in Fig. 1. In the present context only the most important implications of the newer palaeomagnetic data are summarized. Experimental details and further interpretational aspects of the results will be dealt with in a forthcoming paper.

## 2. RESULTS

As a general rule the magnetization of the plutonic and volcanic rocks, and to some extent also the interbasaltic sediments, is of polyphase structure, consisting of sub-components of opposite polarity and which may cause widely dispersed directions of the resultant magnetization vectors. However, including the results from the Brumunddal lavas [11], a total of five different magnetization axes have now been retrieved. The palaeomagnetic component which is interpreted as the oldest one is termed the A-component (cf. Fig. 2). This magnetization is found in the southern and western areas of the rift zone and constitutes the principal magnetization in the investigated plutonic and volcanic rocks. Fig. 3 demonstrates the magnetization build-up in one of these sites. The A-pole (see Fig. 4) plots in close agreement with the important European Palaeozoic pole position at ca. 20N, 160E and which seems to have been stationary from at least Ordovician to around Middle Carboniferous [8]. The subsequent polar shift into the Permo-Carboniferous pole position (pole 2 of Fig. 4) seems to correlate with the onset of the Kiaman geomagnetic interval (reversed) which is dated to ca. 300 m.y. B.P. [3]. This signifies that at least in the southern and western areas of the rift considerable igneous activity must have occurred prior to 300 m.y. B.P.

On a whole, the investigated interbasaltic sediments form a separate group of palaeomagnetic directions which is termed the B-component (cf. Fig. 2). The angular difference between the A and B magnetizations is nearly 40 degrees and statistically the two mean directions are significantly different. There is some evidence that the two sedimentary sites in the southern districts (cf. Fig. 1) contain the pre-Kiaman axis of magnetization along with a strong Kiaman overprint but this seems not to be the case for the other sedimentary sites which come from the districts around Oslo. After applying tectonic correction the B-pole plots in extremely good agreement with the Permo-Carboniferous pole position (pole 2 of Fig. 4). This suggests that at least the Oslo district was swept by volcanism in late Carboniferous or Permian times. Also, the structural tilt must be post-volcanic, a conclusion which is supported by geological evidence [4]. Unfortunately,

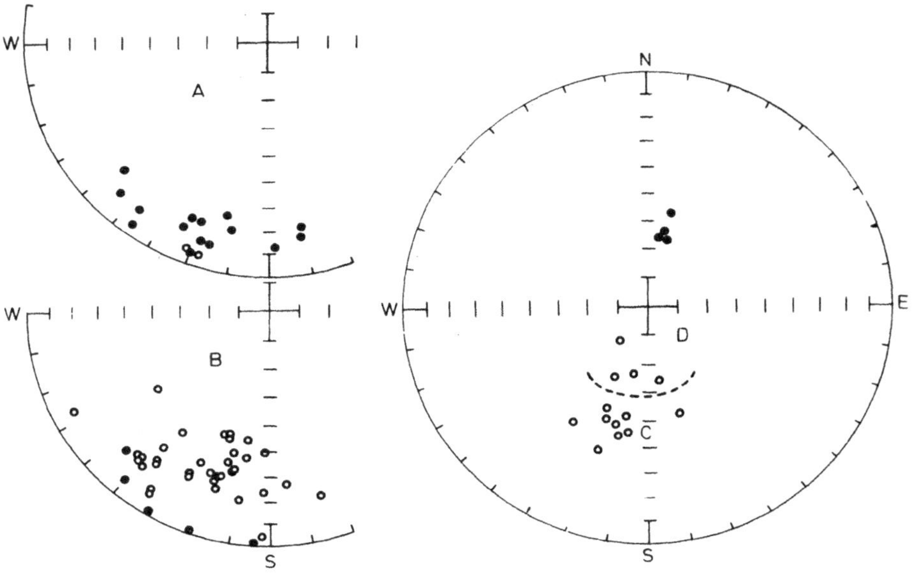


Fig. 2. Palaeomagnetic components from the Oslo rift zone. The A- and B-magnetizations are from plutonic/volcanic rocks and interbasaltic baked sediments respectively. These two components are represented by specimen directions. Magnetizations C and D are from dike rocks in the Oslo and Kragerø districts (cf. Fig. 1). Each dike shows a very good internal consistency of its remanence directions and is believed to constitute a 'point' in time (original). Components C and D are therefore represented by dike mean directions. Note the practically perfect antiparallelism for dike mean directions of the D-component. Data are without tectonic correction. Open (closed) symbols: upward (downward) pointing magnetization vectors. Projection is equal area.

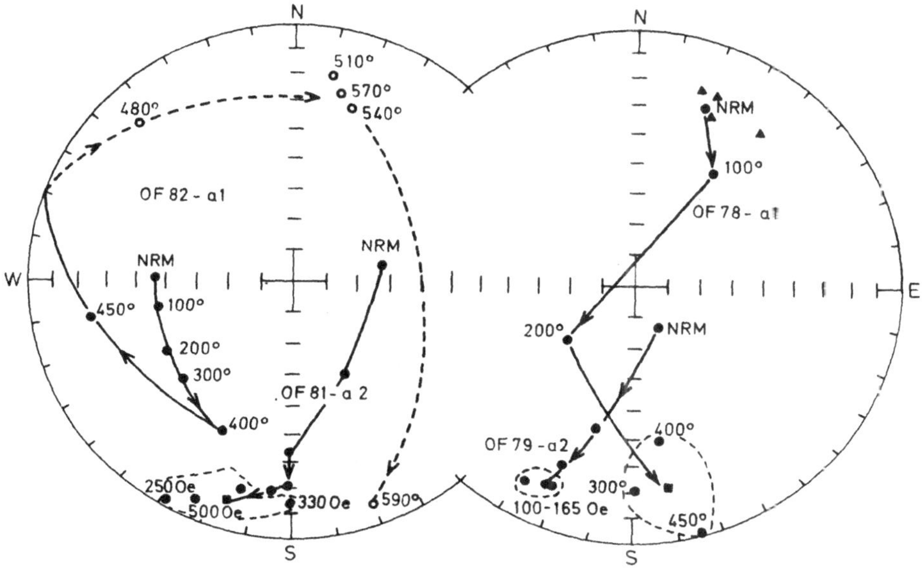


Fig. 3. Directional pattern versus progressive demagnetization for specimens from the ekerite. Note the two-polarity magnetization build-up in specimen OF 82-a1: extended great circle paths through the temperature ranges of NRM-400°C and from 540°-590°C respectively would intersect in a position that agree with stable end-point data from the same site (cf. OF 78-a1, OF 79-a2 and OF 81-a2) and would be closely antiparallel to the normal magnetization directions for the interval 510°-570°C. The triangles represent various vector subtraction estimates for specimen OF 78-a1, defining the erasing vector within the temperature range NRM-300°C. Squares denote the mean high temperature directions (shallow reversed) for individual specimens. Diagram conventions as for Fig. 2.

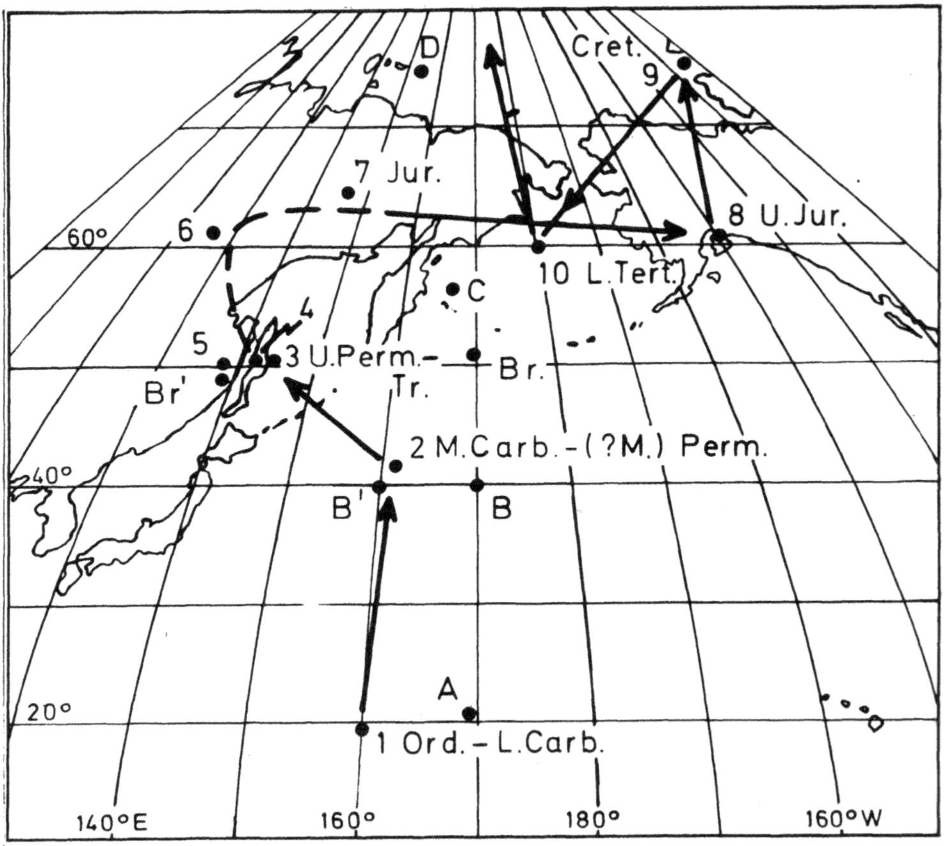


Fig. 4. The Oslo region palaeomagnetic poles seen in conjunction with the suggested European polar wander path for part of Phanerozoic time. Poles 1, 2, 3 and 7 are mean poles from the following sources: 1 [8]; 2, 3 and 7 [5] (other sources give similar results). Individual poles are as follows: 4 [15]; 5 [12]; 6 [13]; 8 [1]; 9 [6]; 10 [10]. A-D refers to the magnetization components of Fig. 2. Br denote the Brumunddal pole [10]. Poles after tectonic correction are shown by dashed symbols (B' and Br'). The Jurassic polar path is very uncertain. With reference to the poles of Fig. 4 an alternative path may be: 6(7)-D-C-8.

at present there are no results from the lavas immediately overlying the investigated sediments, but the available palaeomagnetic evidence from lavas and sediments indicates that the various RP lava units may be of much more local extent than previously suggested [4].

The results from the Brumunddal lava series [11] were rather unexpected in that the Kiaman geomagnetic field (reversed) was only very weakly developed in these rocks while, on the other hand, the Permo-Triassic field axis (both polarities) was very pronounced. The lavas were severly oxidized and in view of a supposedly Lower Permian age for all rhomb porphyries in the Oslo rift zone one was forced to postulate that some unknown process had delayed the oxidation process over a time span of the order of at least 40 m.y. However, by considering the very porous character of the overlying Brumunddal Sandstone (supposed to be a direct continuation of the sedimentary horizons between the lavas) this explanation hardly seems tenable now: a much more straight forward explanation of the data is that the Brumunddal volcanism took place at around the termination of the Kiaman geomagnetic epoch (at ca. 240-230 m.y. B.P.) so that the major oxidation of these rocks occurred in the Permo-Triassic. A late Permian age for the Brumunddal lavas also lines up with the recent suggestion that the Brumunddal Sandstone may be of Triassic age [7]. As seen in Fig. 4 the tectonically corrected Brumunddal pole is in excellent agreement with the late Permian-Triassic palaeomagnetic pole. This implies that the tectonic disturbance, probably connected with the subsidence of the rift, is of post-Permian age.

Several of the dikes collected in the Oslo district are unstably magnetized and reliable palaeomagnetic directions for such rocks have not been obtained. 8 dikes, however, have given very well defined palaeomagnetic results and these are in complete agreement with previously published data from 9 dikes in the Kragerø district (cf. Fig. 1) [9]. The results tend to be grouped along two different axes of magnetization. The combined Oslo and Kragerø results are shown in Fig. 2 as the C and D magnetizations respectively. The C and D poles plot in Mesozoic-Tertiary positions (Fig. 4). A further palaeomagnetic age interpretation of these results is difficult, primarily because of poor knowledge of the Jurassic polar path for Europe. Recent data [2] suggest, however, that pole D may be of Jurassic age and this age may well apply for pole C too. It is therefore tentatively suggested that major dike swarms in the Oslo rift were injected at the time the Central North Sea graben was formed.

Major conclusions:

I Magmatism in the Oslo rift zone initiated prior to 300 m.y. B.P.

II Volcanism covered a time span from pre 300 m.y. B.P. to ~ 240 m.y. B.P.

III In pre-Mesozoic times the igneous centres appear to have propagated northward.

IV Tectonic tilt is post-volcanic and most likely of Mesozoic age.

V Two generation of dikes were injected in Mesozoic times.

REFERENCES


1. E. Briseid and E. Halvorsen, Phys. Earth Planet. Int., 9, 45, 1974.
2. F. Heller, J. Geophys., 42, 475, 1977.
3. M.W. McElhinny, Palaeomagnetism and plate tectonics, Cambridge Univ. Press, Cambridge, 1973.
4. Chr. Oftedahl, Permian rocks and structures of the Oslo region, Norges Geol. Unders., 212, 298, 1960.
5. J.D. Phillips and D. Forsyth, Geol. Soc. Am. Bull. , 83, 1579, 1972.
6. S.T. Sandal and E. Halvorsen, Phys. Earth Planet Int., 7, 125, 1973.
7. N. Spjeldnæs, Abstract - Nordisk geol. vintermøte, Oslo, 1972.
8. K.M. Storetvedt, Earth Planet. Sci. Lett., 3, 444, 1967.
9. K.M. Storetvedt, Norsk Geol. Tidsskr., 48, 154, 1968.
10. K.M. Storetvedt, Phys. Earth Planet. Int., 1977 (in press).
11. K.M. Storetvedt and N. Petersen, Z. Geophys., 36, 569, 1970.
12. K.M. Storetvedt and R. Steel, Scott, J. Geol., 1977 (in press)
13. K.M. Storetvedt, C.M. Carmichael, A. Hayatsu and H.C. Palmer, Phys. Earth Planet. Int., 1977 (in press).
14. R.O. Van Everdingen, Skrifter - Det Norske Videnskapsakademi, Mat-Nat. kl., No. 1, Oslo, 1960.
15. J.D.A. Zijderveld, Paleomagnetism of the Esterel rocks, Thesis (unpublished), Utrecht, 1975.

# SEISMIC MAPPING OF THE FENNOSCANDIAN LITHOSPHERE AND ASTHENOSPHERE WITH SPECIAL REFERENCE TO THE OSLO GRABEN REGION

E.S. Husebye and F. Ringdal

NTNF/NORSAR, P.O.Box 51, N-2007 Kjeller, Norway

ABSTRACT. This paper presents a review of past and present seismological research efforts aimed at mapping lateral inhomogeneities in the lower lithosphere and asthenosphere in Fennoscandia and adjacent areas. Most of the results obtained here are derived from high-quality data from the large aperture NORSAR array which overlies the northern part of the Oslo Graben. The array siting area has been mapped seismically in great detail using novel P-wave travel time and amplitude analysing techniques. In the Oslo Graben area the seismic anomaly pattern does not exhibit an obvious correlation with the graben surface contours nor with the associated gravity anomalies. An integral feature of the Fennoscandian lithosphere appears to be seismic velocity perturbations of order a few per cent. There are also some indications that the lithosphere beneath the Caledonides is relatively complex as compared to that of the Baltic Shield. Interestingly, the strongest seismic velocity anomalies are found in the lower lithosphere of southern Scandinavia which has an estimated thickness of around 250 km. Lateral variations in the upper mantle down to about 500 km have been observed over wave paths between NORSAR - North Atlantic, NORSAR - Western Russia and NORSAR/UK - Southern Europe.

## 1. INTRODUCTION

The abundant geological evidence on variability of earth material properties and in petrological and tectonic processes has been known to geophysicists for a long time, but is nevertheless mostly ignored in the structural models used in the interpretation of geophysical observations, notably seismic profiling data. There

*I. B. Ramberg and E.-R. Neumann (eds.), Tectonics and Geophysics of Continental Rifts, 297-311.*

are several reasons for this, one being that these effects are often restricted to second order terms, another perhaps more important, is that the proper handling of such effects requires access to large amounts of high quality data. In this respect the Oslo Graben is no exception; this area has been subjected to comprehensive geological and gravimetric surveys (e.g., see Dons and Larsen [1] and Ramberg [2]), while its status as a geologically anomalous province is not reflected in interpretations of traversing seismic profiles [3,4].

Modern hypotheses on taphrogenesis suggest that intracontinental rifting occurs as the incipient stage of an evolutionary process normally leading to mid-oceanic ridge features and subsequent creation of new lithosphere. The continued existence of some old continental paleorifts, however, testifies that this evolutionary process is sometimes aborted, as is the case for the Oslo Rift. The rifting is usually tied to the rise of hot, astenospheric material and the associated diapir mechanism is explainable in terms of gravitational instability in the low-viscosity asthenosphere [5]. At least-during the initial stages of magma ascent, this process must make an imprint on the surrounding lithosphere, that is, result in geophysically detectable anomalous structures.

Concerning the Oslo Paleorift, which is of Permian age, the once hot magmatic masses have now cooled and recrystallized. Therefore, heat flow anomalies originally characterizing the Oslo Rift, and consequently other geophysical anomalies associated with the initial stages of the rifting process [6], may also have been lost. We must here differentiate between the upper part of the lithosphere (the crust of thickness around 35 km) and the lower lithosphere (covering the upper mantle down to around 250 km depth). In the case of the crust, the existence of pronounced gravity and magnetic anomalies tied to the surface contours of the Oslo Graben has been amply demonstrated [1,7], while the form of any corresponding seismic anomalies is less clear as discussed by Husebye, England and Ramberg [8, this volume].

The topic of this study is a presentation and discussion of lower lithosphere seismic velocity anomalies reported in the general Oslo Rift area, and in this respect we will restrict ourselves to results derived from analysis of the excellent seismic wave recordings from the large aperture seismic array NORSAR, overlying the northern part of the Oslo Graben. We will not discuss magnetic and gravimetric observations as these types of observations do not have sufficient resolving power for structural heterogeneities of sufficiently small wavelength in the lower lithosphere to be of interest for this study. In addition to studies in the immediate region of the rift, we will also discuss large-scale upper mantle heterogeneities in the Fennoscandian region.

## 2. DATA AND METHOD OF ANALYSIS

Most of the results to be presented here are derived from analysis of short and long period seismic wave recordings from the NORSAR array overlying the northern part of the Oslo Graben (see Fig. 1). The array comprises 22 subarrays which each encompasses 6 short period vertical seismometers and one 3-component long period seismometer. The respective sampling rates are 20 and 1 Hz. The digital sampling and associated magnetic tape storage greatly facilitate analysis of large amounts of earthquake recordings. Details on NORSAR hardware and software systems are given in [9,10]. Additional data come from the International Seismological Centre (ISC) bulletin tapes for 1971-74, comprising P-wave residuals from all the reporting stations in Fennoscandia.

The complexities of seismic P-wave signal variations across large aperture arrays like NORSAR (and LASA in Montana) were novel to seismologists when data from these installations became available a decade ago. Notably, this type of observation was clearly incompatible with standard spherical symmetric earth models with uniform material properties within layers. Particularly disturbing features were travel time anomalies as large as ± 0.5 sec for seismometer separations of only 50-100 km and amplitude variations of teleseismic signals that often could reach a factor of 10 or more across NORSAR [10,11]. The geological significance of these anomalies was clear, namely, that dense sampling of an area offered by an array of diameter around 100 km exhibited differences in seismic wave parameters at depth which traditionally had been attributed to structural variations only over much larger distance. Although at that time the proper physical understanding of the anomalous features mentioned above was inadequate, their potentially adverse effect on array operation could fairly easily be offset by introducing special corrections and flexible signal analysing techniques [12,13,14,15,16]. This ability to remove undesirable residuals to some extent also removed the need to interpret them geophysically, and it is only comparatively recently that these data bases have become extensively used in interpretation of the NORSAR subsurface structural heterogeneities. In general, we distinguish between two principal methods in interpretation of seismic data. One of these is the forward problem of seismology, that is, to compute the wave response for a given structural model and then compare the 'theoretical' results with the actual observations. Usually some sort of perturbation scheme is used here in order to ensure a reasonable fit to the observational data, as the initial model may be too coarse. The second approach, often denoted the inverse problem of seismology, is to derive a structural model directly from the observational data. In practice, the observational data are frequently very complex due to, e.g., wave scattering effects. In such cases, the <u>deterministic</u> approaches of the types mentioned above, at

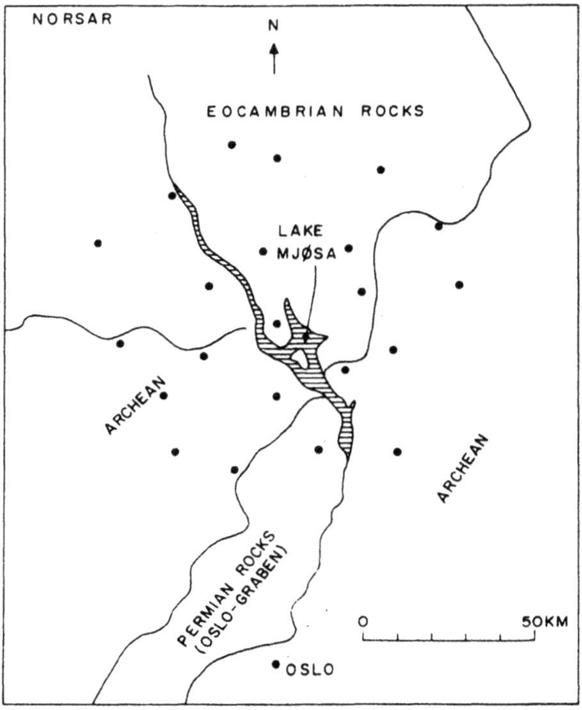


Fig. 1. The NORSAR array configuration comprising the 22 subarray centres. The Oslo Graben igneous rock series are outlined.

least in the initial stages of analysis, are often replaced by random medium models where generalized, statistical descriptions of medium heterogeneities are sought. All of the above approaches have their merits as will be demonstrated in the following sections, though the final goal is, of course, a deterministic mapping of the subsurface structures.

## 3. SEISMIC MAPPING OF THE LITHOSPHERE IN THE CENTRAL OSLO GRABEN AREA

The 'antipode' to conventional deterministic modelling of the lithosphere is to assume that its structure is so complex with respect to seismic waves that it behaves as a medium with random inhomogeneities. This particular kind of modelling has successfully been introduced in the interpretation of NORSAR observed P-wave travel time and amplitude anomalies by using the Chernov [17] and Tartarski [18] theories for acoustic wave propagation in random media. These theories are based on a statistical description of the medium in terms of the mean square of fractional fluctuation of velocity, the correlation distance (i.e., the distance between two points in the medium at which velocity fluctuation becomes uncorrelated) and the linear extent of the inhomogeneous region. For example, Berteussen et al. [19] and Dahle et al. [20] were able to explain about 60% of the variance in NORSAR observed time and amplitude anomalies by such an approach. The correlation distance was of the order of 30 km and tentatively interpreted as a lower limit of the wavelength of significant inhomogeneities in the lithosphere in the NORSAR siting area. The refractive index or rms-velocity perturbations were of the order of 3% for a medium thickness of around 150 km. It is important to note that in this approach to interpreting seismic data or other types of geophysical observations it is very difficult to differentiate between a thin medium with large inhomogeneities and one thicker but less inhomogeneous. Another interesting result of the above random medium experiments was that seismic velocity measurements are seen to be unstable unless the aperture of the array or profiling lengths greatly exceed the wavelength of the lithospheric heterogeneities and these are typically of crustal dimensions or greater [20]. For example, the first attempt to explain NORSAR time anomalies was in terms of a corrugated Moho-interface. The best fitting surface in a least squares sense could only account for approx. 20% of the observed time variance [21], while the Kanestrøm and Haugland [22] model for Moho beneath NORSAR did not result in any time variance reduction.

From this random medium approach it was obvious that significant lateral inhomogeneities exist in the lithosphere but from a geological or geophysical point of view it is more important to actually locate these anomalies than have a statistical estimate of their size. It is possible to do so because a large aperture array like NORSAR represents a form of 2-dimensional sampling in principle similar to that used in photography and holography, albeit in the latter cases the sampling is far more dense. A successful method for solving the problem of inverting array data has recently been demonstrated by Aki, Christoffersson and Husebye [23]. The essence of their approach, as applied to the NORSAR siting area, was to divide the lithosphere into 5 layers which in turn were subdivided into 9 x 9 blocks of size 20 x 20 km. The problem was as formulated in a least squares sense, to estimate the velocity perturbations with each block relative to the average layer velocity. The input data were teleseismic travel time residuals at the 132 short period instruments in the array and good azimuthal and distance coverage were essential for optimum cross-sampling of the individual blocks. Some of their results for the lower lithosphere, depth range 66-126 km, are shown in Figs. 2 and 3. (The results bearing on the upper part of the lithosphere are shown elsewhere in this volume [8].) The dominant feature in these figures is high velocities to the east,

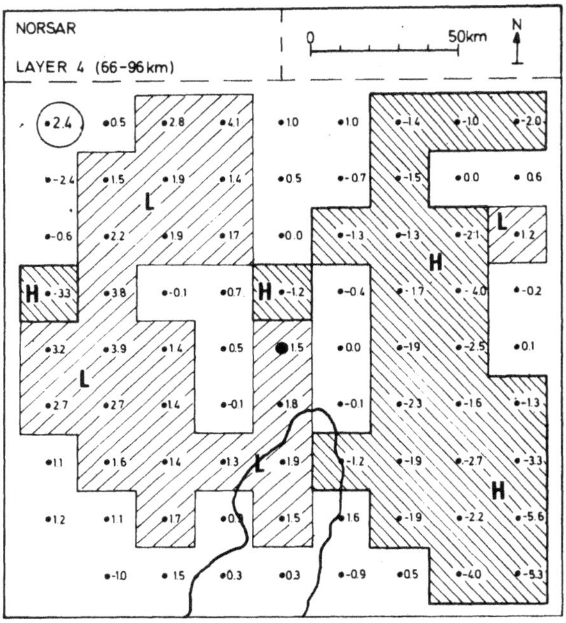


Fig. 2. Seismic velocity anomalies in the lower lithosphere, depth range 66-96 km, beneath the NORSAR siting area. The Oslo Graben igneous rock series are outlined. The numbers indicate the fractional velocity perturbation in per cent of the average velocity of 8.2 km $s^{-1}$. The shaded areas correspond to the magnitude of solution greater than 1 time the standard error. The letters L and H refer to low and high velocity anomalies, respectively. The central subarray is marked by a larger solid circle. The figure corresponds to the generalized inverse solution for Layer 4 in the Aki, Christoffersson, Husebye paper [23]. The seismic anomalies in the upper lithosphere layers are presented elsewhere in this volume [8].

low velocities to the west and no obvious correlation with the Oslo Graben surface contours.

Considerable research efforts have been invested on constraining earth models in general and lithospheric models in particular by incorporating P-wave amplitude in analysis. Most of such work appears to have been only moderately successful, and the main reason for this seems to be that the rather shorth 'wavelength' of the observational data is not matched by sufficiently detailed structural models for which amplitude responses can be calculated. To our knowledge, only the random medium approach and certain types of multivariate statistical analysis have proved satisfactory in quantitative modelling of amplitude anomalies [15,19,20]. Recently, however, Haddon and Husebye [24] have undertaken a comprehensive study of NORSAR P-wave anomalies; their approach is briefly outlined here.

Their starting point is to approximate the subcrustal anomalous structures by a two-dimensional thin lens which is obtained by projecting travel time residuals to the bottom of the lithosphere. This is chosen to be at various depths ranging from 50-250 km. The time residuals are then converted to seismic velocity anomalies and the thin lens models thus constructed have many features similar to the velocity anomaly patterns shown in Figs. 2 and 3. Then Claerbout's [25] wave equation approach, which has found wide application in seismic oil prospecting, is used to calculate the corresponding P-wave amplitude pattern across NORSAR. A satisfactory agreement between observed and theoretical amplitude values has been obtained particularly when the thin lens is regarded as being between 150 and 200 km depths. In addition, Haddon and Husebye have used a variant of Poisson's equation for 'reversing' the above investigations scheme, that is, calculating time anomalies from amplitude anomalies. In this particular case very good agreement was obtained between observed and 'theoretical' values.

The main conclusions of the study of amplitude residuals were: a) the same structural anomalies are responsible for both P-wave time and amplitude anomalies, b) the amplitude anomalies are very sensitive to velocity gradients so proper modelling here is difficult; this is the reason why the 3-D model of Aki, Christoffersson and Husebye [23] was replaced by the 2-D thin lens model, c) the lower part of the lithosphere appears to be relatively heterogeneous in comparison to its upper part, and d) the thickness of the lithosphere beneath NORSAR appears to be of the order of 200 km, and is better constrained than that using time data.

A fundamental concept in many geoscience branches is the lithosphere which by classical definition is that shell of the earth which can sustain long-term nonhydrostatic stresses, whereas the underlying asthenosphere is the region where the flows associated with isostatic adjustments take place [26]. The latter definition appears unduly restrictive in view of very recent gravity analysis, where 'isostatic flow' is hypothesized to take place also within the lithosphere [27]. Seismologically, however, the lithosphere-asthenosphere boundary is not well defined, and the reason for this is at least partly implicit in the schemes presented above. In addition, it may well be that the low velocity zone for P-waves is not widely well-developed, but in view of the significant viscosity decrease probably associated with this transition zone; it may be 'seen' by seismic S-waves. For example,

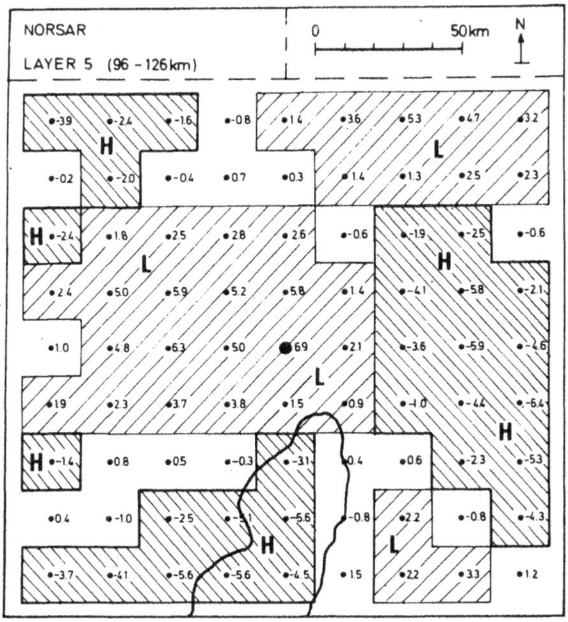


Fig. 3. Seismic velocity anomalies in the lower lithosphere, depth range 96-126 km, beneath the NORSAR siting area. The Oslo Graben igneous rock series are outlined. The figure corresponds to the generalized inverse solution for Layer 5 in the Aki, Christoffersson, Husebye paper [23]. For explanation of symbols and average layer velocity see Fig. 2.

in the lithosphere-asthenosphere boundary is sharp enough, S-waves incident on this discontinuity may give rise to detectable, refracted P-waves. Comprehensive analysis of teleseismic S-wave recordings at NORSAR, distance range 72-82 deg, has detected arrivals which may be interpreted in this sense, and from the differential P-S travel times a lithosphere thickness of around 250 km is inferred in the general NORSAR area assuming a horizontal interface [28].

From the NORSAR studies presented above, it is obvious that significant heterogeneities exist in the lithosphere in the NORSAR siting area and, of course, beneath the general Oslo Graben area. Although the correlation between the seismic results and the surface expression of the graben structures is not quite obvious (see also [8]), even so the observed seismic anomalies may reflect the evolution of the graben or some other tectonic process in the history of the region. An alternative, perhaps extreme, hypothesis is that the above seismic anomalies are typical for lithospheric structures anywhere and that lower lithospheric structures bear no relation to surface tectonic features except in regions of active tectonism. Under this hypothesis, any seismic anomalies in the lower lithosphere resulting from the formation of the Oslo Graben were either purely thermal, and have decayed so that they are now undetectable, or have been blurred or removed by secular deformation of the lower lithosphere. In the next section we will present results which will shed some light on this kind of problem, mainly with regard to velocity structures in Fennoscandia, encompassing parts of the Caledonides and the western extension of the Baltic Shield.

## 4. LATERAL VARIATIONS IN THE FENNOSCANDIAN UPPER MANTLE

NORSAR recordings have also been used in studies of the gross structure of the upper mantle beneath southern Europe, western Russia and the North Atlantic [29,30,31]. The basic data used were absolute P-wave travel times as a function of epicenter distance, in combination with the well-known Herglotz- Weichert and tau inversion schemes [32,33]. The essence of the southern Europe and western Russia studies is shown in Fig. 4, implying different upper mantle differences down to depths of around 500 km. This is mainly based on differences in triplications of the travel time curves which are reflected less clearly in the external bounds on the velocity depth profiles. The velocity distribution beneath the N. Atlantic is, not unexpectedly, significantly different from that of western Russia down to around 200 km but seems to be substantially the same below around 300 km [31]; these results emphasize the velocity contrast to be expected in the upper 200 km of the earth, but suggest that anomalies at depths greater than this are exceptional, since the Russian platform and the North Atlantic would appear to represent extremes of tectonic activity. Furthermore, using P-to-S conversion observations, Vinnik [34] has located the major upper mantle discontinuities at depths around 420 and 670 km, respectively, in agreement with the regional studies of [29] and [31].

The above work reflects large-scale regional differences; the young, oceanic lithosphere in the North Atlantic, the old, stable shield area of western Russia and the continent collision zone of southern Europe. We are mainly interested in relatively small-scale heterogeneities in the general Fennoscandian lithosphere, that is, seismic subsurface mapping in detail roughly equivalent to that obtained for NORSAR. A major obstacle for such ventures is lack of adequate data; in this case this has been circumvented by using data from the conventional Fennoscandian seismograph network shown in Fig. 5. The observational data comprise about 45,000 teleseismic P-wave travel time readings from the Fennoscandian seismograph network, while the novel

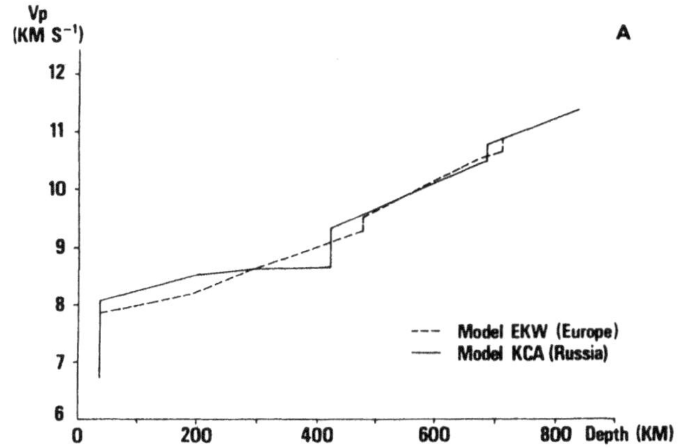


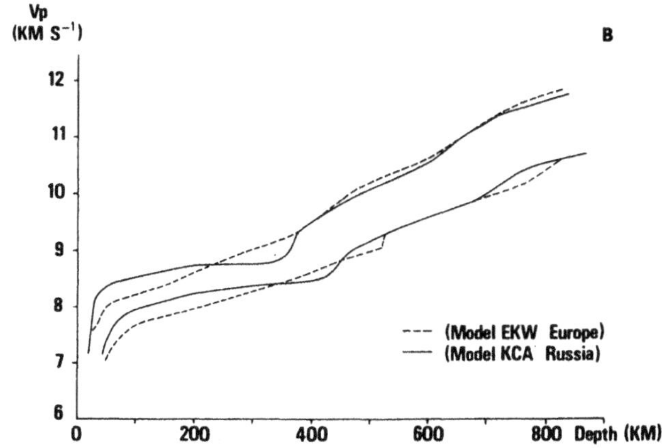


Fig. 4a. Upper mantle P-velocity models for Fennoscandia and adjacent regions resulting primarily from NORSAR array and travel time data. The model KCA is that of King and Calcagnile [29] for western Russia, while the model EKW is that of England, Worthington and King [30] for southern Europe. In the latter case, supplementary Eskdalemuir array data were used. Fig. 4b gives extremal bounds on the KCA and EKW velocity models. Figure redrawn from England et al. [30].

analysing techniques represent a blend of those introduced by Aki et al. [23], Haddon and Husebye [24] and Ringdal [35]. The results obtained so far (see Fig. 5), although preliminary, clearly indicate that the deeper parts of the lithosphere in the general Fennoscandia region are not homogeneous. There are also

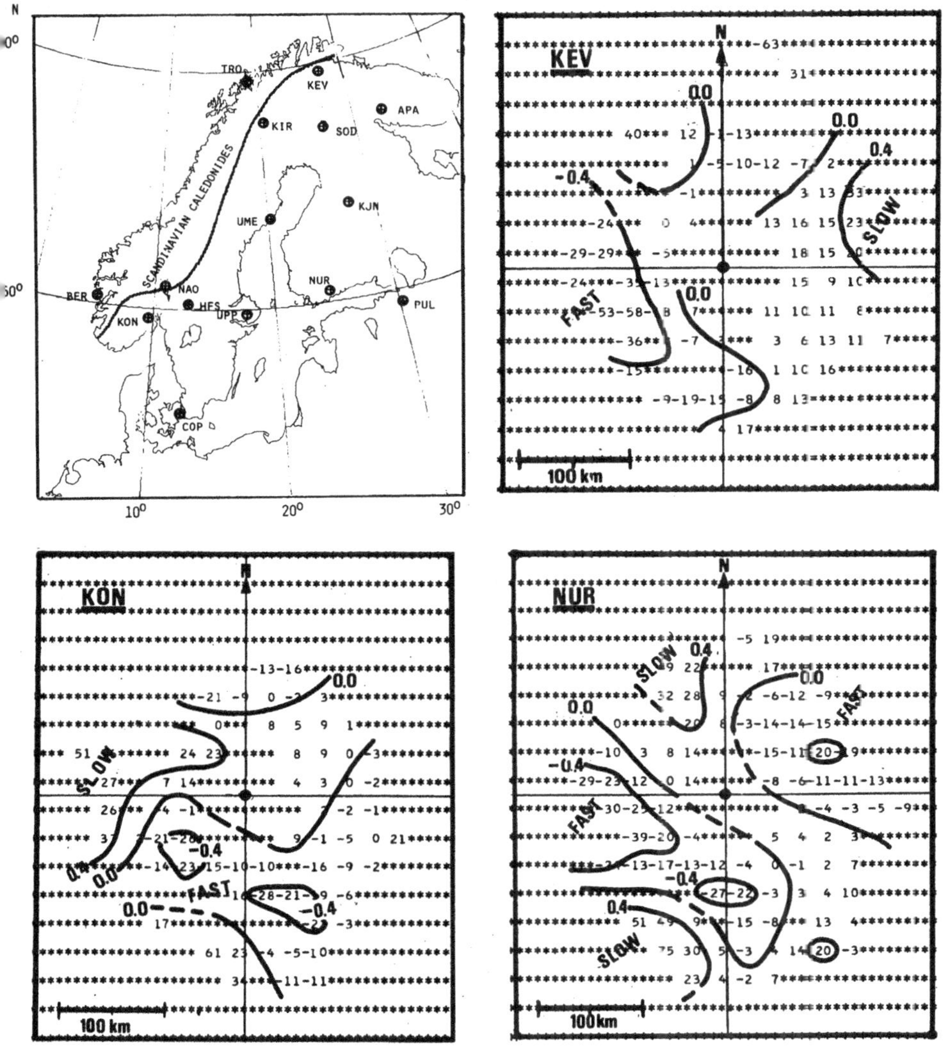


Fig. 5a. The Fennoscandian seismograph network, (upper left corner) used in analysis of P-wave travel time residuals as reported by the International Seismological Centre (ISC) for the period 1971-1974. The western extension of the Baltic Shield, the Caledonides, is outlined. The Kevo (KEV), Kongsberg (KON) and Nurmijärvi (NUR) P-wave travel time residuals are shown in the remaining part of the figure. The given numbers are in units of 1/50 sec, except for the time residuals associated with the contours which are specified in seconds. The time anomaly patterns presented with areas of slow and fast velocity structure indicated are derived from absolute time observations.

some indications that seismograph stations located within the Caledonides generally exhibit a more complex time residual pattern than stations located within the Baltic Shield. We note in passing that the new analysing method introduced here extracts significantly more information from observational travel time data than traditional techniques do. For further details on this work and various tests introduced to check the reliability of the final results, we refer to [36].

The results presented above indicate that seismic velocity fluctuations of the order of a few per cent are an integral part of the Fennoscandian lithosphere. However, this does not appear to be a unique feature for this particular region in view of the so-called precursor waves to the seismic phases like PP and P'P' observed at seismograph stations anywhere. Comprehensive studies of these waves strongly favor a generating mechanism tied to scattering from small heterogeneities in the general lithosphere [37,38]. Further evidence of a general 'complex' lithosphere also comes from analysis of data from seismic networks in America and Asia using the Aki, Christoffersson, Husebye-inversion techniques; for references see [39,40,41,41]. The only exception so far appears to be the Garibindaur array located in the central parts of the Indian subcontinent [43]. Nevertheless, the correspondence of deep-seated seismic anomalies with regions of Caledonian and Permian tectonic activity suggests that, in Fennoscandia at least, anomalies of this sort do reflect tectonic activity, but certainly in a complex fashion.

## 5. DISCUSSION

In the previous sections we have presented the results of comprehensive seismological studies of the Fennoscandian upper mantle, with emphasis on its uppermost part, the lithosphere. Furthermore, the NORSAR siting area overlying the northern part of the Oslo Graben, has been subjected to particularly detailed analysis. In the latter case, it is noteworthy that although widely different types of seismic wave recordings and interpretational techniques have been used, the final results are in very good agreement except for an indeterminacy in the proper depths of the detected heterogeneities. With the sampling offered by the NORSAR array, it is clear that the bulk of the observed seismic anomalies originate in the lower lithosphere, but whether this should be around 150 km or 200-250 km we cannot say. In the latter case, improved resolvability requires data from a network having a significantly larger aperture.

In the context of dominant tectonic features in Fennoscandia, their correlation with seismic results appear to be complex as pointed out previously. An important factor here is the occurrence of velocity anomalies of the order of a few per cent which appears to be an integral part of the lithosphere in this region. Furthermore, the lithosphere thickness is likely to be around 250 km [28], a result in line with gross estimates for stable, shield regions characterized by modest heat flow [44].

The important feature of the above results is the consistent evidence of relatively strong seismic velocity perturbations in the lower lithosphere, which appear to be at least of the same order as those reported for the crust-Moho regime. The relation of these residuals to tectonic features suggests that inhomogeneities may be 'frozen' into the lower lithosphere over 100's of m.y. at least; this is not in line with recent suggestions [27] that the lithosphere behaves rigidly perhaps only to depths of 50 km. Another explanation is that the lithosphere-asthenosphere decoupling is far from perfect, and that localized heat and/or mass interchange can occur between them, in particular beneath the dominant continental plates in view of their relatively slow movements of a few centimeters a year. It is possible that the asthenosphere could become gravitationally unstable below such effective thermal blankets and that hot asthenospheric material is, to various extents, 'injected' into the lower lithosphere. Processes of this kind have been simulated in laboratories under realistic rheological conditions [5], though specific geophysical evidence for such processes is lacking. The fact that we found relatively stronger seismic anomalies beneath the Caledonides is attributed to lithospheric thickening, structural heterogeneities and stress differences typical for continent-to-continent collision processes. There are well-documented geological evidence for stress anomalies with well-determined ages that exceed 100 m.y. [45] in the crust, but it is more likely that the inhomogeneities in the lower lithosphere are chemical or thermal in view of the expected low rigidity there.

## REFERENCES


1. J.A. Dons and B.T. Larsen (eds.), The Oslo Paleorift, Norg. geol. Unders., 1978 (in press).
2. I.B. Ramberg, Norg. geol. Unders., 325, 1, 1976.
3. J. Tryti and M.A. Sellevoll, Report on a reconnaissance seismic crustal study of the Oslo Rift, in: The Norwegian Geotraverse Project Final Report, ed. by K.S. Heier, Oslo Univ., 1977.
4. R. Kanestrøm and K. Haugland, Profile Section 3-4, in: Deep Seismic Sounding, ed. by A. Vogel, Uppsala Univ., Sweden, 1971.
5. H. Ramberg, Lithos, 4, 259, 1971.
6. R.J. Bridwell (this volume).
7. E.S. Husebye and I.B. Ramberg, Geophysical investigations, in: The Oslo Paleorift, ed. by J.A. Dons and B.T. Larsen, Norg. geol. Unders., 1978 (in press).

8. E.S. Husebye, P.C. England and I.B. Ramberg, (this volume).
9. H. Bungum, E.S. Husebye and F. Ringdal, Geophys. J. Roy. Astron. Soc., 25, 115, 1971.
10. H. Gjøystdal, E.S. Husebye and D. Rieber-Mohn, Seismol. Soc. Amer., Bull., 63, 549, 1973.
11. E.S. Husebye, A. Dahle and K.-A. Berteussen, J. Geophys. Res., 79, 2967, 1974.
12. K.-A. Berteussen, NORSAR location calibrations and time delay corrections, Sci. Report 2-73/74, NTNT/NORSAR, Kjeller, Norway, 1974.
13. K.-A. Berteussen and E.S. Husebye, Amplitude pattern effects on NORSAR P-wave detectability, Sci. Report 2-74/75, NTNF/ NORSAR, Kjeller, Norway.
14. A. Christoffersson and E.S. Husebye, Geophys. J. Roy. Astron. Soc., 38, 525, 1974.
15. F. Ringdal, E.S. Husebye and A. Dahle, P-wave envelope representation in event detection using array data, in: Exploitation of Seismograph Networks, ed. by. R.G. Beauchamp, Nordhoff-Leiden, The Netherlands, 1975.
16. J. Fyen, E.S. Husebye and A. Christoffersson, Geophys. J. Roy. Astron. Soc., 42, 529, 1975.
17. L.A. Chernov, Wave Propagation in a Random Medium, McGraw-Hill, New York, 1960.
18. V.I. Tartarski, Wave Propagation in a Turbulent Medium, McGraw-Hill, New York, 1961.
19. K.-A. Berteussen, A. Christoffersson, E.S. Husebye and A. Dahle, Geophys. J. Roy. Astron. Soc., 42, 403, 1975.
20. A. Dahle, E.S. Husebye, K.-A. Berteussen and A. Christoffersson, Wave scattering effects and seismic velocity measurements, in: Exploitation of Seismograph Networks, ed. by K.G. Beauchamp, Nordhoff-Leiden, The Netherlands, 1975.
21. K.-A. Berteussen, J. Geophys., 41, 71, 1975.
22. R. Kanestrøm, PAGEOPH, 105, 729, 1973.
23. K. Aki, A. Christoffersson and E.S. Husebye, J. Geophys. Res., 82, 277, 1977.
24. R.A.W. Haddon and E.S. Husebye, (in prep.)
25. J.F. Claerbout, Fundamentals of Geophysical Data Processing, McGraw-Hill, New York, 1976.
26. J. Barrel, J. Geol., 22, 655, 1914.
27. R.J. Banks, R.L. Parker and S.P. Huestic, Geophys. J. Roy. Astron. Soc., 51, 431, 1977.
28. I.S. Sacks, A. Snokes and E.S. Husebye, Lithosphere thickness beneath the Baltic Shield, Tectonophysics, 1978 (in press).
29. D.W. King and G. Calcagnile, Geophys. J. Roy. Astron. Soc., 46, 407, 1976.
30. P.C. England, M.H. Worthington and D.W. King, Geophys. J. Roy. Astron. Soc., 48, 71, 1977.
31. P.C. England, B.L.N. Kennett and M.H. Worthington, A comparison of the upper mantle structure beneath Eurasia and the North Atlantic, Geophys. J. Roy. Astron. Soc., 1978 (in press).

32. E.N. Bessonova, V.M. Fishman, V.Z. Ryaboyi and G.A. Sitnikova, *Geophys. J. Roy. Astron. Soc.*, **36**, 377, 1974.
33. B.L.N. Kennett, *Geophys. J. Roy. Astron. Soc.*, **44**, 517, 1976.
34. L.P. Vinnik, *Phys. Earth Planet. Inter.*, **15**, 39, 1977.
35. F. Ringdal, *J. Geophys.*, **43**, 611, 1977.
36. F. Rindgal, E.S. Husebye, P.C. England and R.A.W. Haddon, 2-dimensional seismic mapping of the lithosphere using ISC-reported P-wave time residuals, (in prep.).
37. D.W. King, R.A.W. Haddon and E.S. Husebye, *Phys. Earth Planet. Inter.*, **10**, 103, 1975.
38. R.A.W. Haddon, E.S. Husebye and D.W. King, *Phys. Earth Planet. Inter.*, **14**, 41, 1977.
39. K. Aki, A. Christoffersson and E.S. Husebye, *Seismol. Soc. Am., Bull.*, **66**, 501, 1976.
40. E.S. Husebye, A. Christoffersson, K. Aki and C. Powell, *Geophys. J. Roy. Astron. Soc.*, **46**, 319, 1976.
41. B.J. Mitchell, C.C. Cheng and W. Stauder, *Seismol. Soc. Amer., Bull.*, **67**, 1061, 1977.
42. W.H. Menke, *Seismol. Soc. Amer., Bull.*, **67**, 725, 1977.
43. K.-A. Berteussen, E.S. Husebye, R.F. Mereu and A. Ram, Quantitative assessment of the crust-upper mantle heterogeneities beneath the Gauribidinaur seismic array in southern India, *Earth Planet. Sci. Lett.*, 1978 (in press).
44. H.N. Pollack and D.S. Chapman, *Tectonophysics*, **38**, 279, 1977.
45. G.H. Eisbacher and H.U. Bielenstein, *J. Geophys. Res.*, **76**, 2012, 1971.

THE IDEAL-BODY CONCEPT IN INTERPRETATION OF THE OSLO RIFT GRAVITY DATA AND THEIR CORRELATION WITH SEISMIC OBSERVATIONS

E.S. Husebye[1], P.C. England[2] and I.B. Ramberg[3]

[1]NTNF/NORSAR, P.O.Box 51, N-2007 Kjeller, Norway
[2]Dept. of Geol. & Geophys., Univ. of Cambridge, Madingly Rise, Cambridge, U.K.
[3]Dept. of Geology, University of Oslo, P.O.Box 1047, Blindern, Oslo 3, Norway

ABSTRACT. The ideal-body concept of Parker offers a means of putting rigidly determined limits to gravity interpretations. This method has been applied to gravity profiles across the Oslo Graben, where a pronounced axial gravity high has previously been interpreted as the combined effect of crustal thinning and the presence of a dense, mantle-derived paleorift 'cushion' or asthenolith. The existence and approximate location of the dense intrusive body are generally supported by geochemical and petrological data. Interpretation based on Parker's technique favors the model of a dense body in the lower to intermediate crust. However, seismic velocity anomalies in the upper mantle obtained from 3-D inversion of NORSAR array data only exhibit a weak correlation with such anomalies and seem instead to reflect broader velocity variation below the Baltic Shield. There is also qualitative agreement between inversion of travel time and gravity anomalies, but the complex relationships between velocity and density in the intermediate and lower crustal materials allows quantitative agreement only in the depths deduced for the intruded body using the two sets of data.

## 1. INTRODUCTION

The formation of grabens or rifts has long been of interest to geoscientists and there are several conflicting hypotheses about the origins and deep structures of these features. These conflicts may arise from genuine differences between rift systems or may just reflect inadequacies of the geophysical data. In this paper we present a combined gravimetric and seismic approach to the

*I. B. Ramberg and E.-R. Neumann (eds.), Tectonics and Geophysics of Continental Rifts,* 313-327.

problems of the Oslo Graben in an attempt to provide better constraints on its deep structures.

## 2. GEOPHYSICAL INVESTIGATIONS IN THE OSLO REGION

A pronounced axial gravity high associated with the entire Oslo Rift has been interpreted [1-3; see also Ramberg & Spjeldnæs, this vol.] as the combined effect of crustal thinning and the presence of a dense paleorift 'cushion' or asthenolith primarily formed by accumulation and subsequent injection of buoyant mantle-derived fractional melt into the lower crust. It is suggested [1-3] that a mixture of these meta-intrusives and the pre-existing crust forms an inhomogeneous high density body located at depths between ca. 20 km and the base of the crust.

Geochemical and petrological studies [e.g. 4,5] support this contention, but most of the seismic studies do not. Crustal profiling work [6,7] and, in particular independent Moho depth determinations [8] show no convincing correspondence between crustal thickness and the surface expression of the graben; as will be discussed later, deeper studies of the seismic structure [9-11] show no trace of the hypothesized asthenolith in the lower part of the lithosphere. The latter works refer to research tied to body wave recordings from the large aperture array NORSAR [12] overlying the northern part of the Oslo Graben as shown in Fig. 1. For a comprehensive review of this work we refer to [13].

We now investigate those areas in which the seismic and gravity observations do overlap, giving special attention to potensial correlations between these two types of geophysical measurements.

## 3. GRAVITY DATA - THE IDEAL BODY METHOD

Parker [14] describes a method by which relatively few gravity observations may be analysed to yield an optimum bound on certain physical parameters associated with the generating body. In particular, this method may be used to determine a bound on the depth of burial - a bound which is improved if an upper bound on density contrast can be specified or estimated. Although this method can tell us nothing about the shape of the generating body, it does give 'indisputable' information about the buried system, in contrast to more conventional computer modelling, where conclusions may be drawn from accidental features of a particular solution [14].

We use this method to determine the maximum depth of burial of the material giving rise to the gravity high which follows the

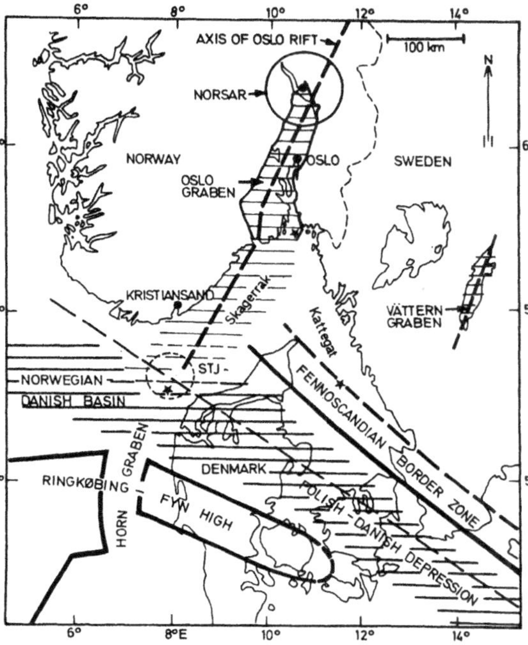


Fig. 1. Tectonic setting of the Oslo Graben - Oslo Rift area. The diameter of the NORSAR array is indicated by a circle, while the presumed locations of the 3 largest recorded earthquakes are marked by stars [19]. STJ- presumed location of the Skagerrak triple junction (the Jutland junction of Burke and Dewey [25].

axis of the Oslo Graben [1]. We take three profiles, one each from the north, the middle and the south of the region (profiles V, VII and IX in Figs. 2 and 3). In each case we assume a two-dimensional generating body, which although not an entirely valid assumption leads to a modest overestimation of the depth bound.

Fig. 4 illustrates the bounds obtained on the maximum depth of burial as a funciton of density contrast for the three profiles. The maximum depth of burial is relatively insensitive to the assumed density contrast above about 100 kg $m^{-3}$, and for a density contrast of about 200 kg $m^{-3}$ (for example, the contrast between basalt of density 2960 kg $m^{-3}$ and intermediate crustal rocks of density 2750 kg $m^{-3}$) the maximum depth of burial varies from 24 km for profile III to 14 km for profile II. These bounds are probably conservative since no attempt was made to use points from the profiles where the anomaly was less than 5 mgal and the bounds obtained with the ideal body method improve as more of the information contained in the profiles is used.

Although we may reasonably expect the depth to the composite instrusive body to vary with position along the graben, the maximum burial depth of 14 km which characterizes profile VII may bethe artefact of a local shallow anomaly - the Tyrifjord high [1]. The depth estimate in profile VII is, therefore, considered less

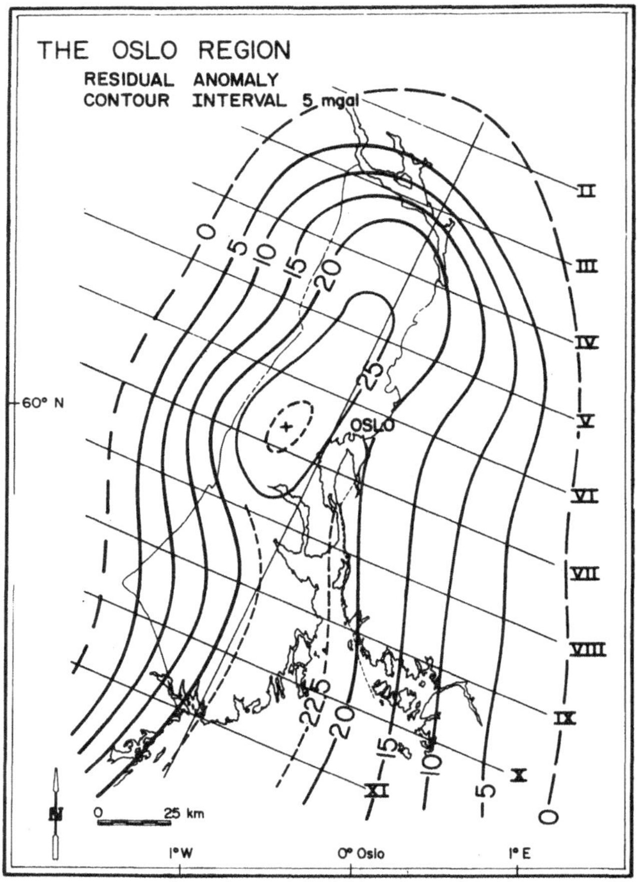


Fig. 2. 'Residual' gravity map of the Oslo Graben obtained by removal of the effects of Moho undulations and of shallow structural features. Figure reproduced from Ramberg [1].

reliable than in the other two profiles, and a closer analysis of the various assumptions and uncertainties originally discussed [1] in connection with the construction of the residual anomaly (Fig. 2) might help decide the accuracy of the depth estimates.

The Ideal Body Method is very time-consuming and so far only the three profiles shown above have been run. A more extensive analysis, including the effect of data errors in the Ideal Body approach will be the subject of a later paper. Despite the preliminary nature of these results, it is clear that this inversion of the gravity data suggests maximum burial depths similar to

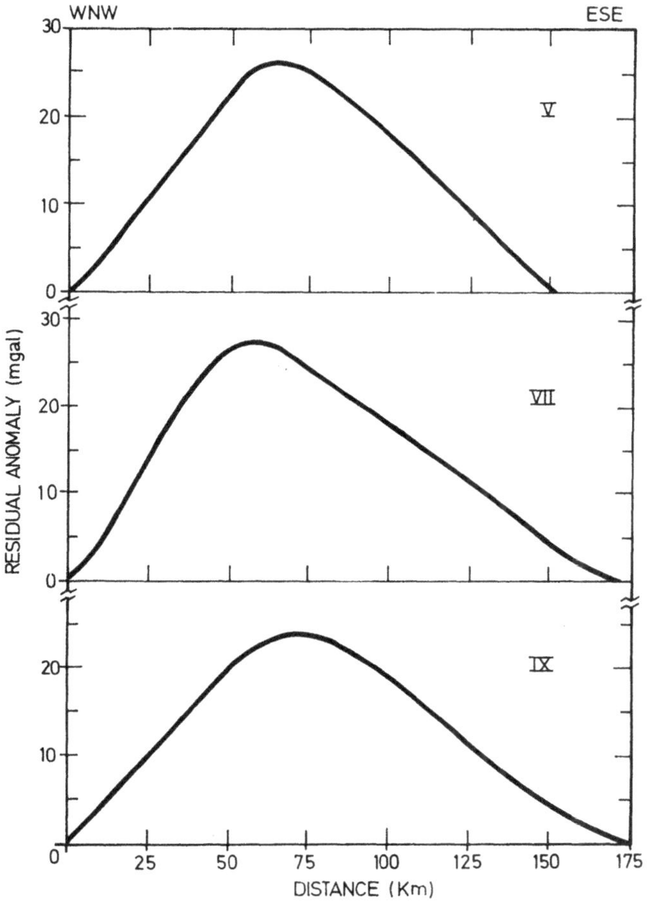


Fig. 3. The three gravity profiles subjected to inversion using Parker's method [14]. The profile numbers compare to those in Fig. 2.

previous estimates [1-3], but possibly at somewhat shallower and undulating levels, particularly since the bounds obtained on depth of burial are conservative.

## 4. SEISMIC INVESTIGATIONS

Unfortunately, the seismic profiling work undertaken in the wider Oslo Graben area [6] is of little help in the present context, since the sensor spacing was too large to detect an inhomogeneity

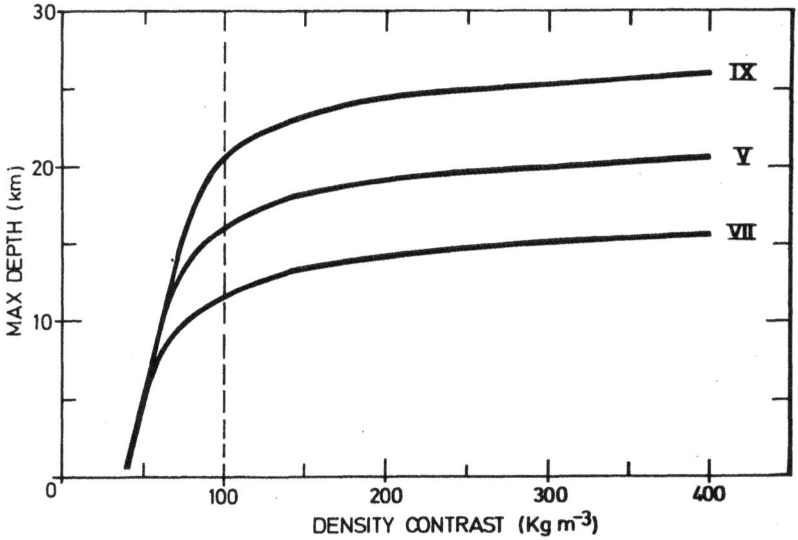


Fig. 4. Estimates of max. depth to intrusive body as a function of presumed density contrast. Notice that for contrasts exceeding 100 kg $m^{-3}$ the max. depth increases very slowly.

of the likely size of the body which generates the axial gravity high. In addition, some unease is introduced by a mismatch of about 4 km in the estimated Moho depths at the anchor point shared between this profile and one crossing southern Sweden [15]. Seismic profiling within the graben proper [7] suggests a Moho depth around 31 km and, interestingly, implies that the crustal structure there is very similar to that reported from widely separate ports of Norway [16].

The large aperture seismic array NORSAR [12] overlies the northern part of the Oslo Graben as shown in Fig. 1. Teleseismic P-wave recordings from this installation have been extensively used for detailed seismic mapping of the lithosphere in the area of the array and also for the upper mantle in Fennoscandia and adjacent areas [9,10,13,17]. Two of these studies are of particular interest to us: the Aki-Christoffersson-Husebye [9] (ACH) technique for 3-dimensional seismic mapping of heterogeneities in the lithosphere by direct inversion of 2-dimensional array travel time residuals, and Berteussen's [8] contouring of Moho-depths beneath the array based on spectral ratio analysis of long-period P wave recordings. A characteristic feature of these approaches is flexibility in modelling when compared with more conventional methods based on the assumption of spherically layered earth models.

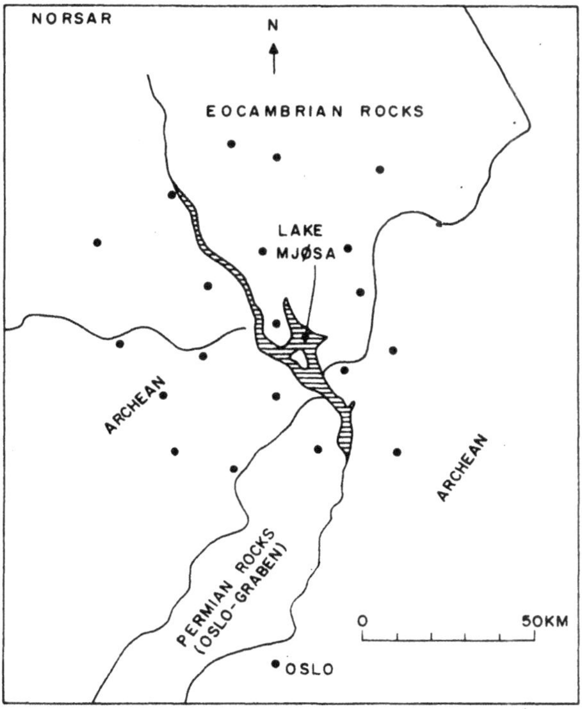


Fig. 5a. Map of NORSAR subarray centres. The Oslo igneous rocks series is outlined.

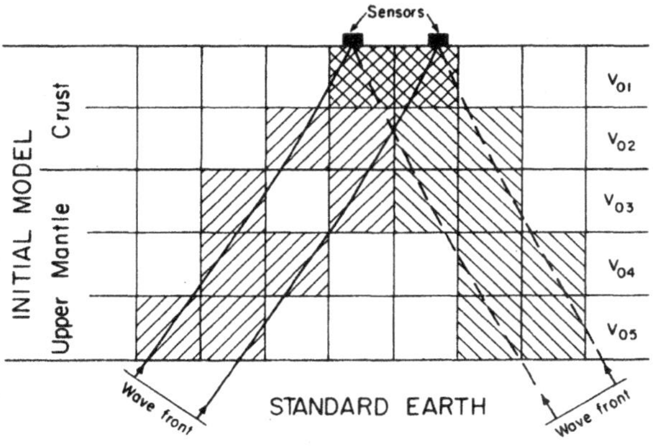


Fig. 5b. The ACH [9] initial layered model with constant thickness and average velocity $V_{oi}$ is divided into many blocks. The problem is to estimate the velocity perturbation $\underline{m}$ for each block using the P travel time data for many teleseismic events observed at the NORSAR receivers above the model.

## 5. THREE-DIMENSIONAL SEISMIC INVERSION

Since our conclusions rest in part on the results of the ACH [9] inversion, a brief summary of their technique is appropriate. In the area to be investigated the lithosphere is divided into a number of layers whose thicknesses are chosen with regard to the resolving power of the method and to the approximate depth of major features such as the Moho; each layer is subdivided into blocks of a size again determined by the resolving power. The aim of the method is to estimate the seismic velocity anomaly within each block relative to the average layer velocity. To do this it is necessary to obtain a large number of high-quality P-wave travel time data sampling the blocks as closely and uniformly as possible. These data are then inverted to give the mean velocity anomalies in each block and hence a representation of the three-dimensional velocity structure in the area. NORSAR provides an ideal site for the acquisition of this kind of data.

In the results illustrated in Figs. 6, 7 and 8, the first two layers correspond to depth intervals 0-17 km (Fig. 6), 17-36 km (Fig. 7) and (i.e., approximately to upper and lower crust, respectively) and three further layers divide the upper mantle down to 126 km: 36-66 km (Fig. 8) and 66-98 km and 96-126 km. Each layer is subdivided into blocks 20 x 20 km. The dominant features in the results are the relative increase of seismic velocity from west to east at depths greater than about 20 km. This feature is common to the layers of Figs. 7 and 8 and to the deeper layers which are not illustrated here (see [9]; it is also insensitive to the choice of block sizes or layer spacing.

The principal feature of Fig. 6, a north or north-west trending band of high velocities, is similarly insensitive to block size or to the thickness assigned to the layer representing the upper crust, provided that this layer is not made too small to be samples adequately nor so large that velocity anomalies from deeper levels dominate the shallow structure. The deeper structure, as depicted in Figs. 7 and 8 and layers 2-5 of [9] is very similar to that proposed by Haddon and Husebye [10] in an extensive analysis of NORSAR P-wave travel time and amplitude observations and their simulation in terms of seismic velocity heterogeneities at depths between 100 and 200 km.

This well-established feature shows no clear relationship with the surface expression of the Oslo Graben and is more likely

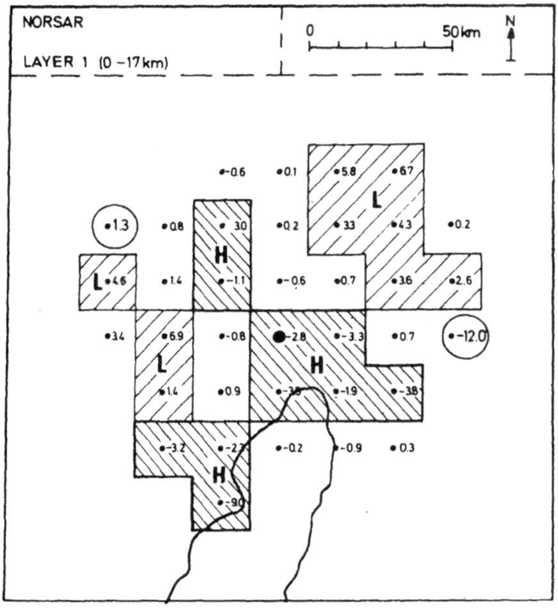


Fig. 6. Generalized inverse solution for Layer 1 (plan view). The numbers show the fractional velocity perturbation in per cent of the average layer velocity of 6.1 km $s^{-1}$. The shaded areas correspond to the magnitude of solution greater than 1 time the standard error. The letter L and H refer to low and high velocity anomalies, respectively. The central subarray is marked by a larger solid circle. Figure redrawn from ACH [9].

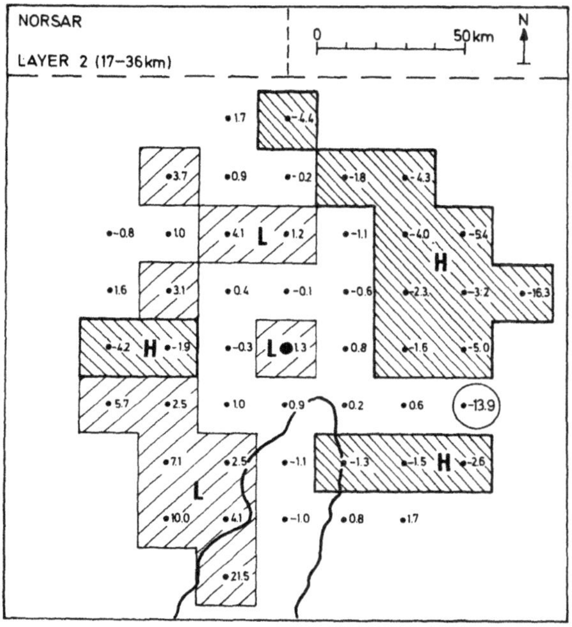


Fig. 7. Generalized inverse solution for Layer 2 (plan view) with average layer velocity of 6.9 km $s^{-1}$. See Fig. 6 for explanation of symbols.

to correspond to the contrast between the rocks of the Baltic Shield and those to the west which have been disturbed by the Caledonian orogeny. The velocity anomalies of Fig. 7 show a clear relation with the variation in Moho depth determined by Berteussen [8] which is reproduced in Fig. 9 - a thinner crust leads to high average velocity for a block in this layer and vice versa.

The only layer in this inversion which shows seismic velocities seemingly related to the surface expression of the Graben is the upper layer (Fig. 6) where the northward-trending high velocities continue the line of the Graben and then turn northwest to follow the line of Lake Mjøsa (Fig. 5a). This feature is coincident with the dominant trend of time and amplitude residuals reported by Gjøystdal and coworkers [18,19] and with a secondary feature in Ramberg's [1] gravity observations. This high velocity feature is coincident in the south of the region studies with the line of the Oslo Graben and in its northwestward extension underlies a surface feature (Lake Mjøsa) which [19] follows the main trend of the down-faulted sedimentary Sparagmite Basin, or Østerdal Aulacogene, of Eocambrian age.

As will be discussed later, there are several reasons to expect a poor correlation between gravity and seismic data; nevertheless, the coincidence of this high velocity feature with

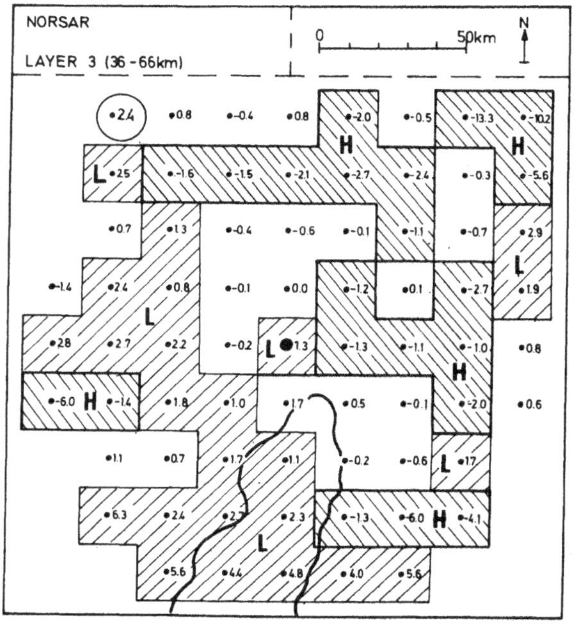


Fig. 8. Generalized inverse solution for Layer 3 (plan view) with average layer velocity of 8.2 km $s^{-1}$.

surficial features and its occurrence in the layer chosen to represent the top half of the crust in the inversion are not considered random phenomena.

## 6. DISCUSSION AND CONCLUSION

Two aspects of this study are of importance: first the extent to which the various types of geophysical observations correlate and secondly the constraints which the combined interpretation of these data can place on the evolution of the Oslo Graben.

It is clear that the gravity contours follow closely the surface expression of the Oslo Graben and associated igneous intrusions [1]; this is also true of the magnetic anomaly data for this area, although no detailed interpretation of these observations has been published so far [20]. Interpretations of the gravity data - both those of Ramberg [1] and those presented here using Parker's Ideal Body technique - favor the intrusion of a dense body into the low to intermediate crust; the latter interpretation may be taken to favour somewhat shallower depths of burial than the former.

Density contrasts of around 100-200 kg $m^{-3}$, which are tentatively chosen for the difference between the deeper crustal layers and the meta-basaltic derivatives, are sufficient to account for these observations, and the total mass, which may be calculated from the residual anomaly profile, amounts to about $1.2 \times 10^5$ $km^3$ [1]. This amounts to a layer 10-18 km thick, depending of density contrast applied, spread uniformly below the area of the surface igneous rocks of the Graben. Geological arguments would clearly favor the smallest density contrast (~100 kg $m^{-3}$), leading to a thick body.

If the estimates of deep to intermediate crustal burial depths for this body are correct, it might consist of a narrower intrusion along the axis of the rift, still extending to the base of the crust, or more plausibly, of an inhomogeneous mixture of crustal and intrusive rocks spread over a larger volume of the middle and lower crust.

In either case, the relatively weak correlation of seismic velocity anomalies with a body of this size requires some discussion: if we take a simplified view of the relation between velocity and density, and assume Birch's law to hold, we would expect a density contrast of 100-200 kg $m^{-3}$ to imply a velocity contrast of around 0.25-0.5 km $s^{-1}$ in the intermediate or lower crust. This amounts to a 3-7% velocity contrast over a zone as much as 40 km wide along the rift axis, and ought to be detectable by the inversion scheme described above.

Indeed, the zone of higher velocities depicted in Fig. 6, is of roughly the right width and some of the blocks exhibit velocity anomalies of thereby supporting the low density contrast alternative. But the magnitude of these anomalies is somewhat sensitive to the smoothing kernels chose in the inversion scheme [9]. Thus while the width of the zone of high velocity is in accord with the gravity data, this do not apply directly to its depth in the upper part of the crust. Let us add that proper depth resolution is the most non-unique part in inversion of seismic data.

The seismic data provide no very clear indication of the contrast in density between this zone and the surrounding region. This is not surprising as data compiled by Press [21] and others, and various laboratory experiments [22] show that considerable variation in velocity is to be expected in intermediate and lower crustal rocks, and that this variation need not be related to

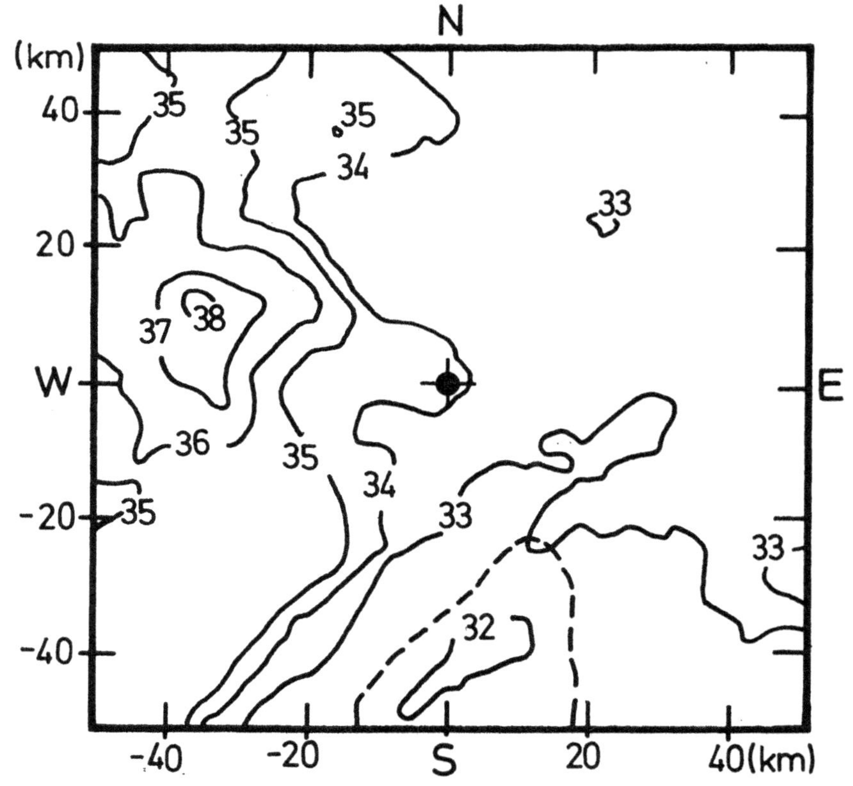


Fig. 9. Moho depth contours based on spectral ratio analysis of NORSAR long-period P-wave recordings. Figure redrawn from Berteussen [8].

density in any simple manner.

Additional difficulties in the positive identification of the high velocity body by seismic means alone come from the facts that velocity anomalies detected outside the graben are at least of the same order as those postulated within it [24]. It should be noted that an equivalent approach to the travel time anomalies at the large aperture seismic array LASA in Montana showed velocity anomalies in the crust and upper mantle of the same magnitude as those discussed here but with a negligible accompanying gravity anomaly [25].

These factors just discussed, and the probable laterally varying nature and depth of the high density body explain the present lack of success of conventional seismic profiling techniques in detecting it. However, an explosion seismic experiment with densely sampled profiles within and outside the graben is now the only practical way to obtain more information on the depth and extension of the high velocity body below the Oslo Graben.

The combined seismological and gravity approach here has enabled us to place somewhat tighter constraints on the deep structure of the region than were previously possible. The gravitational data alone suggested a lower or intermediate crustal depth for the higher density body, and the improved bounds offered by Parker's method may favor the shallower end of this range. The 3-D inversion of the seismic data suggests that an anomalous body lies, at least mainly, in the layer chosen to represent the top section of the crust - in the presentation of Fig. 6, depths 0-17 km. Since the bounds must represent maximum depths to the top of the high density body, and the assumption of two dimensionality in the ideal body technique leads to an overestimation in the bounds, there may exist a qualitative agreement between the seismic and gravity observations. We recommend carefully designed seismic profiling to further explore this hypothesis.

The deeper seismic investigations [9-11] show no sign of a low density body within the lithosphere, although there is a general velocity increase from west to east across the area of the NORSAR array. We regard this as resulting from the compositional contrast between the rocks of the Baltic Shield on the east as compared to those on the west side of the graben. It is reasonable to expect that any asthenolith which existed in Permian times would have cooled long ago; its only remaining physical expression is the large volume of mantle-derived material which in Paleozoic time was intruded into the crust of the region.

REFERENCES

1. I.B. Ramberg, Norg. geol. Unders., 325, 1, 1976.
2. I.B. Ramberg, Nature Phys. Sci., 240, 149, 1972.
3. I.B. Ramberg and S.B. Smithson, Geol. Soc. Am., Bull., 86, 769, 1975.
4. K.S. Heier and W. Compston, Lithos, 2, 133, 1969.
5. E.R. Neumann, Petrology of plutonic rocks, in: The Oslo Paleo-rift, ed. by J.A. Dons and B.T. Larsen, Norg. geol. Unders., 1978 (in press).
6. R. Kanestrøm and K. Haugland, Profile Section 3-4, in: Deep seismic sounding in Northern Europe, ed. by A. Vogel, Uppsala Univ., Sweden, 1971.
7. J. Tryti and M.A. Sellevoll, Report on a reconnaissance seismic crustal study of the Oslo Rift, in: The Norwegian Geotraverse Project Final Report (ed. by K.S. Heier), Oslo Univ., 1977.
8. K.-A. Berteussen, Phys. Earth Planet. Inter., 15, 13, 1977.
9. K. Aki, A. Christoffersson and E.S. Husebye, J. Geophys. Res., 81, 277, 1977.
10. R.A.W. Haddon and E.S. Husebye, Joint interpretation of P-wave travel time and amplitude anomalies in terms of 2-D lithospheric heterogeneities, in preparation, 1978.
11. K.-A. Berteussen, Seism. Soc. Am., Bull., 66, 719, 1977.
12. H. Bungum, E.S. Husebye and F. Ringdal, Geophys. J.R. Astron. Soc., 25, 115, 1971.
13. F. Ringdal and E.S. Husebye, Seismic mapping of heterogeneities in the Fennoscandian lithosphere and asthenosphere, Norg. geol. Unders., 1977 (in press).
14. R.L. Parker, Geophys. J. R. Astron. Soc., 42, 315, 1975.
15. S. Gregersen, Profile section 4-5, in: Deep seismic sounding in Northern Europe, ed. by A. Vogel, Uppsala Univ., Sweden, 1971.
16. R. Kanestrøm, Seismic investigations of the crust and Moho in southern Norway, in: The Norwegian Geotraverse Project - Final Report, ed. by K.S. Heier, Oslo Univ., 1977.
17. P.C. England, B.L.N. Kennett and M.H. Worthington, A comparison of the upper mantle structure beneath Eurasia and the North Atlantic, Geophys. J. R. Astron. Soc., 1977 (in press).
18. H. Gjøystdal, E.S. Husebye and D. Rieber-Mohn, Seism. Soc. Am. Bull., 63, 549, 1973.
19. E.S. Husebye, H. Bungum, J. Fyen and H. Gjøystdal, Earthquake activity in Fennoscandia between 1497 and 1975 and intraplate tectonics, Nor. geol. Tidsskr., 1977 (in press).
20. E.S. Husebye and I.B. Ramberg, Geophysical Investigations, in: The Oslo Paleorift, ed. by J.A. Dons and B.T. Larsen, Norg. geol. Unders., 1978 (in press).
21. F. Press, Science, 160, 1218, 1968.
22. D.H. Chung, J. Geophys. Res., 75, 7353, 1970.
23. A. Dahle, E.S. Husebye, K.-A. Berteussen and A. Christofferson, Wave scattering effects and seismic velocity measurements,

in: Exploitation of Seismograph Networks, ed. by K.G. Beauchamp, Nordshoff-Leiden, The Netherlands, 1975.
24. K. Aki, A. Christoffersson and E.S. Husebye, Seism. Soc. Am. Bull., 66, 501, 1976.
25. K. Burke and J.F. Dewey, J. Geol., 81, 406, 1973.

COMPARATIVE STUDIES

# COMPARISON OF THE EAST AFRICAN RIFT SYSTEM AND THE PERMIAN OSLO RIFT

R.W. Girdler

School of Physics, The University, Newcastle upon Tyne, NE1 7RU, U.K.

## 1. INTRODUCTION

The Oslo palaeorift system and the East African rift system have many features in common. Both are intracontinental, the former being within the European plate and the latter breaking through the Afro-Arabian plate. Both are regions of extension and both have large axial intrusive zones. Both show some lateral spreading, the Oslo graben in the south and the East African rift system in the north.

As might be expected, there are similarities in some geophysical parameters. In particular, the Bouguer gravity anomaly is positive over the Oslo graben and over the axis of the eastern rift in Africa north of Lake Natron. In both cases, this phenomenon has been interpreted as being due to the combined effect of crustal thinning and the intrusion of dense mafic rocks beneath the graben axis [1-5]. In the final stage of this process (where the largest Bouguer anomalies are observed) there has been complete crustal separation with the evolution of oceanic crust as in the Red Sea and the Gulf of Aden [6].

Although there are many similarities, there are some important differences. The first is of age and the second of scale. Whereas the Oslo graben is about 300 million years old, the East African rift system is only about 30 million years. The East African rift system is seismically active whereas the Oslo graben is relatively inactive. The heat flow over the Oslo graben is about normal whereas the heat flow over much of the East African rift system is high or very high. This large difference in age may throw light on processes of rifting. In the East African rift system, recent and

*I. B. Ramberg and E.-R. Neumann (eds.), Tectonics and Geophysics of Continental Rifts,* 329-345.

present day rifting processes may be studied whereas in the Oslo graben, because of several hundred million years of erosion, the more deep seated processes of rifting may be studied.

The Oslo graben is about 35 to 65 km wide and extends for

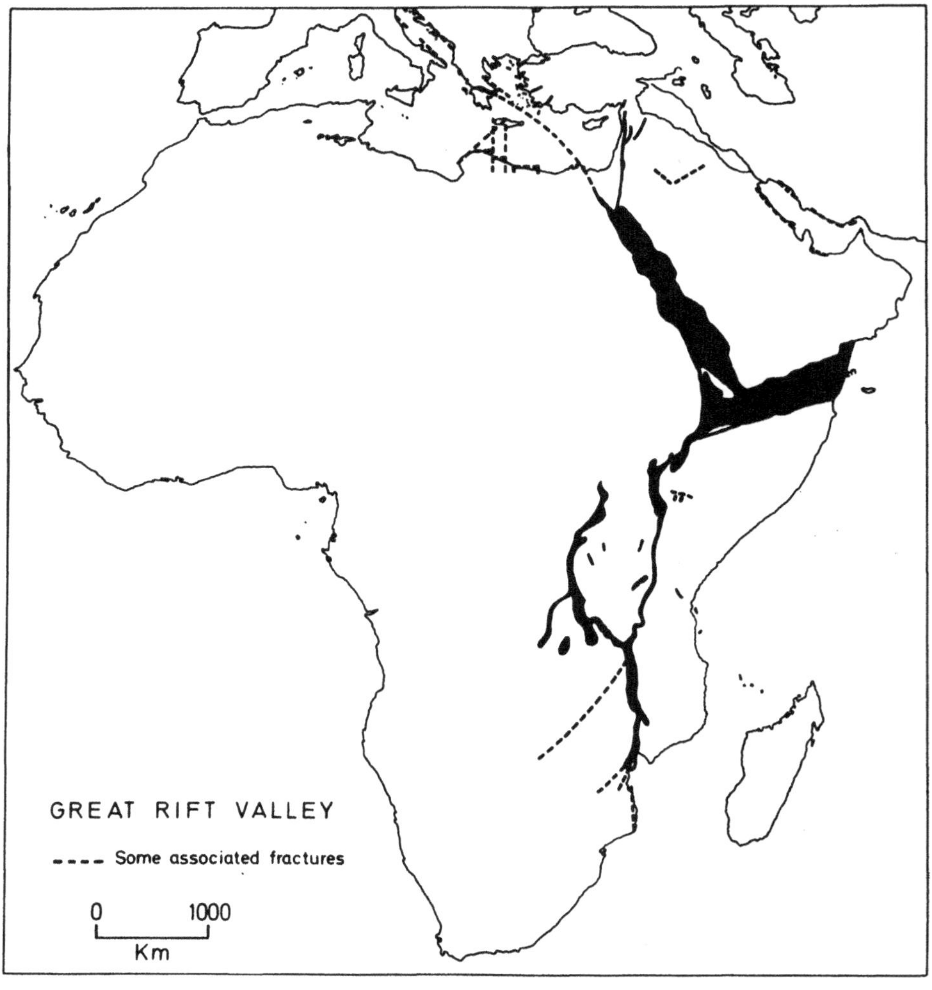


Figure 1. The East African rift system traced from the original map (1896) by J.W. Gregory.

about 200 km. It is usually considered to extend beneath the Skagerrak Sea and even if it is somewhat longer it is small compared with the East African rift system which extends for thousands of kilometres (Fig. 1). However, the vertical displacement on the major faults in the Oslo Graben and throughout the East African rift system is about the same, being 1 to 3 km.

One of the values of studying the East African rift system is that various stages in the evolution of rifting can be seen. In the north-east, the Gulf of Aden connects to the Carlsberg ridge in the Indian ocean and is a small but fully developed ocean. In the north west, the Red Sea divides into the Aqaba-Dead Sea transform system and the Suez rift. The northern Red Sea is a slow spreading centre, the rate of spreading increasing towards the south. In East Africa, there seem to be little or no crustal separation but there is evidence for intrusive zones beneath the rift in Ethiopia and Kenya. In Tanzania, there are normal faulted rifts with little or no volcanism and the rifting dies out in the extreme south. Each of these will be considered in turn and the student of the Oslo palaeorift will be left to decide which is the closest analogue.

## 2. GULF OF ADEN

It is appropriate to start with the Gulf of Aden as here a fully developed oceanic structure may be recognised on a small scale. Bathymetric profiles show a rift along the axis of a "mid-ocean ridge" which is flanked by "deep sea plains" (the main trough) and narrow continental shelves bordering Arabia and Somalia. Such features are difficult to recognise in the Red Sea, which was an enclosed evaporite basin until the end of the Miocene and became filled with salts.

The possibility that the Red Sea and the Gulf of Aden have formed in at least two stages has been the subject of recent studies [7-10]. There now seems little doubt that there have been at least two stages of sea floor spreading but the early phase has proved somewhat difficult to date. To help resolve this, a detailed magnetic survey was carried out in the western Gulf in 1975.

The location of the survey is shown in Fig. 2 which also shows the geology of the neighbouring land areas and the earlier seismic refraction results. Discussion will be limited to the interpretation of magnetic anomalies and an overall summary relating to the structural and geological history of the Gulf.

In general, the largest magnetic anomalies are found over the axial ridge. These are of the order of 2000 nT; in contrast, the anomalies over the main trough are a few hundred nT. Profile

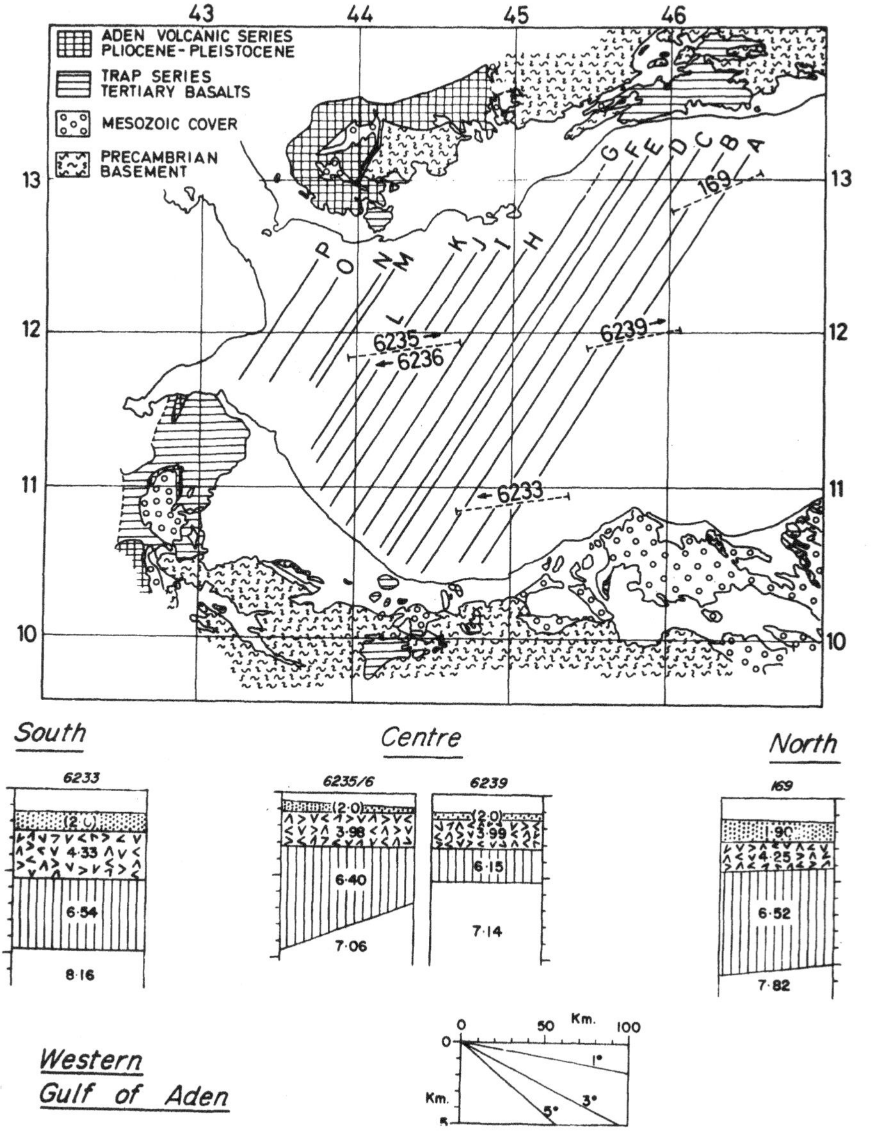


Figure 2. The location of a recent geophysical survey in the western Gulf of Aden (RRS SHACKLETON Legs 2 and 3/1975). Earlier seismic refraction profiles are also shown; all profiles show oceanic structure (after Girdler and Styles [10], in press).

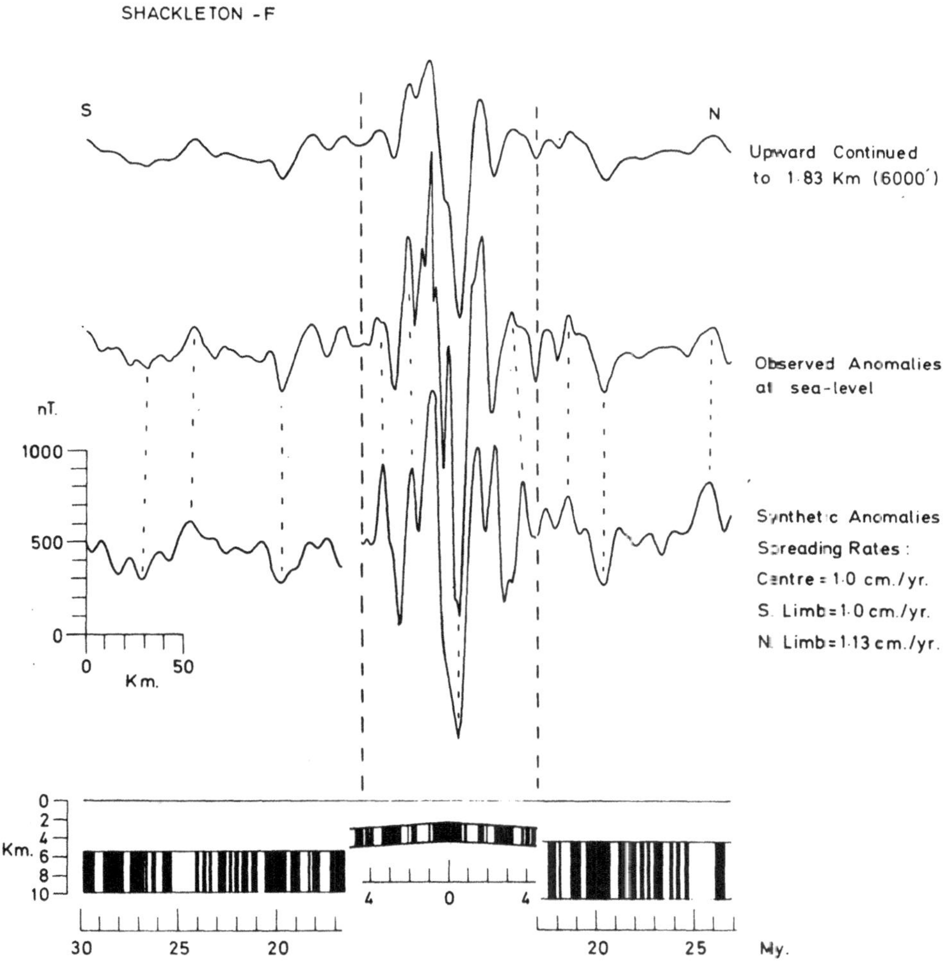


Figure 3. Sea floor spreading model to explain the magnetic anomalies for profile F (see Fig. 2) in the western Gulf of Aden. The magnetization is assumed to reside in layer 2 for the new crust in the centre of the Gulf and in layer 3 for the older crust beneath the main trough. The depth to and thicknesses of these layers are taken from the seismic refraction profiles shown in Fig. 2 (after Girdler and Styles [10], in press).

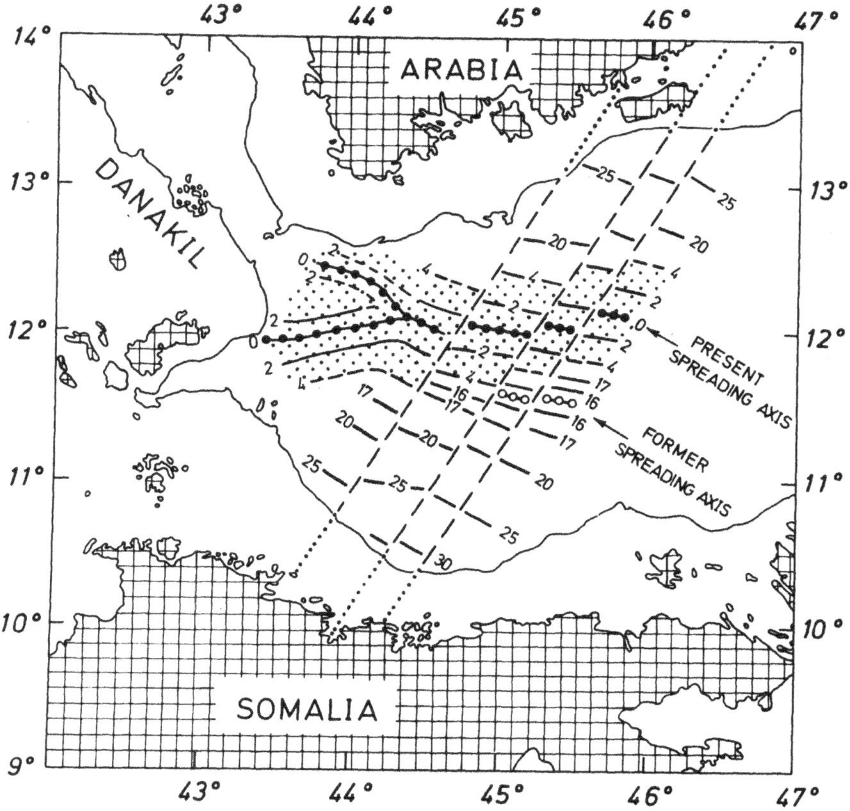


Figure 4. Isochrons (My) deduced from magnetic anomalies for the western Gulf of Aden. The present spreading axis is shown by a chain of solid circles and the former spreading axis by a chain of open circles. Three new transform faults were discovered and these are shown by dashed lines and their possible extensions by dotted lines. Land over 914 m (3000 feet) is shown shaded (after Girdler and Styles [10], in press).

F (Fig. 3) is chosen to illustrate the nature of the anomalies. Using a recent reversal timescale [11], synthetic sea floor spreading magnetic anomalies were generated and compared with the observed. The axial anomalies over the central ridge can readily be explained by assuming spreading over the last 4½ million years. The anomalies over the main trough can be explained best by assuming a sea floor spreading phase from about 30 My to about

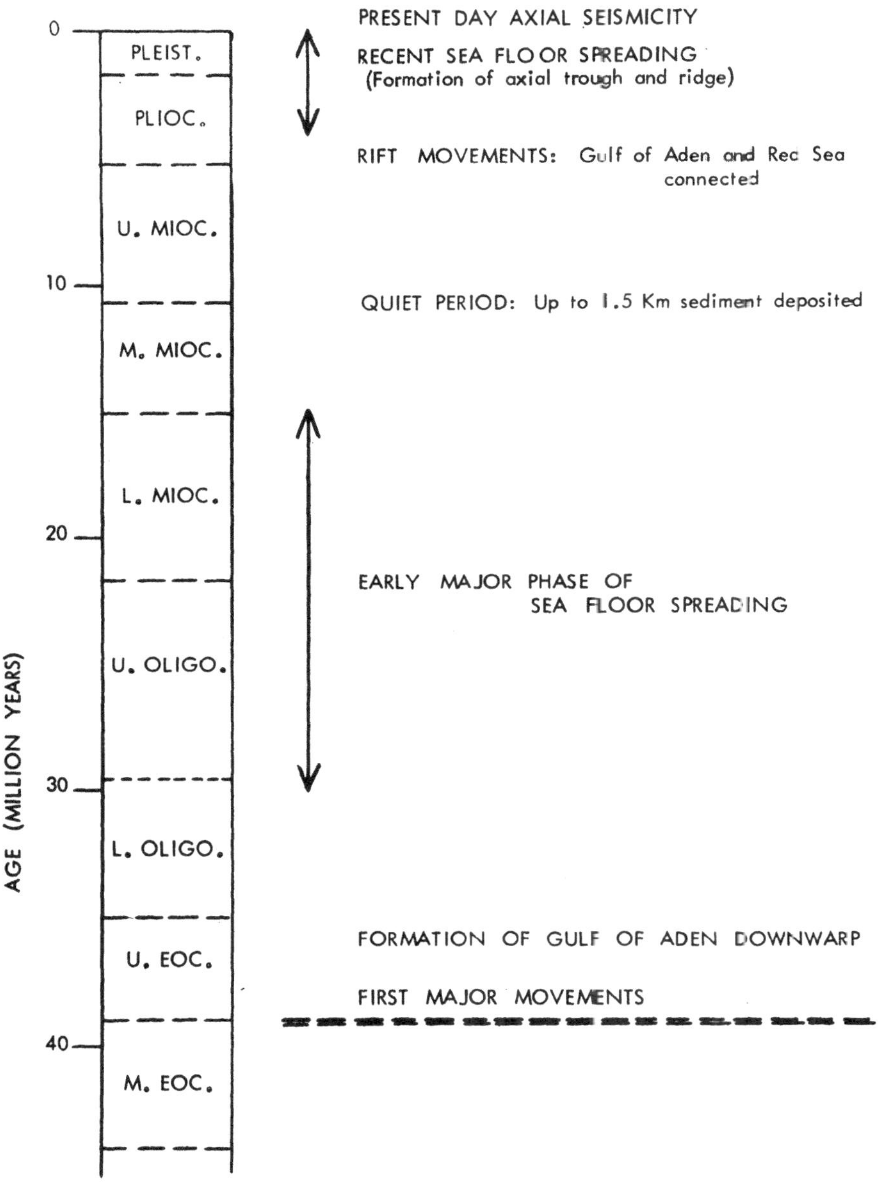


Figure 5. Possible geological history of the Gulf of Aden utilizing the interpretation of the sea floor spreading magnetic anomalies.

17 My. Using this interpretation an isochron map has been prepared for the western Gulf of Aden (Fig. 4). The two distinct phases of spreading are shown, the new phase being distinguished by stippling. A summary of the geological history of the Gulf is given in Fig. 5.

Thus the main results are that nearly all the crust beneath the Gulf is oceanic and the oldest crust began forming about 30 My ago in the upper Oligocene. The Gulf spread in a north easterly direction for about 15 My (to the end of lower Miocene) and this was followed by a hiatus for about 10 My at the end of which a new phase of spreading commenced in the early Pliocene which still continues today.

## 3. THE GULF OF AQABA-DEAD SEA RIFT AND THE NORTHERN RED SEA

We now move some 2000 Km to the northernmost Red Sea. Here the amount of separation may be largely determined by the amount of shear along the Aqaba-Arava-Dead Sea rift (Figs. 6 & 9). In 1958, Quennell gave an ingenious mechanism for the origin of the Dead Sea rift [12]. He suggested that the Dead Sea occurs where the fault is offset some 20 Km to the west; as Arabia moved north with respect to Sinai, the offset opened up in a south-north direction giving the Dead Sea rift. In later work, Freund has established that the shear occurred in two stages with 40 to 45 Km dated as post-Miocene and 60 Km poorly dated as post-Cretaceous to pre-Miocene [13,14]. The evidence for this is impressive and clearly the formation of the northernmost Red Sea must be related to the two-stage transform motion along the Aqaba-Dead Sea rift.

To estimate the total extension in the northernmost Red Sea, the extension across the mouth of the Gulf of Suez has to be estimated and included. This is difficult as little information is available. Pearce [15] in a study of the larger earthquakes which have occurred in the mouth of the Gulf of Suez has discovered a region of low Q. It is therefore likely that there has been some recent lateral spreading in this region. Until more data are available, the total extension has to be assumed to be of the order of a few tens of kilometres.

There is very little geophysical information for the northern Red Sea. A total intensity magnetic profile obtained by CONRAD 9 (location shown in Fig. 6) has been studied by Girdler and Styles [16]. This suggests a fairly slow half spreading rate of 0.5 cm $y^{-1}$ over the last few million years and is consistent with the estimates for the post-Miocene shear along the Aqaba rift and the small extension in the Gulf of Suez.

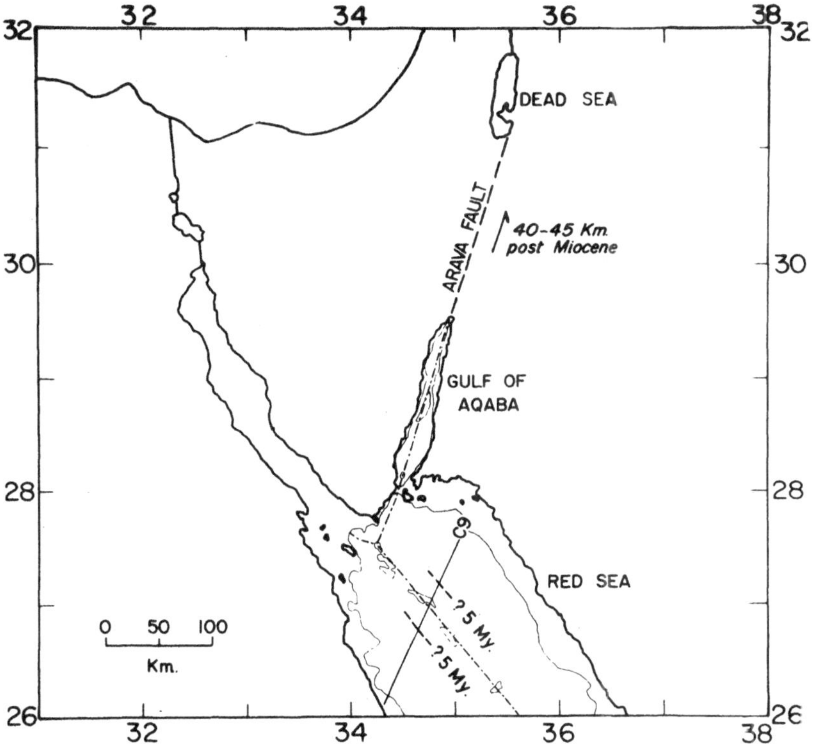


Figure 6. Map of the northern Red Sea, Gulf of Suez and Gulf of Aqaba. The possible present day plate boundaries are shown by dashed lines. The 40-45 km post-Miocene shear motion along the Aqaba-Dead Sea rift is consistent with the opening of the central Red Sea with a half rate of about 0.5 cm $y^{-1}$ over the last 5 My [16].

## 4. CENTRAL AND SOUTHERN RED SEA

There are many more geophysical data sets for the central and southern Red Sea and it should be possible to establish the sea floor spreading history. Ideally, the amounts and times of movements for the Gulf of Aden, Red Sea and the Aqaba-Dead Sea rift should all be consistent unless another degree of freedom is introduced in East Africa.

The last phase of spreading (post Miocene) is fairly easily established from the large magnetic anomalies over the centre of

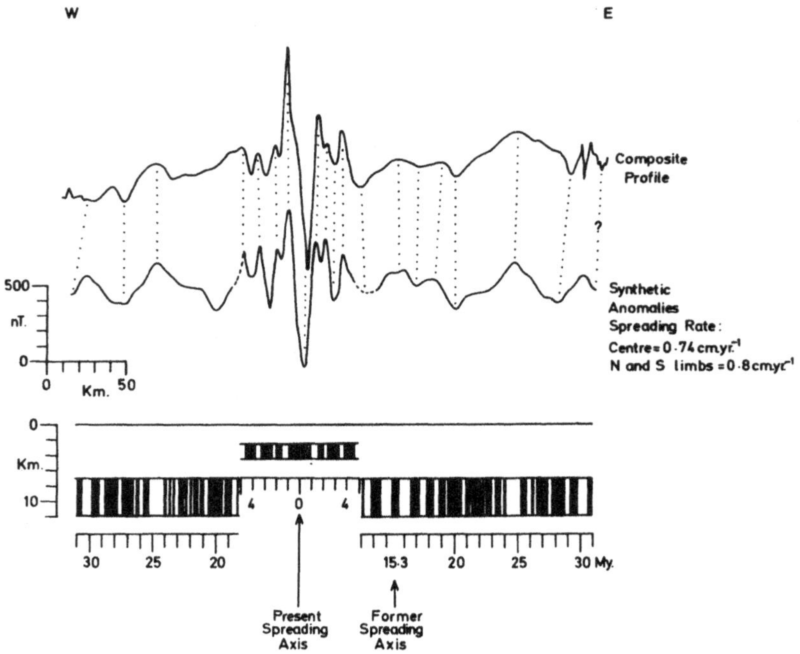


Figure 7. A composite mean total intensity magnetic profile over the Red Sea and its interpretation assuming the ages for the model in the Gulf of Aden (Fig. 3) apply to the Red Sea (by courtesy of Peter Styles).

the Red Sea. The earlier phase is much more difficult to establish due to there being few long profiles from shore to shore. There are several marine surveys over the centre and several air surveys over the margins but these are difficult to assemble. Several attempts have been made at interpreting the anomalies over the sides [8,9,17] and a further discussion is given by Noy [18].

The 1975 RRS SHACKLETON survey in the Gulf of Aden has satellite navigation and extends almost to the shores. It is therefore the best available data set and it is important to establish whether the magnetic anomalies over the Red Sea give an interpretation consistent with that given for the Gulf of Aden in section 2.

Figure 7 shows an attempt at modelling the Red Sea anomalies using the Gulf of Aden model. It is seen that the same ages could apply but the fit is not perfect. Further models are discussed by Noy in this volume [17].

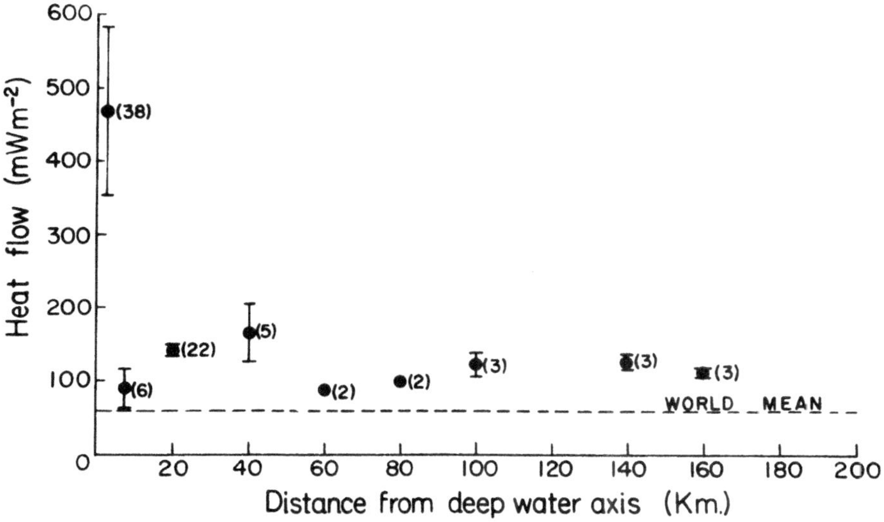


Figure 8. Red Sea heat-flow means (number of observations in parenthesis) as a function of distance from the deep water axis as measured on the bathymetric chart of Laughton [32]. All the means are appreciably higher than the world mean given by Chapman and Pollack [33]. After Girdler and Evans [23].

Space does not permit discussion of other geophysical data such as seismic refraction [18], seismic reflexion [19,20] and gravity [21] which lend support to the view that nearly all the Red Sea is underlain by sea floor which has evolved in at least two stages. Mention is made of the heat flow which is spectacularly high over the centre of the Red Sea and which is about twice the world mean over the margins. It is now established that the high heat flow extends at least as far as the Red Sea coasts [22]. The pattern shown in Fig. 8 is consistent with the presence of relatively recent oceanic crust beneath the whole of the Red Sea.

## 5. PLATE GEOMETRY FOR THE GULF OF ADEN, RED SEA AND EAST AFRICA

As more data have accumulated for the Gulf of Aden and Red Sea, it has become apparent that the amounts of separation are almost exactly those obtained by fitting the coastlines as originally envisaged in the pioneering works of Alfred Wegener. The pre-drift reconstruction is therefore probably close to that shown in Fig. 9 a. Several workers have noticed that the Gulf of Aden and Red Sea have to be fitted separately using different poles of rotation as reviewed in [24]. This implies that there has been considerable relative movement between the Nubia and Somalia plates forming the rift in East Africa.

A reconstruction can also be attempted for the situation at the end of the Miocene using the geological data indicating 40 to 45 km post Miocene shear along the Aqaba-Dead Sea rift and closing together the Red Sea and Gulf of Aden by the amounts given by the large magnetic anomalies indicating the last post Miocene phase of sea floor spreading. This reconstruction is shown in Fig. 9 b. Finally Fig. 9 c shows the present day situation.

A problem arises when the shoreline fits are used. The reconstruction shows that the total extension in north-east Africa could be as much as 100 km. On first sight this seems excessive as the width of the rift in Ethiopia is only about 80 km and it is not known what underlies the volcanics which cover the rift floor. However, there has been considerable thinning of the continental lithosphere in East Africa but it is difficult to obtain a quantitative estimate from this for the total extension. The problem is discussed further in the next section.

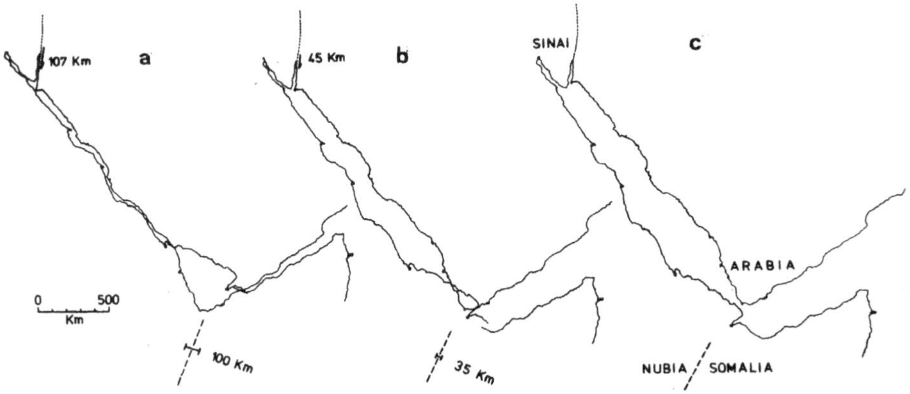


Fig. 9. Reconstructions for the Nubia, Sinai, Arabia and Somalia plates based on geological and geophysical evidence. When the Nubia and Somalia plates are moved back the overlap in Ethiopia becomes 35 km for the end of the Miocene [3] and possibly 100 km at the end of the Cretaceous.

## 6. THE RIFT IN EAST AFRICA

The extent of the rifts in Africa is shown in Fig. 1. An important clue concerning the total extension in East Africa would seem to lie in a careful study of the long wavelength negative Bouguer anomaly over the rift system. The anomaly is very impressive as can be seen from Fig. 10 which is adapted from the Bouguer anomaly map of Africa prepared by the U.S. Defence Mapping Agency Aerospace Centre [25]. The anomaly extends for nearly 5000 km across Africa and has a width greater than 1000 km. This interesting feature is discussed more fully in [26]. It is sufficient to mention that an interpretation in terms of thinning of the lithosphere as suggested by Girdler et al. [27] seems to be the most consistent with the plate tectonic situation described in the last section as it is a corollary of plate tectonics that the lithosphere must decrease to zero thickness before plate separation. Such an interpretation is shown in Fig. 12. More detailed interpretation awaits more detailed seismic information on the structure and thickness of the lithosphere in Africa.

Superimposed on this large negative Bouguer anomaly is a smaller positive Bouguer anomaly. This anomaly is found over the rift floor in Ethiopia and Kenya [28] and dies out near the Kenya-Tanzania border [29]. The anomaly was mapped in considerable detail in Kenya by Searle [28] who found it to be 40 to 80 km wide with an amplitude of 30 to 60 mgal. The anomaly has been explained by the presence of a dense intrusion (s.g. contrast + 0.23) which is about 20 km wide with its upper surface about 3 km below the floor of the rift. Fig. 11 shows how the nature of the intrusion varies and is of particular interest for comparison with similar structures deduced for the Oslo [1-3] and Rio Grande [30] rifts.

When the two interpretations are put together, we arrive at a model for the underlying structure of the rift zones such as shown in Fig. 12. An intrusive zone of the size required in Fig. 11 suggests there has been extreme thinning of the lithosphere beneath the rift valley whilst the long wavelength negative anomaly suggests there has been widespread thinning over the whole region. (The density is positve with respect to the light sialic crust but negative with respect to the deeper and denser upper mantle part of the lithosphere).

In this brief review several problems remain unsolved. Evidence has been accumulating that the Gulf of Aden and Red Sea evolved in at least two major stages but did the East African rift evolve similarly? The East African rift is seismically active [31] and in Ethiopia and Kenya there are hot springs, steam jets, fumaroles, grid faults and microseismicity all

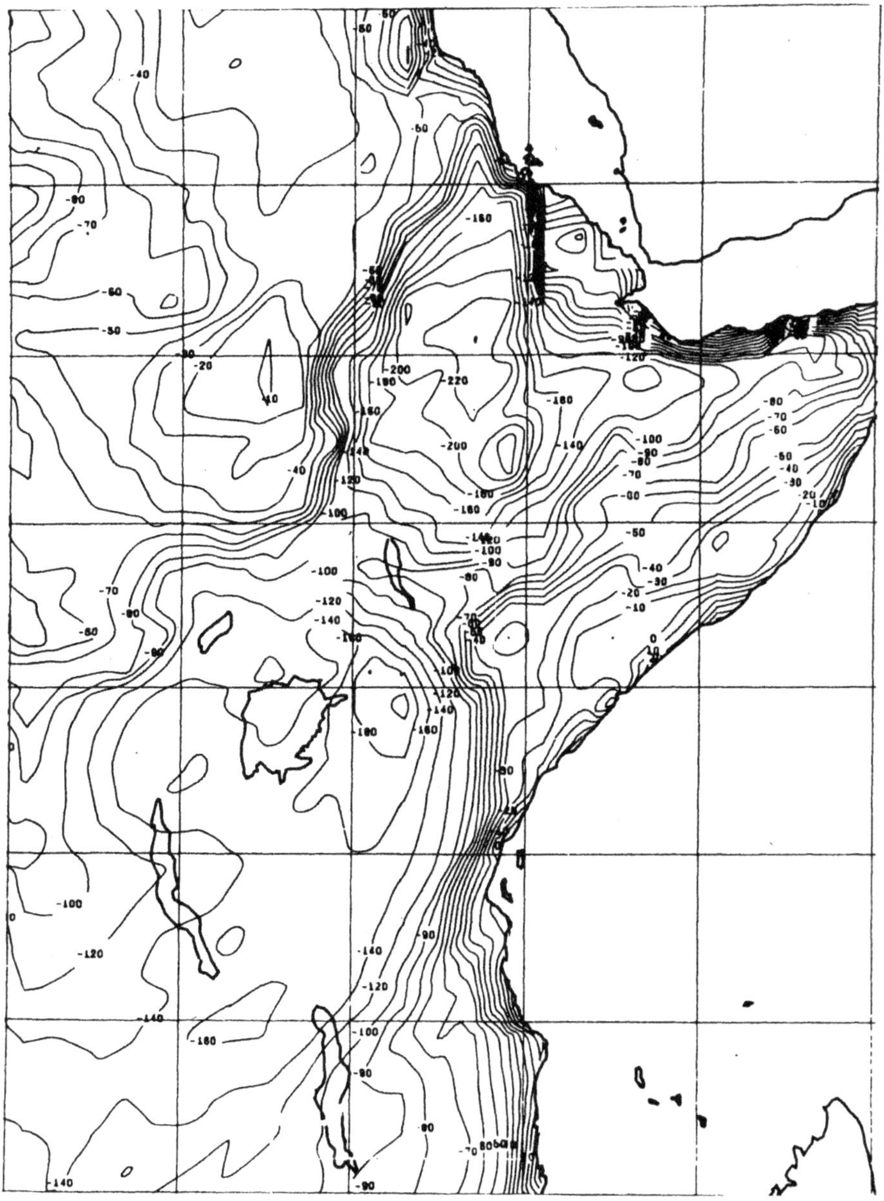

Fig. 10. The large negative Bouguer gravity anomaly over north-east Africa adapted from [25]. The contours are in mgal. The grid is 5° by 5° (Mercator projection).

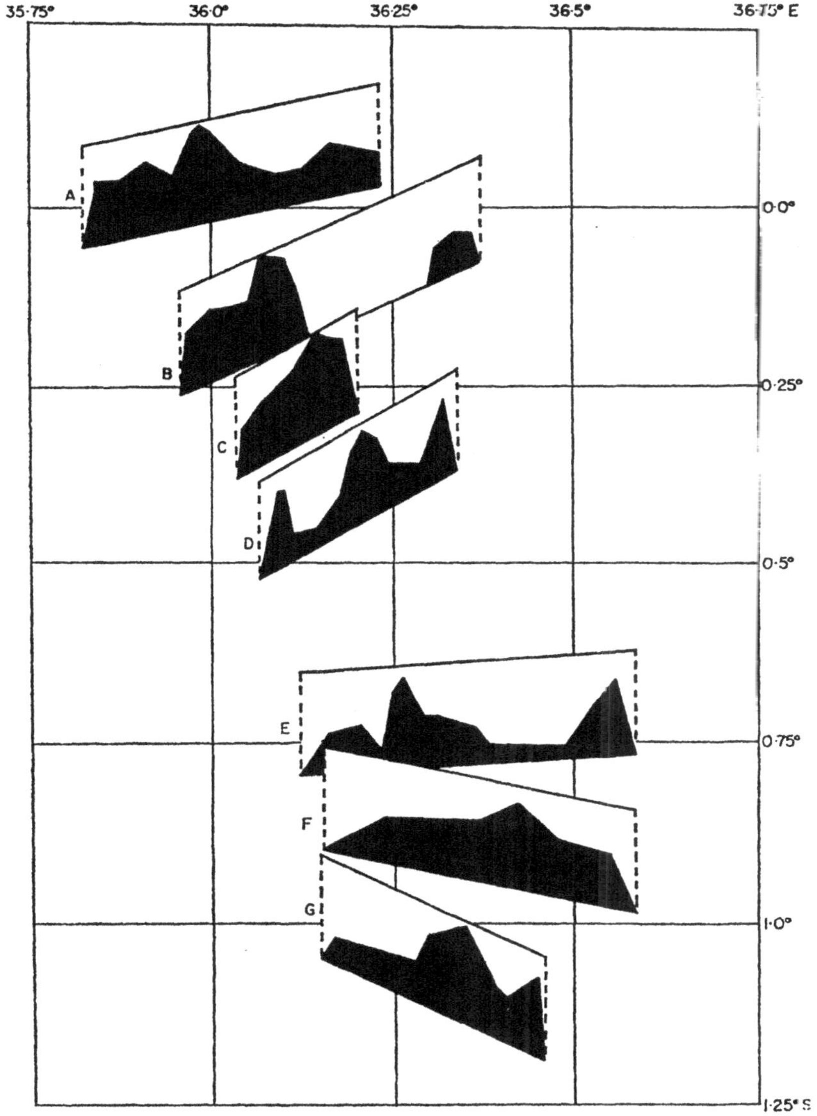


Fig. 11. Models computed for various gravity profiles across the rift in Kenya showing the possible nature of the underlying basic intrusion assuming a uniform specific gravity contrast of + 0.23 for all sections (After Searle [28]).

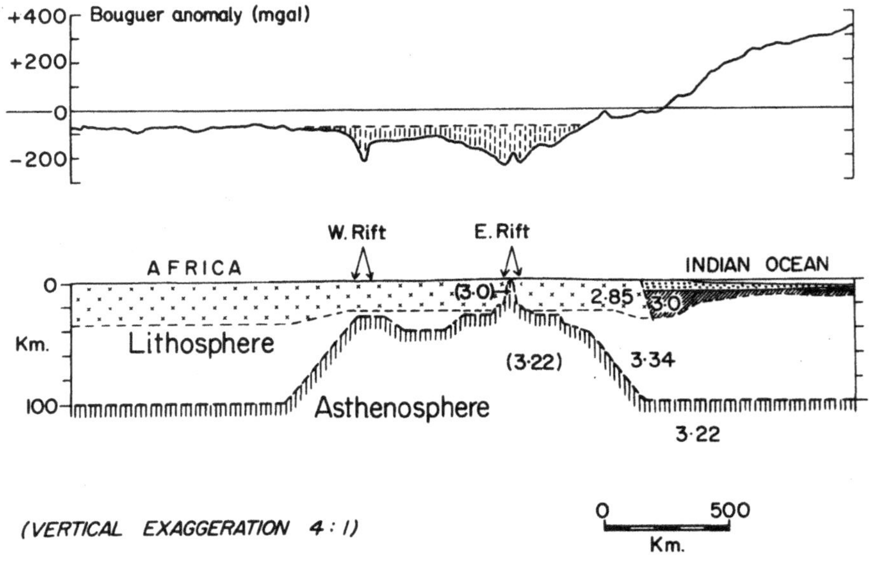


Fig. 12. A possible interpretation of the overall gravity anomaly over East Africa showing thinning of the lithosphere beneath the whole region with extreme thinning beneath the rift floor in Kenya. Based on the model of Girdler et al. [27].

closely related to the axis of the positive Bouguer anomaly. It would seem that these features correspond to the active (post-Miocene) phase of sea floor spreading in the Gulf of Aden and Red Sea. It is next natural to ask whether the large region of lithosphere thinning (Fig. 12) is related to the earlier and longer stage of sea floor spreading in the Gulf of Aden and Red Sea. Unfortunately the measurement of present day potential fields tells us nothing about potential fields in the past. It is perhaps significant that the correspondingly large region of negative gravity for the Oslo palaeorift is not nearly so impressive. Is it possible that this is a feature associated with recent and active rifts? The positive anomalies have been maintained as they are due to basic intrusions which have penetrated the upper crust. The features of Fig. 11 are still active but this stage was reached in the Oslo graben 270 million years ago [2]. It is possible that the deeper and larger features associated with the base of the lithosphere where the temperatures and pressures are so much higher are less permanent.

REFERENCES

1. I.B. Ramberg, Nature Phys. Sci., 240, 149, 1972.
2. I.B. Ramberg, Norges Geol. Unders., 325, 1, 1976.
3. E.S. Husebye and I.B. Ramberg, in: The Oslo Paleorift Part I, A Review, ed. by J.A. Dons and B.T. Larsen, NATO, Oslo 1977.
4. R.W. Girdler, J.D. Fairhead, R.C. Searle and W.T.C. Sowerbutts, Nature, 224, 1178, 1969.
5. R.C. Searle, Geophys. J.R.A.S., 21, 13, 1970.
6. R.W. Girdler, Q.J. Geol. Soc. Lond., 114, 79, 1958.
7. A.S. Laughton, Phil. Trans. Roy. Soc. Lond., A259, 150, 1966.
8. R.W. Girdler and P. Styles, Nature, 247, 7, 1974.
9. R.W. Girdler and P. Styles, in: Afar between Continental and Oceanic Rifting, ed. by A. Pilger and A. Rösler, Schweizerbart'-sche Verlagsbuchhandlung, Stuttgart, 1976.
10. R.W. Girdler and P. Styles, Nature, in press.
11. D.H. Tarling and J. Mitchell, Geology, 4, 133, 1976.
12. A.M. Quennell, Q.J. Geol. Soc. Lond., 114, 1, 1958.
13. R. Freund, Geol. Mag., 102, 189, 1965.
14. R. Freund, Z. Garfunkel, I. Zak, M. Goldberg, T. Weissbrod and B. Derin, Phil. Trans. Roy. Soc. Lond., A267, 107, 1970.
15. R.B. Pearce, personal communication, 1975.
16. R.W. Girdler and P. Styles, Earth Planet. Sci. Lett., 33, 169, 1976.
17. S.A. Hall, PhD. Dissertation, University of Newcastle, 1977.
18. D. Noy, Loc. cit.
19. D. Tramontini and D. Davies, Geophys. J.R.A.S., 17, 225, 1969.
20. S.T. Knott, E.T. Bunce and R.L. Chase, Can. Geol. Surv. Pap., 66-14, 33, 1969.
21. J.D. Phillips and D.A. Ross, Phil. Trans. Roy. Soc. Lond., A267, 143, 1970.
22. M.E. Gettings, EOS, 56, 458, 1975.
23. R.W. Girdler and T.R. Evans, Geophys. J.R.A.S., 51, 245, 1977.
24. R.W. Girdler and B.W. Darracott, Comm. Earth Sci., Geophys., 2, 131, 1972.
25. R.L. Slettene, L.E. Wilcox, R.S. Blouse and J.R. Sanders, Tech. Pap., 73-3 U.S. Defence Agency, 1973.
26. R.W. Girdler, EOS, 56, 516, 1975.
27. R.W. Girdler, J.D. Fairhead, E.C. Searle and W.T.C. Sowerbutts, Nature, 224, 1178, 1969.
28. R.C. Searle, Geophys. J.R.A.S., 21, 13, 1970.
29. B.W. Darracott, J.D. Fairhead and R.W. Girdler, Tectonophysics 15, 131, 1972.
30. J. Bridwell, New Mexico Geo.Soc.Guidebook, 27th Field Con., 1976.
31. R.W. Girdler and J.D. Fairhead, Geophys. J.R.A.S., 24, 271, 1971.
32. A.S. Laughton, Phil. Trans. Roy. Soc. Lond., A267, 21, 1970.
33. D.S. Chapman and H.N. Pollack, Earth Planet. Sci. Lett., 28, 23, 1975.

# A COMPARISON BETWEEN THE OLDER (KARROO) RIFTS AND THE YOUNGER (CENOZOIC) RIFTS OF EASTERN AFRICA

Basil C. King

Department of Geology, Bedford College (University of London), Regent's Park, London NW1 4NS

ABSTRACT. The older rifts or troughs are confined to southern Africa and extend northwards into Zambia and southern Tanzania. They were primarily infilled with Karroo sediments, with a final 'capping' of basaltic volcanics, remarkably similar in sequence to that seen in the main Karroo basins of South Africa and Botswana, despite their mutual isolation. Subsidence by downwarping was contemporaneous with deposition, although, faulting was prominent particularly in late or post-Karroo times. The trends of the troughs are often SW-NE and side trends NW-SE, suggesting a stress system at right angles to this direction; nevertheless, there is a common alignment with older Precambrian 'grains'. They were excavated and partly refilled by Cretaceous deposits and their present relief is the result of erosional exhumation. Volcanism that can be regarded as directly associated with the rifts is rare. The rifts were already 'dead' before the separation of Africa as a continent and, indeed, comparable rifts in west Africa and S. America have been transected by continental separation. The courses of the Zambesi and its tributaries, the Luangwa and Kafue, illustrate not only the typical diversity of trends of the troughs, but also the effects of later exhumation: indeed, in places, as at the Kariba gorge, a block of underlying basement has been reached and transected. The extent of possible marginal uplifts to the troughs cannot now be judged because of later events of uplift and erosion, but it is notable that the Zambesi has maintained its course to the sea despite continued uplift of the eastern side of Africa.

The Cenozoic rifts are also in general infilled by sediments or water; volcanism is the exception rather than the rule and the structures are similar regardless of the nature of the infill. Again they show a correspondence with earlier Precambrian 'grains',

*I. B. Ramberg and E.-R. Neumann (eds.), Tectonics and Geophysics of Continental Rifts, 347-350.*

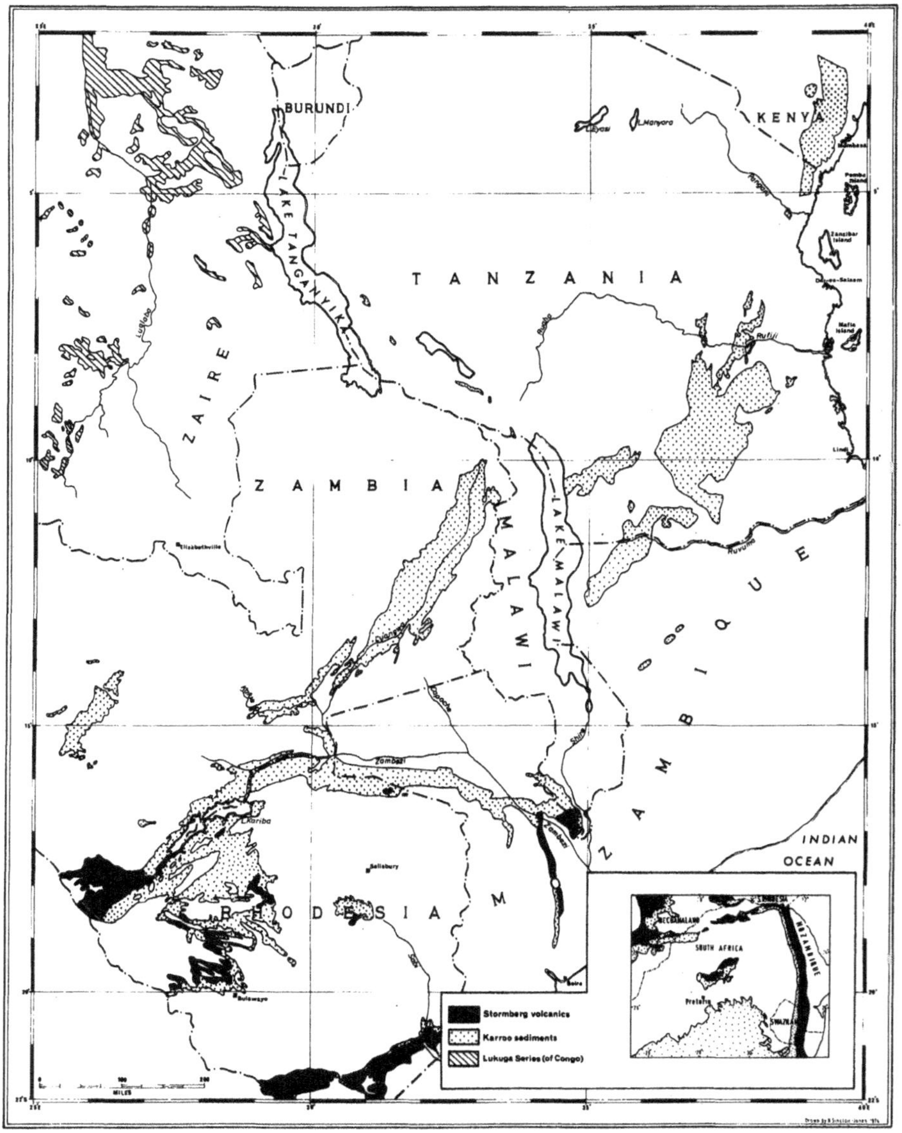


Fig. 1. The Karroo Troughs of Southern Africa.

but being situated farther northand east, they exploit the predominant north-south trends, (e.g. of the Mozambique belt), of these regions. At the same time it is likely that they relate to an east-west stress field, for blocks with a completely traverse east-west grain, such as the Tanganyika 'shield' appear to have been respon-

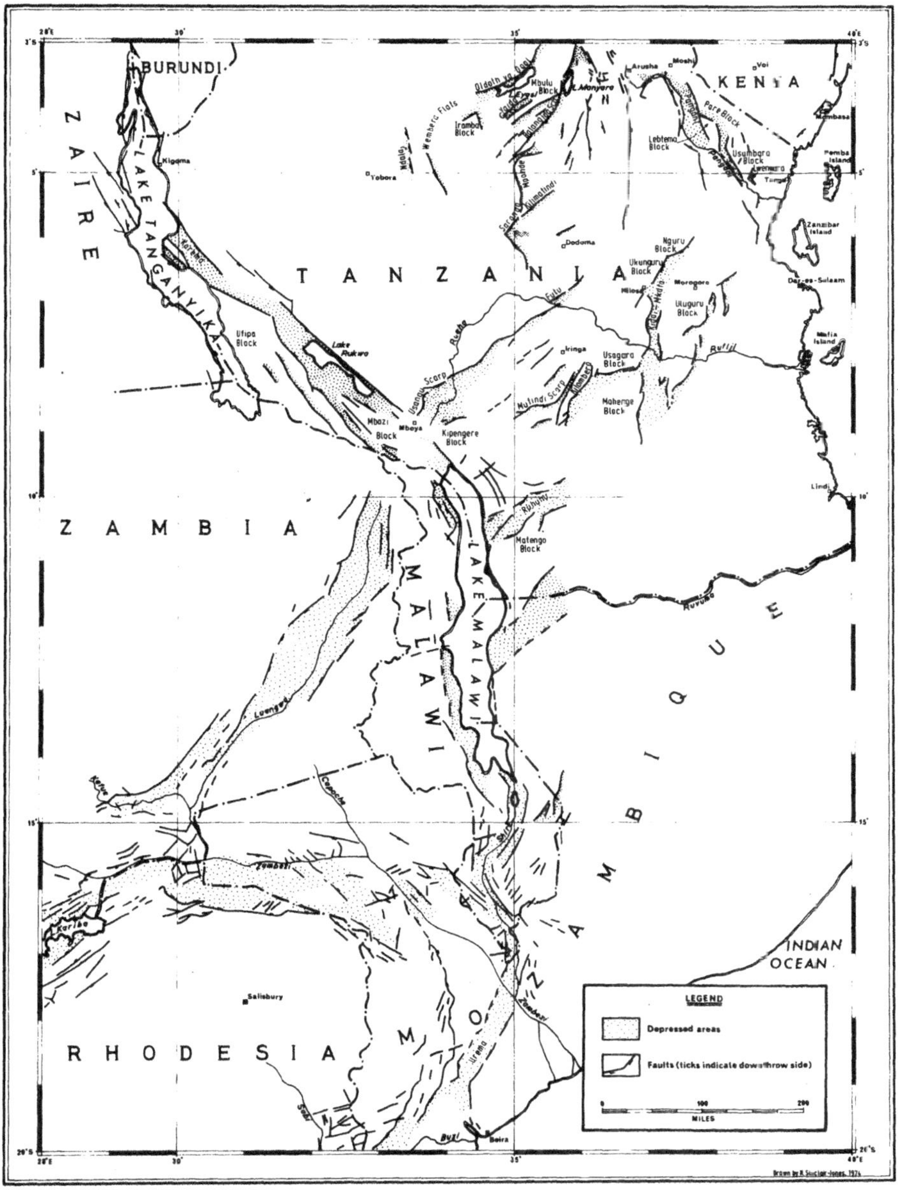


Fig. 2. The structures of the southern part of the East African Rift System.

sible for the division into a western and eastern rift system, while only those of the earlier rifts that were appropriately aligned have been rejuvenated by later movements.

The clearest examples of rejuvenation are shown by the southern troughs of Lake Tanganyika, notably the Rukwa trough. The Lake Malawi rift, although in parts it exploits the older troughs, is largely transverse to them and is essentially a Cenozoic structure.

The repeated uplifts of the eastern margin of Africa, indicated by the elevation of the later Mesozoic surfaces that bevel the Karroo troughs, was evidently a consequence of crustal separation, but clearly shows that, in relation to basement surfaces, the main movements associated with the Cenozoic rifts have been downwarping or downfaulting whereas uplift of the 'shoulders' of the rift has been both local and trivial on a regional scale.

Throughout most of its history the relief of the Gregory rift was insignificant, for the margins have twice been overlapped by sedimentary and volcanic accretion and even the Kenya dome is largely the result of marginal overtopping of the flanks and accumulation in the trough overtaking the rate of subsidence. The present spectacular morphology of the Cenozoic rifts in East Africa is very largely the result of Pleistocene faulting.

The history of all continental rift valleys thus shows a notable similarity of events. They commence with downwarps wider than their present limits, they are 'overtopped' by infill and further downwarped or downfaulted. Movements and infill become concentrated towards the centre at later stages; the present Cenozoic rift system is now in its dying stages, while the earlier southern system is extinct.

In both cases the amount of crustal extension is measurable in terms of hades and throws of faults; it amounts to about 1 in 8 or 1 in 10 over widths of, say, 60 to 100 km. This is comparable with values of extension in many other continental regions which cannot be regarded as related to rift systems.

The notion that continental rifts are 'aborted' oceans should have been tackled before 'conception'.

REFERENCES

1. F. Dixey, Colonial Geol. & Mineral Resources Supp. 1, 77, 1956.
2. Various memoirs and reports of the Geological Surveys of Malawi (Nyosaland), Rhodesia (S. Rhodesia), Tanzania (Tanganyika) and Zambia (N. Rhodesia).

# MAIN FEATURES OF EVOLUTION AND MAGMATISM OF CONTINENTAL RIFT ZONES IN THE CENOZOIC

N.A. Logatchev

Institute of the Earth's Crust
Siberian Branch of the USSR Acad. Sci.
Irkutsk

ABSTRACT. The vertical sequences of volcanic and sedimentary strata from Cenozoic continental rift zones can be divided into two distinct stages, this is true both of zones with strong magmatic activity (Kenya, Ethiopia) and zones with moderate or weak volcanicity (Baikal, Rhine, Western Rift of East Africa). This is due to the fact that a sharp alteration of tectonic movements occurs in the course of rifting. The second stage of the development is always characterized by faulting and taphrogenesis.

Relationships between magmatism and the formation of neotectonic structures in the evolution of continental rift zones are extremely complicated. In rift zones with strong volcanicity (Kenya, Ethiopia), magmatic activity and crustal movements are closely interrelated and interdependent, while in zones with weak volcanism magma formation and volcanicity are essentially independent of the development of rift faults and valleys proper (Baikal, Rhine). Zones of weaker volcanism generally show a more scattered distribution of faults, rift valleys and volcanic fields. An exception is the Western Rift of East Africa which is a zone of weak volcanicity but occupies an intermediate position in terms of the relationship between magmatism and tectonic structures.

All petrochemical series of basaltic rocks are represented in continental rift zones (tholeiites, subalkaline basalts, alkaline olivine basalts, ultra-alkaline nepheline-bearing rocks and carbonatites, kimberlite-like rocks and kimberlites). The petrochemistry of the volcanic series depends on the degree of extension and attenuation of the continental lithosphere during the rifting process as well as on the underlying heterogeneities (cratons, shields, ancient folded belts) which are correlated with heat flow value.

*I. B. Ramberg and E.-R. Neumann (eds.), Tectonics and Geophysics of Continental Rifts,* 361-366.

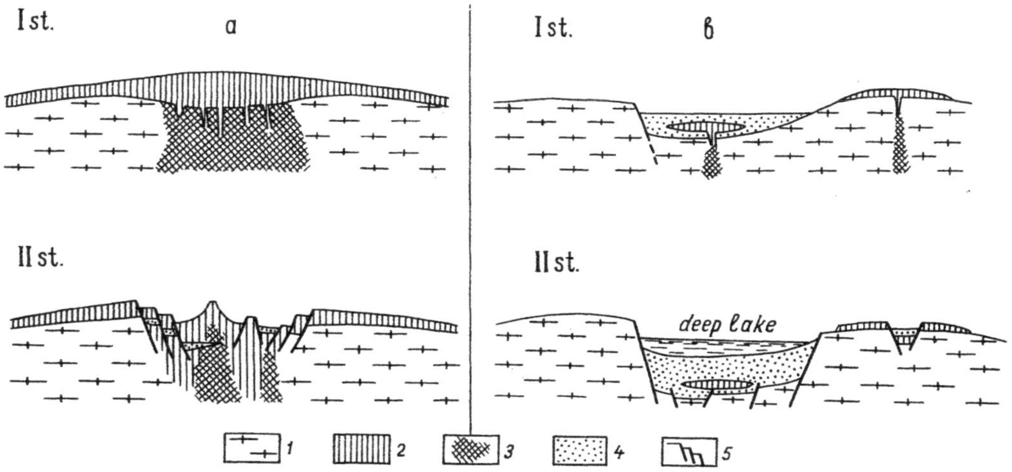


Figure 1. Schematic illustration of the two-stage history of rift zones with strong volcanism (a), and weak volcanism (b). 1-sialic crustal rocks; 2-volcanics, mainly basalts; 3-zones of attenuation and permeability for the upper mantle melts; 4-sediments of various origins 5-faults.

## 1. TWO-STAGE GEOLOGICAL HISTORY

The most universal feature of continental rift zones is the evidence for two-stage formation, which reflects varying regimes of rifting and is independent of geographical position, scale of concomitant volcanicity, removal of deep-seated material and type and age of the basement. Two types of formational complexes are to be distinguished which correspond to the two types of epiplatform continental rift zones, as outlined by Milanowsky [10]. The lower stage of domal volcanic rift zones consists of volcanic layers formed by material erupted from volcanoes of fissure and central types. The largest volumes of endogenous material are erupted during earlier fissure eruptions under considerable extension of the lithosphere. Linear intrusions or voluminous lava outpourings (Fig. 1a) are sufficient to compensate for the extension, so subsidence is a minor factor at that time.

In rift zones with weak volcanicity an extension has no time to be compensated by material rising from the depths, and subsidence in zones of extension commonly finds its morphological expression in the form of elongate depressions with sedimentary infills (marine or continental) depending on the position of the basins relative to the sea (Fig. 1b). Crustal deformation is rather slow at this time and movements along faults are slow or absent and do not produce sharp geomorphic contrasts. In depressions surrounded by low mountain areas terrestrial material, as well as organic and chemical sediments, accumulate. Under marine conditions chemical accumulation leads to formation of thick evaporites in the inner part of the basin (Upper Rhine graben, Suez and Red Sea grabens). The evaporites are replaced by terrigenous or volcanogenic (usually olivine-basalt) series in the outer parts of the graben. The outlines of these primary basins of accumulation may differ considerably from those formed during the second stage of rifting. In the areas involved in volcanicity lava sheets and tuffs are present on uplifts as well as in depressions. In the latter complex sedimentary and volcanogenic intercalations are formed passing into massive basaltic series on the flanks of the depressions. The total volume of the volcanics is tens of times less than in domal volcanic rift zones, though the thickness of sedimentary-volcanic infill forming the lower part of the rift succession with weak volcanicity usually exceeds that in the domal volcanic zones.

Thus, the development of rift zones at an earlier stage is rather variable and depends on the degree of involvement of eruptive material, i.e. ultimately on the intensity of deep-seated processes. In zones with strong volcanicity huge masses of deep-seated material begin to move, the rise of which nullifies the morphological result of extension. In the other case, wide syncline-like basins filled with sediments, locally accompanied by small volumes of lava, form from the very beginning.

In spite of the initial differences in rift zone development, their later evolution becomes almost identical and is characterized by intensification of extension, taphrogenesis and vertical movements. In domal volcanic rift zones this phenomenon eventually manifests itself in the formation of a rift valley, the process being followed by a sharp rise of geoisotherms right under the rift trough and by the formation of a melting zone in the mantle along the entire length of the subsiding area. Initially the melting conditions are more or less uniform, which leads to the outpouring of voluminous basaltic lavas. Differentiation of magmas in the deep-seated chamber leads to formation of trachyte and phonolite. At a certain stage eruption of ignimbrites, along with lavas, begins to play an important role, their deposits extend beyond the outlines of the rift troughs. As the system cools the common magma-generating chamber and a corresponding area of volcanicity are gradually reduced to form separate small chambers and volcanic fields, accompanied by a simultaneous change from fissure eruptions to central eruptions. In these deep residual chambers, or in peripheral chambers, differentiation continues and relatively small amounts of acid melts appear. When these are emptied, calderas in the rift floor collapse and become an additional factor causing rifting.

The declining nature of the thermal regime is not an obstacle to the creation of new magma chambers but the later chambers are situated at much deeper levels than before. Young alkaline carbonatitic centres on the southern extremety of the Kenya rift exemplify this phenomenon very well (e.g. Oldonyo, Lengai and Kerimasi volcanoes). The increasing depth of the chambers may well be related to the penetration of faults and extensional forces deeper into cooling lithosphere.

The second stage in the development of rift zones is characterized by intensification of vertical movements whose rate increases by several times as compared with the previous stage. Vertical displacement prevail along major faults. This stage may be described as the "liberation" of the rift shoulders and is attained by marginal ruptures separating subsiding blocks from adjacent blocks.

In spite of slow tectonic movements the rate of subsidence of old basins at the pre-rift stage exceeds the rate of shoulder uprising. This phenomenon results in subdued topography and almost complete infilling of the basins with sediments and volcanics. At the second stage the rate of block displacements on the "rift-shoulders" increases and becomes uniform. Only in the areas of maximum transverse extension does subsidence prevail, creating great uncompensated depressions such as those of the Baikal, Tanganyika and Nyassa lakes.

At the second stage of development of rift zones with weak volcanicity, changes in thermal regime are similar to those in domal volcanic rift zones. In the Baikal rift zone the conditions of melting and rise of deep-seated material were not improved, on the contrary, at the very beginning of the second stage basalt outpourings in the Tunka depression were reduced and then increased a little again. Sometimes on crustal uplifts (e.g. Udokan, East Sayan, Khamar-Daban, Prikhubsugulye) quite new areas of basalt volcanicity appear. They orginate through local intrusion of heated deep-seated material below areas of domal uplifts. A similar situation probably existed in the Upper Rhine graben where out-of-rift western volcanic fields (Eifel, Siebengebirge) came into being during the second stage of rift zone evolution. They are confined to a domal uplift.

It is interesting to note that in both rift zones of the Western Branch of the Africa-Arabian belt, which are characterized by weak volcanicity and have many features similar to those of the Baikal rift zone, the relations between volcanicity and tectonics differ greatly from that described above. The majority of the volcanic areas originated and developed during the second stage. Except for some rare cases, they are situated at the bottom or on the slopes of rift valleys, usually within sections of the relatively uplifted basement. The general structural position of such fields favours the hypothesis of H. Cloos [3] but they are confined not to marginal faults but to flexure-like crests of the graben (Toro-Ankole) or transverse tectonic volcanic zones involving adjacent rift shoulders (e.g. Virunga field with its 8 major volcanic centres forming a latitudinal volcanic zone). In the latter case the deep-seated magma source is possibly wider than the rift trough. Some minor out-of-rift volcanic fields are situated in close proximity to the graben, being connected with offsets of marginal faults which have not undergone strong vertical displacements (Kamituga, Kahusi, Biega fields and Beni volcano).

Thus it is in the Western branch of the East-African rift system the most complete coincidence of volcanic belts with rift structures can be noted. This branch occupies an intermediate position between zones with weak and strong volcanicity. The scope of the latter is limited but the volcanic areas are confined to rift depressions. Therefore, this branch approaches rift zones of a domal volcanic type. It can be explained by the ratio of arcogenesis to taphrogenesis along the rift zones. The subcrustal stratum below the domal uplifts of the Western branch is close to a critical state, and crustal weakening is sufficient for upper mantle melting and for magma material to be erupted in the rift itself or in the near vicinity of it. In this connection the more "concentrated" character of rifting in the Western branch should be noted (Fig. 2), as compared with rifting in the Baikal

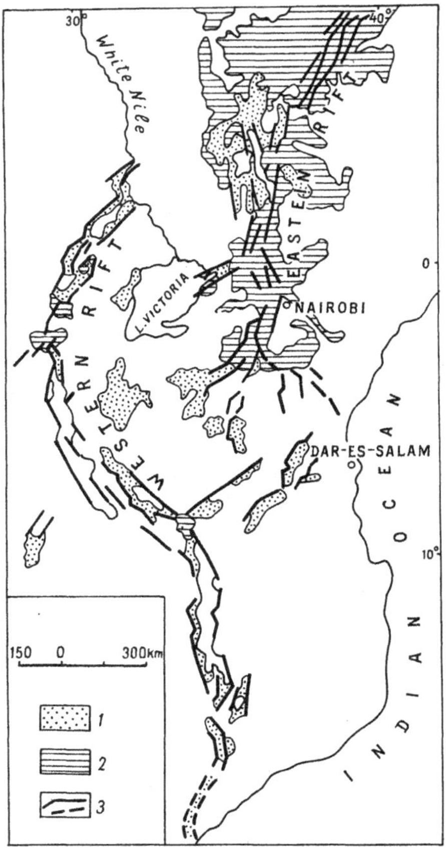


Figure 2. Southern part of the Afro-Arabian rift system characterized by concentrated linear or belt-like groupings of major faults, rift valleys and volcanic provinces. 1-rift valleys and depressions; 2-volcanics; 3-major faults.

rift zone with its scattered depressions (Fig. 3). Except for the section where the Tanganyika and Rukwa-Nyasa zones approach each other in an en echelon manner, there is a single rift trough along the Western branch. In some places it is divided by diagonal or transverse cross-links. Additional tectonic stresses at the dome were sufficient to produce volcanic activity in the rift.

No two continental rift zones are absolutely alike (Figs. 2, 3,4). In spite of their division into two large groups cognate to each other in the character of volcanic events and other important features, each rift zone has its own peculiar features. These differences lie in the time and scale of volcanicity, type of structures created, petrochemistry of products, their distribution relative to faults and rift valleys and conditions of mantle substratum melting. The main reason are the variable nature of the lithosphere, the different potential energy of masses of heated

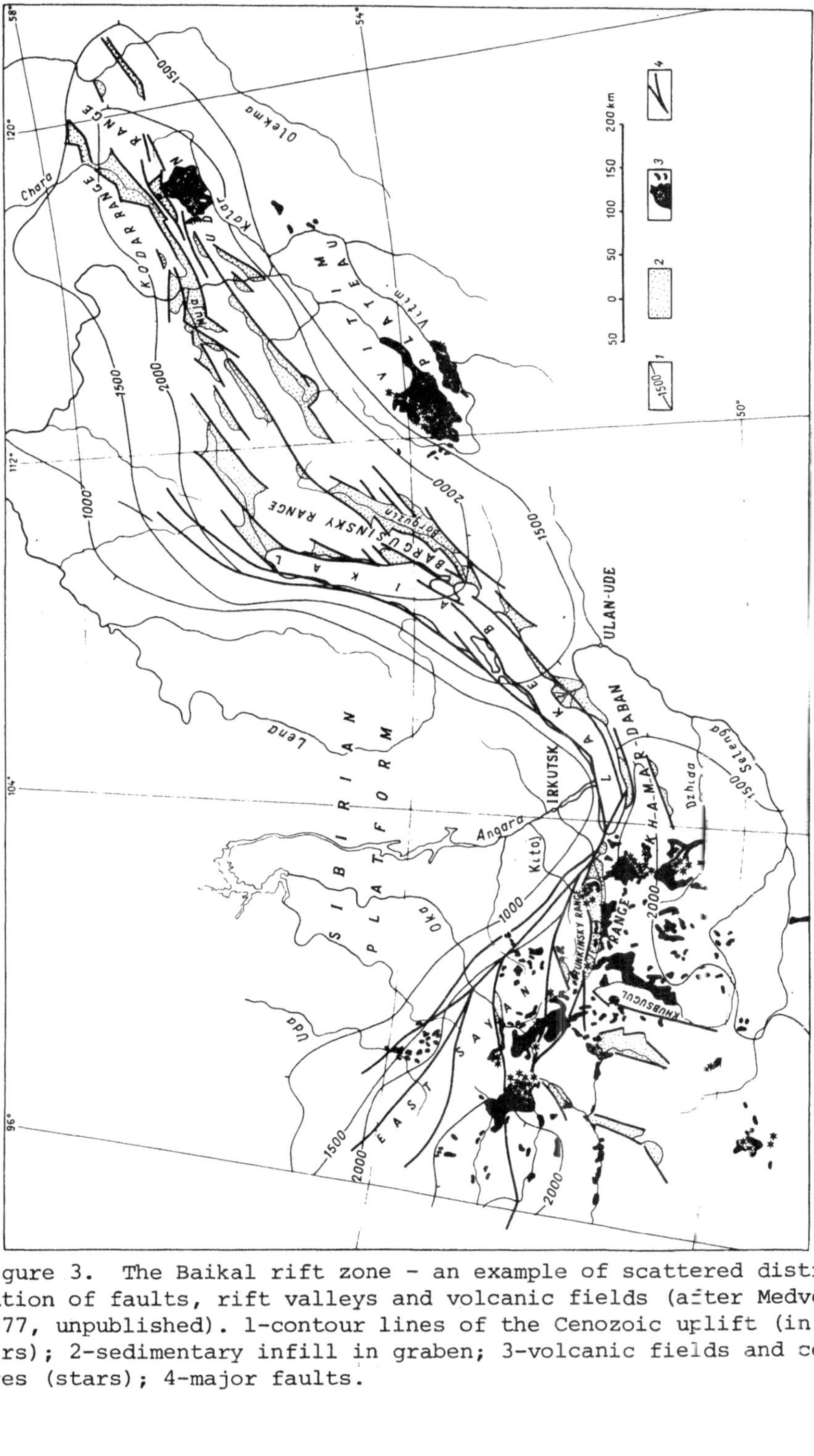


Figure 3. The Baikal rift zone - an example of scattered distribution of faults, rift valleys and volcanic fields (after Medvedev, 1977, unpublished). 1-contour lines of the Cenozoic uplift (in meters); 2-sedimentary infill in graben; 3-volcanic fields and centres (stars); 4-major faults.

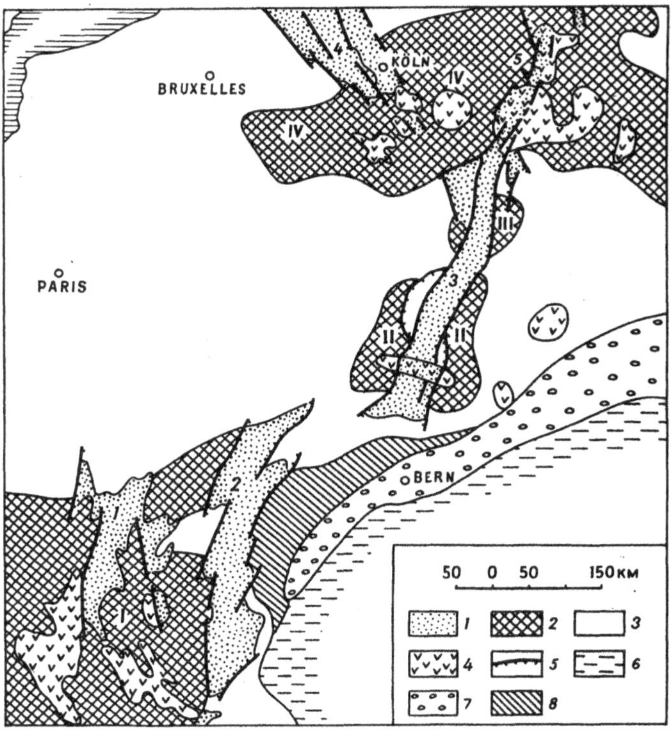


Figure 4. The scattered distribution of faults, rift valleys and volcanic fields in Central Europe. 1-sedimentary infill in graben (1-Limagne, 2-Bresse, 3-Upper Rhine, 4-Lower Rhine, 5-Hessen depression); 2-dome-like uplifts (I-Central French massif, II-Schwarzwald - Vosges, III-Odenwald-Hardt, IV-Ardennes - Schiefergebirge-Thüringenwald); 3-plateaux and plain covered by platform sediments; 4-volcanic fields (generalized); 5-major faults; 6-Alpine orogenic belt; 7-Molasse trough; 8-Jura fold belt.

material rising from the deepest layers of the mantle and extension of the crust varying in time and along the strike of rift zones.

The comparison of rift zones leads us to conclude that their volcanicity is related to the updoming resulting from the heating and partial melting of the upper mantle. In the case under discussion, crustal warping results from phase transitions, in the mantle. In this connection it becomes clear that the volcanic areas of rift zones are confined to domal uplifts. Therefore, we should define paragenetic relations between volcanicity faulting and rift valley subsidence. They are not causes and effects: they are closely connected with each other during rifting [1,8]. Such interaction is well manifested in domal volcanic rift zones.

In the Kenya rift zone further transformation of some deep fissures on a dome crest into marginal faults has been registered [8]. Similar phenomena occur in the Ethiopian rift zone [11]. Finally, complete or partial localization of volcanicity within an axial graben, beginning with its creation, proves the dependence of the distribution of volcanoes and volcanic products on the development of a rift trough. The dynamic conditions under which a rift is initiated intensify melting and magma formation right below the rift block. An axial graben is not the only place where the volcanic process may proceed, judging from the initiation of new areas of volcanicity outside the Kenya rift valley. Nevertheless, volcanicity in the axial graben is in direct genetic relation with the development of the latter, especially with extension and attenuation of the crust which are accompanied by rising of heated mantle material and geoisotherms below the rift valley.

In rift zones of domal volcanic type, the total volume of transported endogenous material is roughly equal to amounts of subsidence (Kenya zone) or even surpasses it (Ethiopian zone). Nevertheless, there are no good reasons to believe the Kenya and Ethiopian rifts to be volcanic tectonic structures. The hypothesis that graben formation on domal volcanic uplifts is a result of collapses over empty magmatic chambers is the preliminary attempt to explain the structural role of massive volcanicity. The influence of volcanicity on taphrogenesis may then be classified by special considerations. In the Kenya rift zone, collapses were not preceded by fissure eruptions of phonolites, but by central-type eruptions lasting for 5-7 m.y., during which extension was weaker than during the previous stage.

If we postulate that rift subsidences resulting from volcanic events are of a subordinate nature, difficulties arise in explaining relations between volcanicity at the pre-rift stage and young volcanicity in the rift itself. Variations in types of eruptions and the petrochemistry of its products during a transition to the rift stage in the Kenya and Ethiopia zones are so sharp that one should not speak of magma formation below the rift valley as a residual phenomenon inherited from the previous stage. In addition we should note that formation of subsidences on domal volcanic uplifts began simultaneously with intensification of depression formation (usually in the Middle Pliocene) and the elevation of rift shoulders in areas with weak volcanicity. This coincidence cannot be fortuitous. It proves the unity of mechanisms of subsidence in areas with sharply differing scales of volcanic activity.

Analysing sequences in the formation of assemblages of volcanic rocks from the Kenya rift zone, the author has come to the conclusion that complex approaches should be used for evaluation of the localization and role of volcanicity under rift tectonic conditions which cause the formation of large graben zones. Such an approach implies an analysis of large tectono-volcanic or tectono-thermal systems [4]. It is based on a knowledge of the successions of main structure-forming and volcanic events, the nature of the substratum involved and its influence on the development of newly formed structures, as well as the types of volcanoes and the petrochemistry of their products. It is most desirable to have data from geophysical investigations providing necessary detailed studies should be available. Unfortunately, while so different in their features, continental rift zones have been studied to a different extent by geologists and geophysicists, the East African rift system having received little attention from geophysicists so far.

## 2. PETROCHEMISTRY OF VOLCANICS

On the basis of analysis of temporal and spatial changes in the volcanic petrochemistry of the Kenya rift [9], as well as on data from experimental studies of basalt compositions under high pressure [5] the author came to the following conclusions: 1) volcanics of the two main petrochemical series (strongly alkaline with carbonatites and mildly alkaline) were formed synchronously; 2) the formation of the parent melts of these two series are separated by a narrow range of P-T conditions. Within this range melts of intermediate composition appear. As a result, the transitional series of volcanics is formed, consisting of a mixture of rocks of two series.

There is a petrochemical asymmetry in the volcanic cover of the Kenya rift since rocks of alkaline-ultramafic affinity are confined to the WSW periphery of the 200 km wide area of conjugation of the Archean Tanganyika shield with the Mosambique belt. (Fig. 5). The distance of penetration of known alkaline-carbonatitic centres into the edge of the Tanganyika shield does not exceed 100 to 120 km (Lower Miocene). Further investigations will prove whether such a distance is a maximum.

As a general conclusion, it follows that lateral replacement of volcanic associations of mild alkalinity by those with strong alkaline affinity (including carbonatites) is stipulated by the primary nature of the parental melts, rather than by differentiation of a single magma of olivine basaltic composition, i.e. it is connected with lateral changes of melting conditions in the transitional zone from the Mosambique belt to the Tanganyika shield. The general inclination of geoisotherms is likely to be a

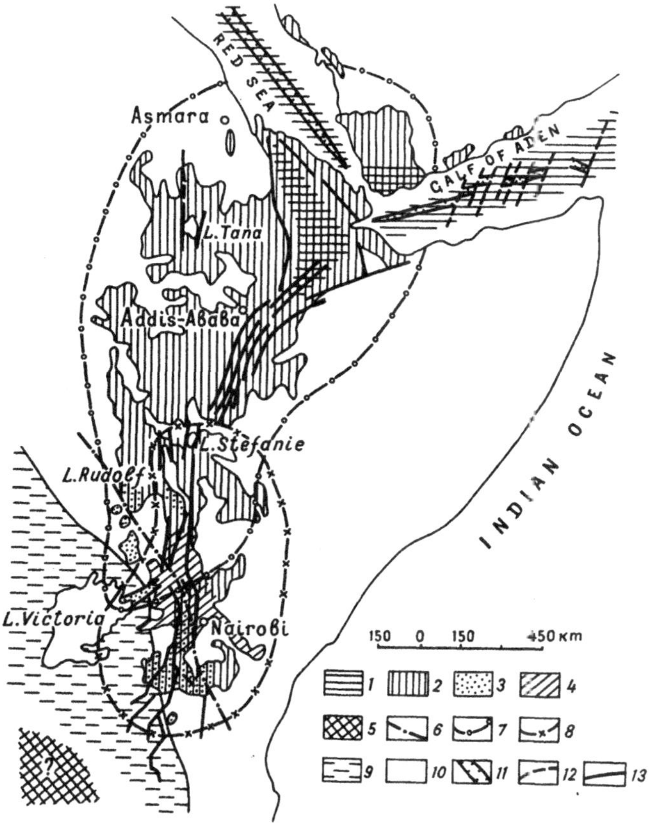


Figure 5. Volcanic rocks' distribution along the Eastern rift of Africa. Major petrochemical series: 1-tholeiites and "transitional" basalts; 2-alkaline olivine basalts (mildly alkaline series); 3-strongly alkaline nepheline-bearing rocks with carbonatites; 4-phonolites; 5-area of supposed kimberlite occurences; 6-north-eastern limit of alkaline-carbonatite centres running almost parallel to the Tanganyika shield margin. Tectono-thermal systems: 7-Ethiopian; 8-Kenyan; 9-Tanganyika shield; 10-Mosambique fold belt rejuvenated by Pan-African orogeny in Late Proterozoic-Early Paleozoic; 11-axial troughs in the Red Sea and Gulf of Aden; 12-transform-like offsets of axial troughs; 13-major faults.

governing mechanism of such changes and the thickness of the lithosphere increases from the Mosambique belt towards the Tanganyika shield (Fig. 6). Melting depths may vary by tens of

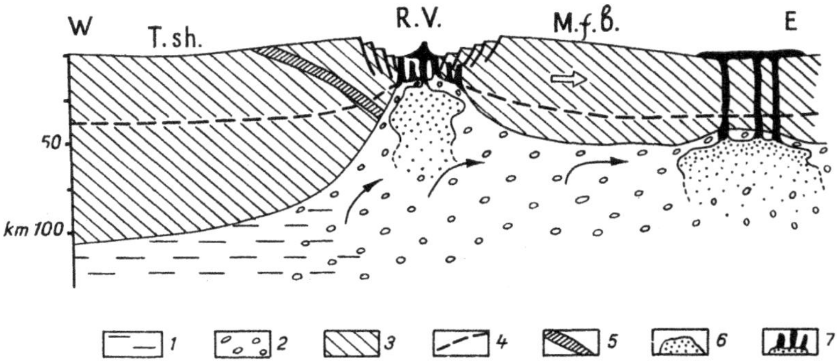


Figure 6. Hypothetical model of deep structure and magma generation in the Kenya rift (central part). 1-asthenosphere; 2-anomalous upper mantle; 3-lithosphere; 4-the base of the Earth's crust; 5-suture in the lithosphere between structural units of the Tanganyika shield (T.sh.) and Mosambique fold belt (M.f.b.); 6-deep magma sources; 7-magmatic feeders and dyke swarms. Solid arrows show the anomalous upper mantle convective displacement towards the east. Open arrow shows the direction of the lithosphere extension.

kilometres. The inclination of the geoisotherms should amount to 15 to 20$^{o}$; partial melting of the upper mantle proceeds at the same temperature (~1250$^{o}$). It is possible that compositional differences in the parental melts are defined by a vertical lamination of the upper mantle, as is nowadays acknowledged by many petrologists and tectonists. Less probable, and leading to a dead end, is the supposition that the nature of parental melts was defined by horizontal inhomogeneity of the mantle composition (remember the co-existence of volcanic associations of strongly and mildly alkaline series, divided by compositional gaps, in the same areas). The postulation that chemical properties of parental melts depend on differences in melting degrees at the same depth in the upper mantle (in particular on the possibility that the degree of partial melting decreases towards the margins of the area of magma generation) is groundless as alkaline carbonatitic rims are present only along the WSW margin of the Kenya volcanic area and have a very definite structural position. In the Ethiopian volcanic area they are absent.

Thus, on the basis of data from the Kenya region which has been considered as a typical area of alkaline magmatism, we have established criteria justifying the recongnition of an individual series of strongly alkaline rocks. The general connections of this series with alkaline rocks of the olivine-basalt series are rather complex and are difficult to explain, especially in areas of ancient intense basaltic volcanicity where volcanic piles are strongly denudated. The relations between the two alkaline series are, however, as important as relations between tholeiitic and olivine basaltic series, though they are present in lesser lava volumes and on a world scale are rare. It is doubtful whether the distinct relationships between volcanics of both alkaline series found in the Kenya volcanic area [1,8] are unique. Similar ones will probably be detected in other alkaline provinces of the world, even those exposing the root system of the volcanoes. But this problem is beyond the scope of the present report. The main condition of simultaneous formation of two series has been provided by tectono-magmatic forces in two large lithospheric blocks. Synchronous formation of the two series in structurally homogenous lithospheric blocks is less probable as the necessary deep conditions for formation are not present. Conditions such as these should be favourable for the development of petrochemically more or less uniform series of rocks of basaltic composition.

One more conclusion should be emphasized which is important for magmatic geology. It appears that the changes of deep conditions which cause changes in regime and the depth of melting of the upper mantle and replacement in time and in the same areas of mildly alkaline basaltic rocks by ultra-alkaline series, or vice versa, take place within a short period. From a few hundred thousands to less than 1 m.y. is sufficient for this process. This conclusion is based on direct geological data. These rapid changes in regimes and depths of melting are in particular, likely to be characteristic of the volcanicity of rift zones. Nevertheless, the question arises: is there anything similar in other alkaline provinces? In the Gulinskaya igneous complex (northern part of East Siberia) alternation of suites of moderate and highly alkaline volcanic rocks occurred, and this is an argument against those who deny that melts of alkaline ultramafic composition can be independent by derived and who explain the variety of rocks by differentiation of alkali-olivine basalt magma.

In spite of the series of alkaline ultramafic rocks and carbonatites being confined to the zone of conjugation of the Tanganyika shield and the Mosambique belt, the distributional limits of rocks from the above mentioned series and alkali-olivine basalt series are variable in time and are characterized by high mobility which is defined by the evolution of the tectono-thermal system of the Kenya rift zone. The depth of spatial inter-penetration of both series along the southern extremity of the Kenya rift amounts to 160 km. The main role in the emergence of ultramafic alkaline series and carbonatites in the Kenya volcanic province belongs to the structural heterogeneity of the ancient

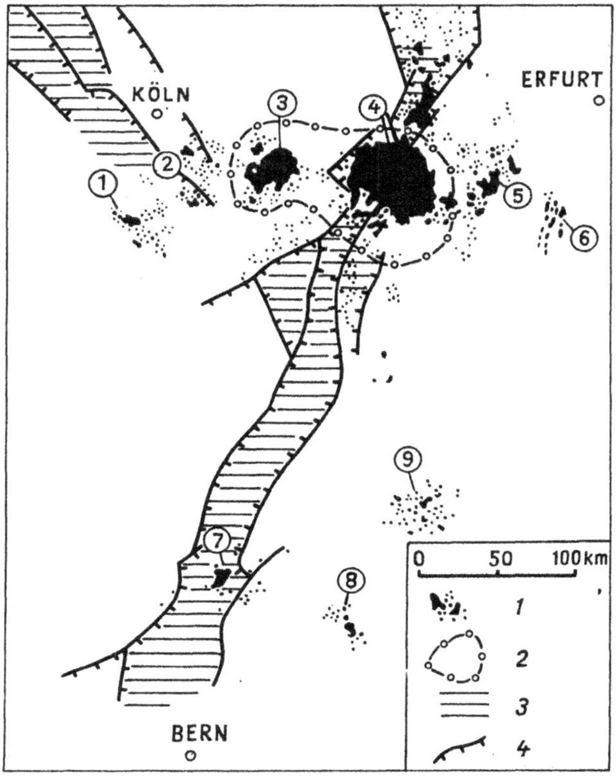


Figure 7. Central part of the Rhine rift zone (mainly after Illies [6] and Wimmenauer [12]. Petrochemical zoning of volcanic rocks of this region remains to be explained. 1-volcanic rocks; 2-contour of primary distribution of olivine alkaline and subalkaline basalts and tholeiites beyond the limit of distribution of rocks from strongly alkaline series (with carbonatites in the massif Kaiserstuhl); 3-sedimentary infill in graben; 4-major rift faults. Numbered circles are volcanic fields: 1-Eifel, 2-Neuwied and Laachern Sea, 3-Westerwald; 4-Vogelsberg; 5-Rhön, 6-Oberpfalz, 7-Kaiserstuhl, 8-Hegau, 9-Urach.

African platform rather than to the rift trough or rift faults proper. But for this heterogeneity the volcanic evolution might have taken place within alkali-olivine basaltic series. The comparison of the Baikal, Rhine and Africa-Arabian rift zones proves that the associated volcanicity is not limited to a narrow petrochemical range, but embraces all basalt series from tholeiites and subalkaline basalts in the axial grabens of the Red Sea and Afar depression to ultramafic rocks with carbonatites in the Kenya,

Tanganyika and Rhine rift zones (Fig. 7). On the whole, igneous rocks of continental rifts are as alkaline as in other areas of magmatic activity in stable zones of continental platforms. The widespread distribution of alkaline carbonatitic centres is due to their position in the lithosphere under unique conditions. The emergence of alkaline carbonatitic complexes depends on different thermo-dynamic conditions which may be indirectly evaluated on the basis of an analysis of structural inhomogeneity within the upper crust.

## 3. CONCLUSION

The two-stage arrangement of the sequence of volcanic and sedimentary rocks accompanying the formation of continental rift zones is defined by a two-stage evolution of the tectonic process. It should be taken into account that one is concerned here with the major stages, each involving a series of smaller tectonic events. Since a two-stage geological history is typical for rift zones with different extents of volcanicity, it must be taken as the most general and peculiar feature of continental rifting as a whole.

The nature of the relations between magmatism and tectonics in rifting is still not clear enough and therefore must be an inspiration for further study. However, along with genetically simple relationships between structure formation and volcanicity in a series of regions, the volcanic fields and rift structures (troughs, faults) are associated with each other only paragenetically, i.e. magmatism and tectonic movements in the upper crust proceed more or less independently of one another, though they have the same abyssal source.

Present geological and experimental data confirm that the chemical composition of initial melts and the degree of partial melting of the upper mantle are determined on the one hand by the degree of extension and attenuation of the lithosphere under the influence of rifting, and on the other by its structural and thermal heterogeneities which existed prior to rifting.

## ACKNOWLEDGEMENTS

The present work and its basic conclusions was possible due to intensive investigations in East Africa by B. Baker, B. King, L. Williams, P. Mohr, I. Gass, D. Bailey, A. Pilger, A. Rösler and E. Milanovsky, in Europe - by H. Illies, W. Wimmenauer and Doeble F., in East Siberia by N. Florensov, V. Solonenko and Yu. Zorin.

REFERENCES


1. D.K. Bailey, Continental rifting and alkaline magmatism, in: The Alkaline Rocks ed. by H.S. Sørensen, A Wiley - Intersci. Publ., 148, 1974.
2. B.H. Baker, P.A. Mohr and L.A.I. Williams, Geol.Soc.Am.Spec. Pap., 136, 67, 1972.
3. H. Cloos, Geol. Rundsch., 30, 401, 1939.
4. J.G. Gass, Phil. Trans. Roy. Soc. London, 267, 369, 1970.
5. D.H. Green, Phys. Earth Planet. Int., 3, 221, 1970.
6. I.H. Illies, Taphrogenesis and plate tectonics, in: Approaches to taphrogenesis, ed. by I.H. Illies and K. Fuchs, 1974.
7. B.C. King and G.R. Chapman, Phil. Trans. Roy. Soc. London, 271, 185, 1972.
8. N.A. Logatchev, Geotectonics, 3, pp. 74, 1974 (in Russian).
9. N.A. Logatchev, Magmatism-tectonics relationship during the Kenya rift development and the prime cause of strongly and mildly alkaline volcanic suite formation, in: Afar between Continental and Oceanic Rifting, ed. by A. Pilger and A. Rösler, E. Schweizerbart'sche Verlagsbh., Stuttgart, 1976.
10. E.E. Milanovsky, Main types of the continental rift zones. Proceed. of Moscow St. Univers., Geology, 2, 13, 1970.
11. P.A. Mohr, J. Geophys. Res., 76, 1967, 1971.
12. W. Wimmenauer, The alkaline province of Central Europe and France, in: The Alkaline Rocks, ed. by H. Sørensen, A Wiley-Intersci. Publ., 1974.

# THE ST. LAWRENCE PALEO-RIFT SYSTEM: A COMPARATIVE STUDY

P.S. Kumarapeli

Department of Geology, Concordia University
Montreal, Canada

ABSTRACT. The concept of a St. Lawrence rift system was first proposed 11 years ago. The rifts of the water-covered eastern half of this tectonic feature are still largely hypothetical, although the basis for their existence is stronger today than it was 11 years ago. The rift zones of the western half of the rift system are reasonably well-authenticated. This part of the rift system alone is about 1100 km long.

Many features of the St. Lawrence rift system compare closely with those of the active rift systems in the East African, Baikal and Rhine regions. However, unlike the latter rifts, the St. Lawrence rift system has little or no geophysical expression; its fault troughs are shallower and have only small amounts of sediments and volcanics which can be regarded as rift related and its geomorphic expression is relatively subdued. Although these differences are thought to be largely a result of its long inactivity, some of the differences may just be inherent features of the rift system.

In its present form, the St. Lawrence rift system appears to have formed in the mid-Mesozoic, possibly as a result of tensional stresses related to the opening of the Atlantic Ocean. However, there are indications that the pattern of mid-Mesozoic rifting was largely superimposed on pre-existing structures, also of tensional origin and possibly of Hadrynian age. These older structures are identified as aulacogens and other structural elements inherited from the rifting of a single North American-African continental plate, as envisaged in recent models of Appalachian evolution.

*I. B. Ramberg and E.-R. Neumann (eds.), Tectonics and Geophysics of Continental Rifts,* 367-384.

## 1. INTRODUCTION

In the southeastern parts of the Canadian Shield and in the adjacent St. Lawrence platform, a few seemingly isolated graben structures were known since the early part of this century [1]. These structures were considered as forming parts of a large continental rift system, the St. Lawrence rift system (Fig. 1), in 1966 [2]. The basis for this synthesis is given in earlier papers [2,3] and will not be repeated here.

The St. Lawrence rift system, as shown in Fig. 1, is about 2200 km long, but only its western half can be considered as sufficiently well authenticated. Work done on this part since 1966 [4], has shown that the rift system is a paleo-structure and not an active one as was originally suggested. The eastern half of the rift system, being largely water-covered, affords us only minimal opportunities for direct study. Geophysics may not be of much help either, because even the better known parts of the rift system are not associated with conspicuous geophysical anomalies. Therefore, the eastern half of the rift system remains hypothetical, although work done on this part since 1966 has enhanced the viability of the rift hypothesis for much of this area [3]. The greatest uncertainty is with respect to the proposed branch of the rift system beneath the outer part of the Laurentian Channel. Recent studies have provided no evidence for a graben structure (in the strict sense) [5,6], but have indicated that this part of the channel may be underlain by some type of structural discontinuity [7,8]. In the light of these studies, it is proposed that this branch of the rift system may exist as a feature, which in the present state of knowledge, may best be described as a fracture zone. This fracture zone may have formed (at the same time as the rest of the rift system) during a mid-Mesozoic reactivation of an older line of structural weakness.

In many parts of the rift system, rift faults clearly cut and displace Paleozoic strata, but in no part of the rift system has it been possible to bracket closely the ages of this rifting by direct methods. Magmatic events whose products range in age from 90 to 150 m.y. occured in widely separated parts of the rift system (Fig. 5). It is entirely likely that the emplacement of these igneous rocks and the greater part of rifting that involved Paleozoic strata may have been mutually connected events [9]. A mid-Mesozoic phase of rifting, therefore, is considered to have been responsible for the present form of the rift system. In places, there are indications that the apparently mid-Mesozoic rifts are superimposed on older aulacogens (e.g. Ottawa Graben) graben-like structures and fracture zones. These <u>Ancestral Structures</u> may have formed in the Hadrynian, synchronously with the emplacement of diabase dyke swarms in the St. Lawrence Region (Fig. 5). This early rifting seems to have taken place at about

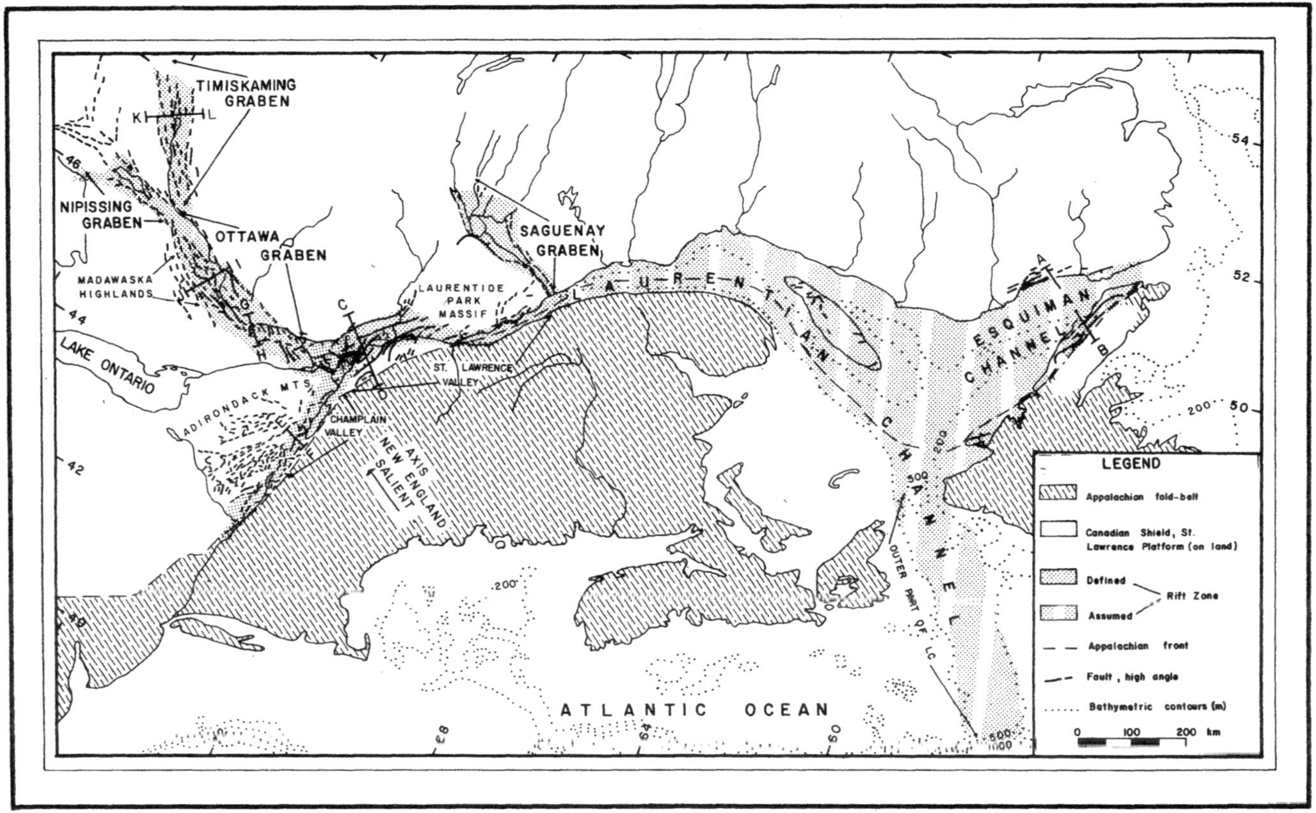


Fig. 1. Rift zones of the St. Lawrence system. AB, CD, etc. are lines of sections shown in Fig. 3.

the same time as the beginning of the stratigraphic record in the Appalachians.

For convenience of reference the following names will be applied throughout this paper: Marginal Segment to the Z-shaped part of the rift system which is set marginally with respect to the Appalachians; Shield Segments to those parts of the rift system set in the Canadian Shield; Appalachian Region to the part of the Appalachian fold-belt rimmed by the Marginal Segment; Shield Region to the general area of the Canadian Shield in which the Shield Segments are set.

Large rift systems have many common characteristics. However, each rift system can be expected to have its own set of distinctive characteristics. The writer proposes to focus attention on the characteristics of the St. Lawrence rift system by comparing them with those of better known rifts, particularly the active rifts of the three classical areas. Comparisons are brief because of space considerations.

## 2. GEOMORPHIC CHARACTERISTICS

Rift zones of the St. Lawrence Region show negative topography typical of rift valleys, but compared with the rift valleys of the classical areas those of the St. Lawrence Region are shallower (commonly less than 300 m deep) and often poorly defined. Fault-line scarps of the St. Lawrence Region are deeply dissected and rarely rise to more than 500 m above the floors of the rift valleys. Four large lakes; Timiskaming, Nipissing, Champlain, and St. Jean, occupy parts of rift valleys. Except lake Champlain, the other three lakes occur in areas where rift faults splay out to form wide structurally depressed areas. Three large elevated blocks of Precambrian rocks occur in close association with the rift system: Adirondack Mountains (maximum elevation 1,639 m) and the Laurentide Park massif (1,166 m) occupy the acute angles between branching rifts and Madawaska Highlands border the Ottawa graben (Fig. 1).

## 3. POSSIBLE RELATION TO A CRUSTAL UPWARP

A crustal swell similar to those that form the habitats of the classical continental rift systems cannot be discerned from a study of the present erosion surface of the St. Lawrence Region. However, if such a feature did exist during the development of the rift system and has since been lost by erosion, then by analogy with the situation in other areas, the crustal swell may have been elliptical in outline with the trunk rift zone along its long axis and with about one half of it involving the Shield

Region and the rest involving the Appalachian Region. The following features of the Shield Region are compatible with the concept of a deeply eroded crustal swell as speculated above: (1) The Shield Region has undergone exceptionally deep erosion to expose rocks formed in the catazonal environment possibly at depths of 15 to 20 km [10]. (2) The main drainage lines of the Shield Region are deeply entrenched and appear to be fracture controlled, and they form a rather distinctive pattern resembling a lateral fracture pattern which develops on an oval uplift. (3) The deep drainage lines of the Shield Region have their sources to the north of the present height of land but flow soutwards through canyons and gorges against the regional slope for considerable distances.

Apparently, the deep valleys were established and structure adjusted in pre-Ordovician times [11], and the deep erosion of the Shield Region, which is largely coextensive with the Grenville structural province, was complete before the beginning of the Paleozoic [11]. Therefore, the indicated updoming may have occurred in connection with the development of the Ancestral Structures rather than the mid-Mesozoic rifting. The block mountains mentioned earlier may be parts of this crustal swell, resurrected during the mid-Mesozoic rifting. The generally southeast sloping surface of the Appalachian Region may also be a product of block tilting related to the mid-Mesozoic rifting.

## 4. THE PATTERN OF THE ST. LAWRENCE RIFT SYSTEM

According to the hypothesis of origin presented later in this paper, the St. Lawrence rift system is the product of superimposition of two similar, but only partly overlapping patterns of rifting. Despite this, the pattern of the St. Lawrence rift system is similar to the rather distinctive patterns of large continental rift systems (Fig. 2). The fault pattern of the western half of the rift system resembles closely the fracture pattern produced by Cloos [12] by swelling an oval-shaped hot water bottle coated with clay.

## 5. DIMENSIONS

Table 1 shows the similarity of widths of some well known continental grabens and the various parts of the St. Lawrence rift system. Values quoted for the St. Lawrence rift system are for the medial parts of the respective rift zones; some of the rifts are narrower at one end and wider at the other. The widths of the Marginal Segment rifts are difficult to ascertain because to the south and east, structural elements of the rifts become intermingled with the structures of the Appalachian fold-belt.

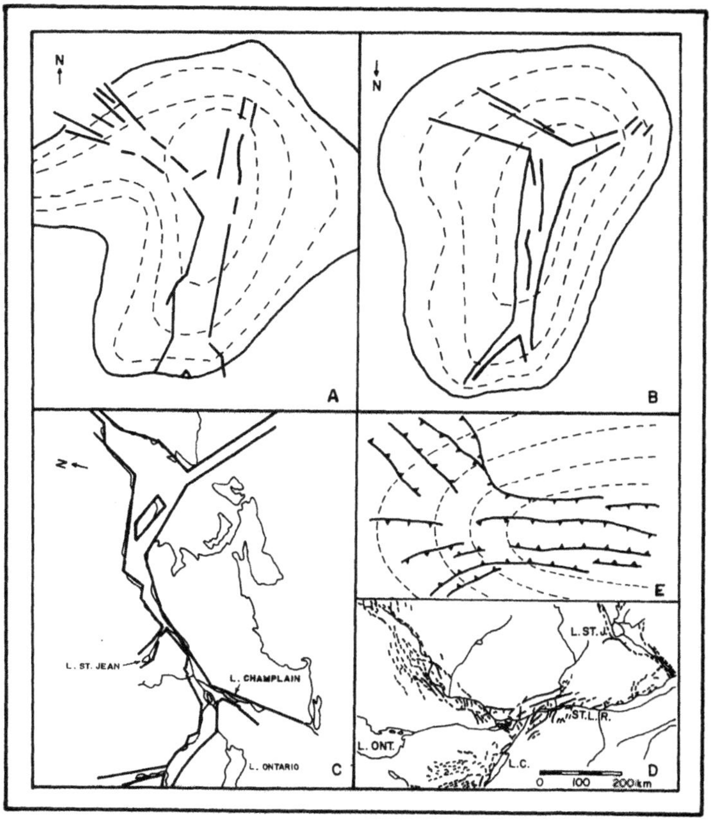


Fig. 2. Diagram showing the similarity of patterns of the Rhine graben (A), Eritrean graben (B), the St. Lawrence Rift system (C), the fault pattern at the south end of the St. Lawrence Valley (D) with the fracture pattern produced by Cloos [12] by swelling a hot water bottle coated with clay (E).

The termination of the rifts in the St. Lawrence Region are poorly known. However, the rift system, as shown in Fig. 1 is about 2200 km long.

Table I: Common widths of Continental Rifts

| Rift | Width in km |
|---|---|
| Gulf Aquaba [x] | 50 |
| Dead Sea [x] | 35 |
| Gulf of Suez[x] | 35 |
| East African (common) [x] | 30 |
| Rhine [x] | 40 |
| Baikal [x] | 50 |
| Ottawa Graben | 55 |
| Nipissing Graben | 30 |
| Timiskaming Graben | 40 |
| Saguenay Graben | 55 |
| Marginal Segment | |
| St. Lawrence Valley | 30? |
| Champlain Valley | 35? |

[x] Data from Beloussov [13]

6. TECTONIC SETTING AND TREND RELATIONS

Excepting the possible branch structure along the outer part of the Laurentian Channel, the rest of the St. Lawrence rift system is set on craton. The large continental rifts of the three classical areas are also superimposed largely on cratons [13]. The craton in the St. Lawrence Region consists mainly of the Grenville structural province of the Canadian Shield and the narrow discontinuous platform, the St. Lawrence platform (not shown separately in Fig. 1) , which lies between the Appalachian fold-belt and the Canadian Shield. The last orogeny in the Grenville structural province took place in the Helikian (ca. 1000 m.y.) and the tectono-stratigraphic evolution of the Northern Appalachians took place from Hadrynian to Permian. The cover rocks of the St. Lawrence platform are dominantly carbonates of Early Paleozoic age. Some segments of the rift system are floored by these rocks.

The gross trend of the East African rift system is N-S, and the Precambrian basement in the area is also characterized by the same general trend. Despite this overall conformity of trends, notable discordancies between the strikes of the Precambrian rocks and of the rifts are observed when they are considered in detail. Trend relations are essentially similar in the Baikal region [13]. The trend relations of the St. Lawrence rift system are similar to those of the East African and Baikal rift system in that:

(i) the gross trend of the St. Lawrence and Baikal rift system conforms with that of the Grenville belt on the one hand and that of the Appalachian belt on the other, (ii) while the Z-shaped Marginal Segment closely follows the boundary of the Appalachian fold-belt, the Shield Segments have trends nearly at right angles to the fold-belt and are superimposed on older rocks discordantly. The trend relations of the Shield Segments are similar to those of the Rhine graben which is discordantly superimposed on Paleozoic and Precambrian structures of the Rhine region.

The Ottawa graben extends westwards from the axis of a salient - the New England Salient - of the Appalachian fold-belt. The axis of the New England Salient (Fig. 1) is nearly coincident with a transverse geosynclinal trough, which first appeared late in the Hadrynian and persisted as a major downwarp throughout the evolution of the fold-belt. The stratigraphic section in the trough is as much as 25 km thick but 50 km northeast and southwest of the axis of the salient the section converges to less than 15 km [14]. The space relations of the trough and the Ottawa graben are such that they appear to have been parts of a single structure, presumably a rift zone, which in the Hadrynian extended into the axial region of the Appalachian fold-belt. The postulated branch of the rift system along the outer part of the Laurentian Channel also extends from a prominent re-entrant of the Appalachian fold-belt.

## 7. STRUCTURE

Six structure sections across selected parts of the St. Lawrence rift system are shown in Fig. 3. The displacement pattern along the Marginal Segment is essentially one of step faulting, down from the shield side towards the Appalachian fold-belt. The Shield Segments consist mainly of complex grabens, commonly asymmetrical. Longitudinally, the grabens are commonly cradle-shaped. In several parts of the rift system, there are indications that the directions of rift faults were controlled by a pre-existing fracture pattern, oriented prevalently northeast and northwest (Fig. 4). The rift faults and the rift zones themselves, for the most part, are oriented or tend to be oriented in one or the other of these two directions. Detailed descriptions of all the structural features of the rifts are not planned in this section, but as a generalization it may be said that almost all the structural variants displayed by the rifts in the three classical areas have their counterparts in the St. Lawrence Region.

The maximum structural subsidence along the St. Lawrence rift system cannot be determined with any certainty, but the indications are that the mid-Mesozoic rifting has not brought about more than 500 m of structural subsidence which is several orders

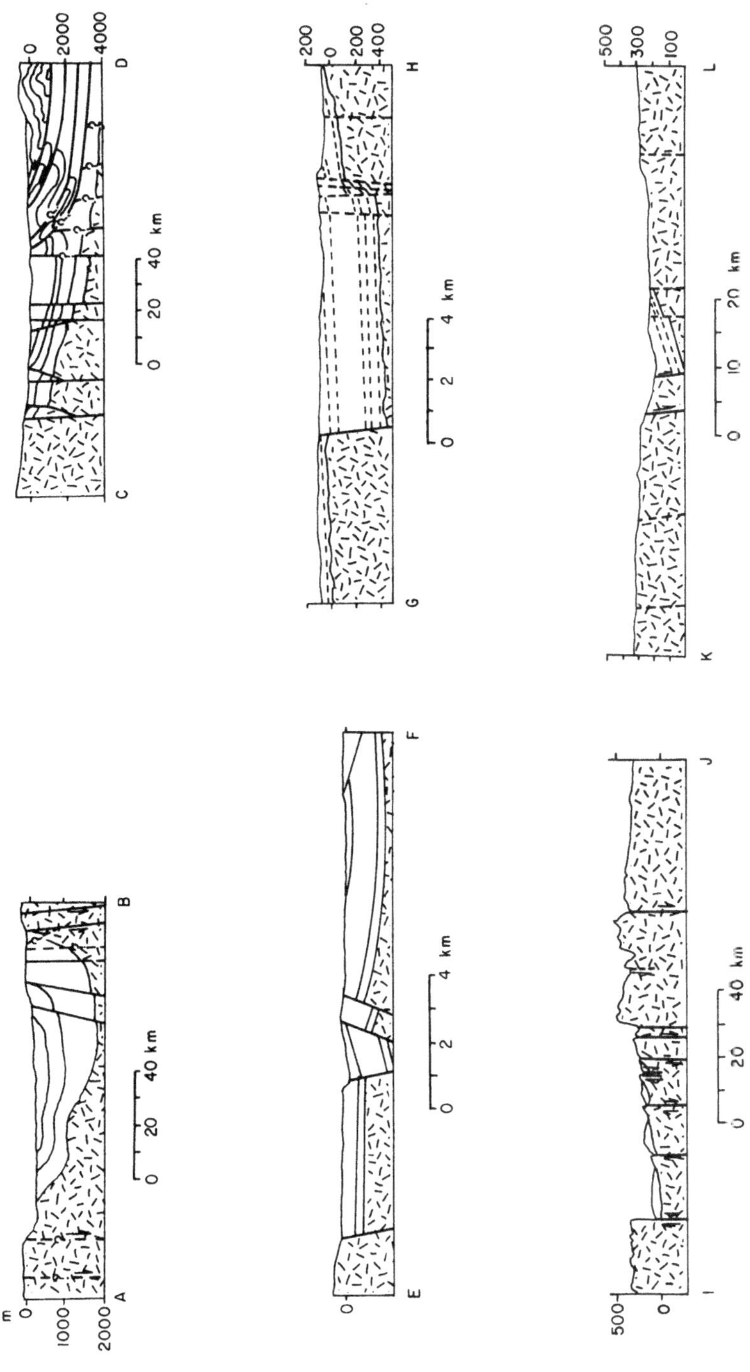


Fig. 3. Structure sections across selected parts of the St. Lawrence rift system. EF after Quinn [17], GH after Wilson [15], IJ after Kay [16].

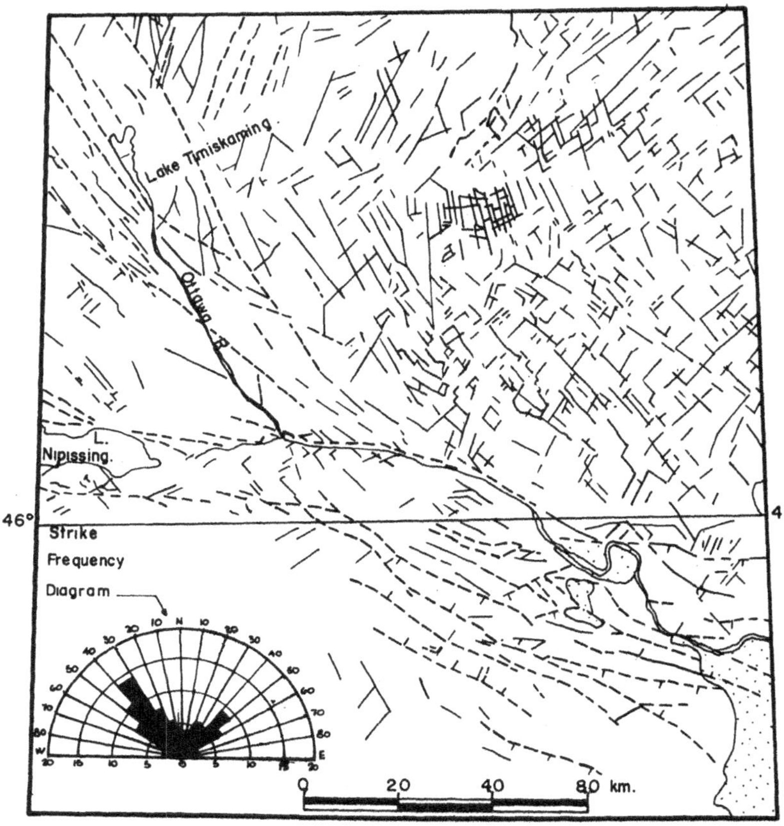


Fig. 4. Fracture-controlled airphoto lineaments of a part of the Shield Region. Thin lines: Airphoto lineaments. Thick broken lines: Mapped and interpreted high angle faults. Stipples: Areas of Early Paleozoic platformal rocks.

of magnitude smaller than the structural subsidence along the classical rift zones. Greater structural subsidence may have taken place during the evolution of the Ancestral structures related to the St. Lawrence rift system, but the magnitude of these movements cannot be estimated.

## 8. SEDIMENTARY-VOLCANIC FILL

Unlike the fault troughs of the classical rift zones, those

of the St. Lawrence rift system do not contain substantial amounts of rift-related clastics and/or volcanics. In fact, the latter have no rift fill that can be regarded as related to the mid-Mesozoic rifting. However, immature clastics in some parts of the Marginal Segment [18,19] and minor flood basalts in the Belle Isle area [20], also in the Marginal Segment (Fig. 5), may be rift-related products. If this should be the case, then they are probably connected with the formation of the Ancestral Structures.

## 9. IGNEOUS ACTIVITY

One rather outstanding characteristic of igneous activity associated with the East African rifts is the unusually large scale development of alkaline plutonic and volcanic rocks often with feldspathoid-bearing differentiates. Petrochemically similar rocks are also associated with the Rhine graben. The most common plutonic bodies of the alkaline series are central ring complexes of ultrabasic rocks and carbonatite. Compared with the igneous rocks of the East African and Rhine rifts, those of the Baikal region are less alkaline. Greatly undersaturated feldspathoid-bearing rocks which occur not infrequently in the East African and Rhine lava series, are unknown in the Baikal region, and carbonatite and ultrabasic alkaline rocks seem to be altogether absent [21]. Olivine basalts, both normal and mildly alkaline, form part of the volcanic products of the East African and Rhine rifts. Furthermore, large volumes of tholeiitic basalts also occur in the East African system [22] but seem to be restricted to the Afar depression. From the description to follow, it can be seen that the igneous products associated with the St. Lawrence rift system resemble those of the classical rifts, in particular those of the East African and Rhine regions, with regard to petrochemistry, tectonic setting and age relations.

Igneous products associated with the St. Lawrence rift system consist of: (a) Tholeiitic and transitional types (between tholeiitic and alkaline basalts) of probable Hadrynian age, represented by dyke swarms and minor flood basalts, (b) alkaline complexes including several alkaline-carbonatite complexes which yield K/Ar ranging from early Cambrian to Cretaceous (Fig. 5).

Tholeiitic and transitional types occur in two main areas: (i) As an E-W trending diabase dyke swarm - Grenville dyke swarm - of tholeiitic character [23], extending for a distance of about 650 km in close association with the Ottawa and Nipissing grabens (Fig. 5), (ii) as a dyke swarm related minor flood basalts of transitional character in the easternmost part of Marginal Segment in Belle Isle area [20]. In both areas, the dyke swarms and graben structures show a considerable unity of trends and patterns. These relationships are interpreted as signifying a genetic relationship between graben formation and dyke intrusion. The dyke swarms constitute the earliest (but post-Grenville) and most pervasive igneous activity along the St. Lawrence rift system. The flood basalts in the Belle Isle area may be the remnants of once extensive plateau basalt fields. Whether some of the dykes of the Grenville swarm were feeders for flood basalt eruptions is an interesting speculation. The precise ages of the above dyke rocks and volcanics are uncertain. Dykes in the northern part of Long Range Mountains (Fig. 5), with which Belle Isle dykes and flood basalts may be comagmatic, have been dated ($Ar^{40/39}$) at 600 $\pm$ 6 m.y. [24]. K/Ar age dating of the dykes of the Grenville swarm has not yielded concordant results, presumably due to argon loss. An age of 700 m.y. has been suggested on paleomagnetic grounds [25]. An anorthosite-gabbro-granite complex, located at Sept-Iles on the northern boundary of the Marginal Segment (Fig. 5) has recently been found to yield a Rb-Sr age of about 540 m.y. [28].

Twenty one central complexes, numerous dykes, sills and plugs of alkaline and related rocks whose K/Ar ages range from 580 m.y. to 90 m.y. occur in close association with the St. Lawrence rift system (Fig. 5). Quantitatively (considering the number of central complexes, dykes, sills, plugs, breccia pipes and the extent of hydrothermal alteration), however, the bulk of this magmatism seems to have occurred in two widely separated periods: one about the beginning of the Cambrian and the other in late Jurassic - early Cretaceous times. The products of the earlier period include rocks of the Lake Nipissing alkaline province [26] and the alkaline-carbonatite complexes in the Saguenay graben and those of the later period include rocks of the Monteregian alkaline province [27]. The early Cambrian alkaline magmatism and the probable Hadrynian diabase dyking and flood basalt eruptions may be aspects of a single phase of igneous activity probably related to the formation of the Ancestral Structures related to the St. Lawrence rift system. The mid-Mesozoic phase of alkaline magmatism which is widespread along the rift system, is considered as coeval with the events which led to the development of the rift system, essentially to its present form.

The space relations between alkaline centres and graben structures of the St. Lawrence, East African and Rhine rift zones are also similar in some respects. The typical setting of the early Cambrian alkaline-carbonatite complexes of the St. Lawrence region is where the Shield Segments begin splaying out. The Rigaud and Chatham-Grenville stocks (Nos. 6a & b in Fig. 5) and the Monteregian intrusions are largely localized at a major bifurcation of the rift system. The Monteregian line of intrusions extends across the rifts to the outlying areas in the manner of lateral volcanic lines in the East African and Rhine setting. On the other hand, this line of intrusions may simply be set along the eastward continuation of the Ottawa graben faults which,

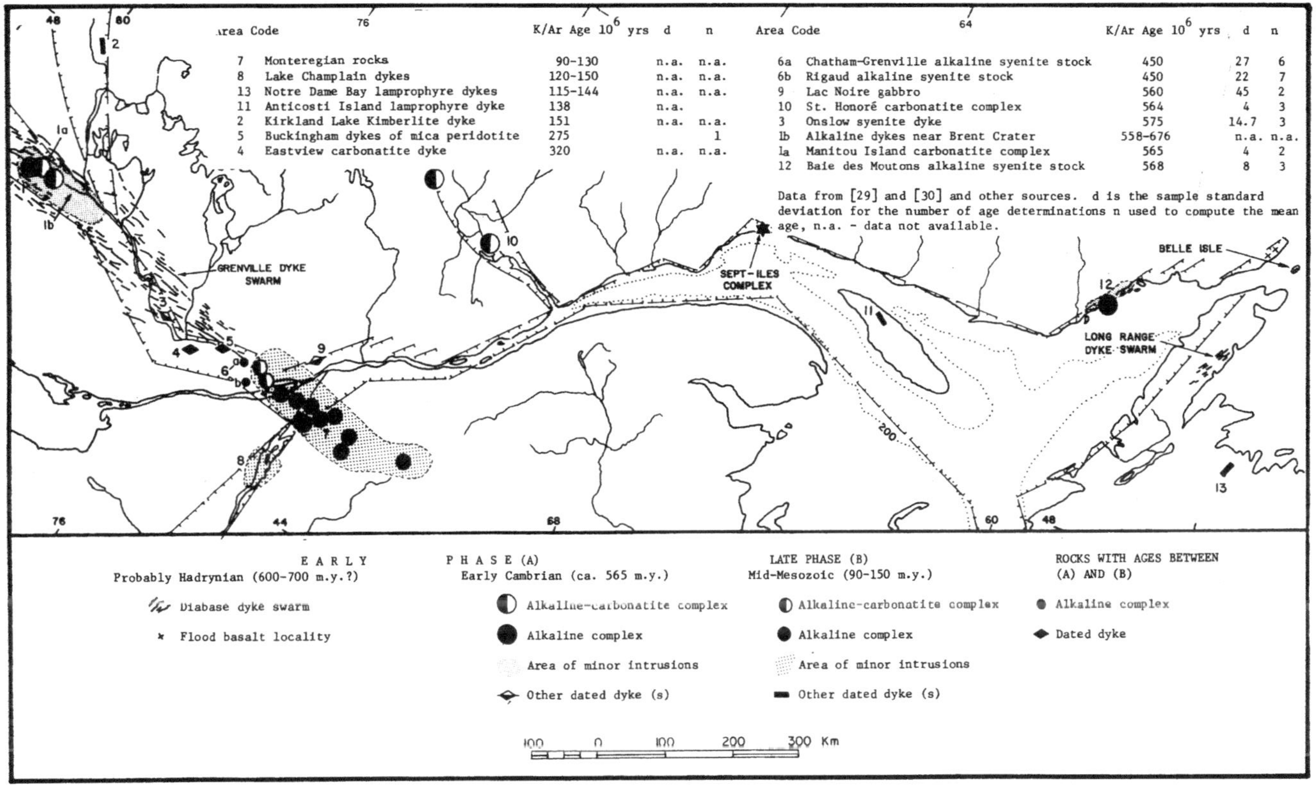


| Area Code | | K/Ar Age $10^6$ yrs | d | n |
|---|---|---|---|---|
| 7 | Monteregian rocks | 90-130 | n.a. | n.a. |
| 8 | Lake Champlain dykes | 120-150 | n.a. | n.a. |
| 13 | Notre Dame Bay lamprophyre dykes | 115-144 | n.a. | n.a. |
| 11 | Anticosti Island lamprophyre dyke | 138 | n.a. | |
| 2 | Kirkland Lake Kimberlite dyke | 151 | n.a. | n.a. |
| 5 | Buckingham dykes of mica peridotite | 275 | | 1 |
| 4 | Eastview carbonatite dyke | 320 | n.a. | n.a. |

| Area Code | | K/Ar Age $10^6$ yrs | d | n |
|---|---|---|---|---|
| 6a | Chatham-Grenville alkaline syenite stock | 450 | 27 | 6 |
| 6b | Rigaud alkaline syenite stock | 450 | 22 | 7 |
| 9 | Lac Noire gabbro | 560 | 45 | 2 |
| 10 | St. Honoré carbonatite complex | 564 | 4 | 3 |
| 3 | Onslow syenite dyke | 575 | 14.7 | 3 |
| 1b | Alkaline dykes near Brent Crater | 558-676 | n.a. | n.a. |
| 1a | Manitou Island carbonatite complex | 565 | 4 | 2 |
| 12 | Baie des Moutons alkaline syenite stock | 568 | 8 | 3 |

Data from [29] and [30] and other sources. d is the sample standard deviation for the number of age determinations n used to compute the mean age, n.a. - data not available.

Fig. 5. Probably rift related, post-Grenville igneous rocks of the St. Lawrence Region and their ages.

as mentioned earlier, probably extend across the Marginal Segment into the Appalachian fold-belt.

The wide scatter of ages of alkaline-carbonatite activity is characteristic not only of the St. Lawrence Region but also of the East African situation [31]. The various igneous events probably reflect episodes of tectonic activity, the two main phases of magmatism in the St. Lawrence Region reflecting tectonic movements of a more general nature along the rift system.

## 10. GEOPHYSICAL CHARACTERISTICS

There are no conspicuous gravity anomalies which can be correlated with the rift zones of the St. Lawrence system, although low amplitude (< 15 mgals) axial lows are present along parts of the Marginal Segment [32] and along the Saguenay graben [33]. The lows in the Marginal Segment can be explained in terms of density characteristics and thickness of platform sedimentary rocks within the rifts, and the gravity low along the Saguenay graben has been explained as resulting from down-faulting of denser rock units within the crust [33]. A regional gravity high similar to the one associated with the Oslo graben [34] is also not apparent in the St. Lawrence Region. Consistent with their long inactivity, the St. Lawrence rifts are not characterized by above average heat flow. Mild seismicity in the St. Lawrence Region appears to be unrelated to rifting processes. The contemporary stresses in the area appear to be horizontal and compressional with the maximum principle stress direction oriented dominantly E to NE [35]. The seismicity is probably a result of release of these stresses along the many faults in the area. Experiments have been conducted to study the parameters of the crust and the upper mantle beneath parts of the St. Lawrence rift system [36,37], but the results of these experiments have not provided a basis for any firm conclusions. Electrical conductivities of the Marginal Segment have revealed that for part of its length, it is the locus of current concentration [38], which is consistent with the deep faulting along this zone.

The main points of the foregoing comparative study are summarized in Table II.

## 11. ORIGIN

Although there is general agreement that rifts form as a response of the crust to tensional stresses, the deep seated causes which give rise to the tensional stresses are far from being understood. These causes may have been different, at least in detail, for different rifted areas. Therefore, the comparative study presen-

TABLE II. Comparison of some selected features of the rift systems of East Africa, the Baikal, the Rhine and of the St. Lawrence

| FEATURE COMPARED | EAST AFRICAN RIFT SYSTEM | BAIKAL RIFT SYSTEM | RHINE GRABEN | ST. LAWRENCE RIFT SYSTEM |
|---|---|---|---|---|
| Length in km | 6500 | 2500 | 300 | > 2200 |
| Common widths of rifts, km | 50 | 50 | 30 | 30-55 |
| Geomorphic expression | Commonly form deep broad valleys with steep edges | | | Shallow (<300 m) valleys and often poorly defined. |
| Age | Tertiary (Miocene or later ?) | | | Formed in two main stages: Ancestral Structures probably in the Hadrynian. The rift system in its present form in Mid-Mesozoic. |
| Tectonic Setting | On crustal swell involving craton. | On crustal swell involving partly craton and partly foldbelt (Transbaikalian) but the rift mainly on craton. | On crustal swell involving craton. | Largely on craton. A crustal swell cannot be discerned. One may have existed during the formation of Ancestral Structures. |
| Gross Trend | N-S, conforms with the general trend of the Precambrian basement. | NE-SW, conforms with the general trend of the Precambrian structures. | N-S. Discordantly superimposed on the Paleozoic and Precambrian structures. | NE-SW. Conforms with the general trends of the Precambrian and Paleozoic structures. |
| Structure | Complex grabens bounded for the most part by step faults and en echelon fault zones, commonly asymmetrical in cross-section and cradle-shaped longitudinally. Vertical displacements up to several thousand metres, largest in the Baikal ( >7000 m). Younger faults in the Baikal and East African regions controlled by Precambrian structures. | | | Shield Segments underlain by complex grabens, Marginal Segment characterized by zones of step-faulting. Rift faults controlled by a NE and NW trending fracture pattern. Mid-Mesozoic structural subsidence is probably no more than 500 m. Total subsidence may be substantially greater. |
| Sedimentary-volcanic fill | Fault troughs partly filled with rift-related (Cenozoic) sediments and volcanics, which in the deeper parts of the troughs are several thousands of metres thick, thickest in the Baikal, 2000 to 6000 m. | | | No rift-related sediments and/or volcanics of Mid-Mesozoic age. Probably some related to the development of Ancestral Structures. |
| Associated igneous rocks | Olivine basalts normal and mildly alkaline; tholeiitic basalts. Also plutonic and volcanic equivalents of ultrabasic alkaline rocks and carbonatites. | Similar to the igneous rocks associated with the East African rift system but greatly undersaturated alkaline rocks and carbonatites are unknown. | Igneous rock assemblages similar to those of the East African rift system. | Tholeiitic and transitional dike swarms, minor flood basalts. Ultramafic alkaline complexes and alkaline-carbonatite complexes. |
| Geophysical Characteristics | Rift zones commonly characterized by negative Bouguer anomalies (maximum about 50 milligals) and sharp axial gravity highs, above average heat flow and shallow focus seismicity. | | | Appears to have little or no geophysical expression except for mild shallow focus seismicity. |
| Contemporary Stresses | Horizontal tension normal and horizontal compression parallel to the rift zones. | | Horizontal compressional stresses of regional extent with the greatest principal stress direction northwest-southeast. | Horizontal compressional stresses of regional extent with the greatest principal stress direction east to northeast. |

ted in this paper would be incomplete unless a statement regarding the origin of the St. Lawrence rift system is included.

Any genetic model of the origin of the St. Lawrence rift system must provide a coherent rationale for the following features of the rift system: (i) The Marginal Segment follows closely the northwestern edge of the Appalachian fold-belt; (ii) during the Hadrynian, the (ancestral) Ottawa graben appears to have extended eastwards to the axial region of the Appalachian fold-belt; (iii) the postulated branch of the rift system along the outer part of the Laurentian Channel extends from a prominent re-entrant of the fold-belt; (v) the first appearance of the Ancestral Structures and the beginning of the Appalachian stratigraphic sequences are nearly contemporaneous (note that the Belle Isle volcanics occur at the base of the Appalachian stratigraphic sequence in the Belle Isle area); (v) the rift system appears to have developed, essentially to its present form, in the mid-Mesozoic.

According to the time relationship listed in (v) above, the present form of the rift system is largely the result of a tectonic event which took place after the Appalachians ceased to be a mobile belt. However, the other relationships are such that they suggest a causal relationship between the initiation of the mobile belt and the first appearance of the Ancestral Structures of the rift system. All recent models of Appalachian evolution contain the idea that geosynclinal activity began in the Hadrynian when a rift system developed somewhere along the present site of the Appalachian eugeosynclinal belt. This eo-Appalachian rift system [3] was set in a continent consisting of North America and Africa then joined together. The rift later widened to form the Iapetos ocean (e.g. Bird-Dewey model [39] or a downwarped zone occupied by a broad inland sea (similar to the present Mediterranean) with a dominantly sialic floor (Chidester-Cady model [40]). Whichever, the case might be, the writer proposes that the Ancestral Structures of the St. Lawrence system can be understood as features inherited from the eo-Appalachian rift system.

The crustal upwarping and distension related to the formation of the eo-Appalachian rift system and further crustal distension related to subsequent widening of the rift led to extensive block faulting of the crustal plates on either side of the main rift. To the N and W of the rift, the block faulting extended as fas as the present limit of the Marginal Segment. Much of this block-faulted terrain was subsequently incorporated into the fold-belt except along the Marginal Segment, where high angle faults remained largely unhealed and susceptible to later reactivation. The eo-Appalachian rift system contained two major triple junctions (Fig. 6) (and perhaps a minor one) which formed over local mantle upwellings or plumes of hot material [41]. An arm of one of the triple junctions extended along the axis of the New England

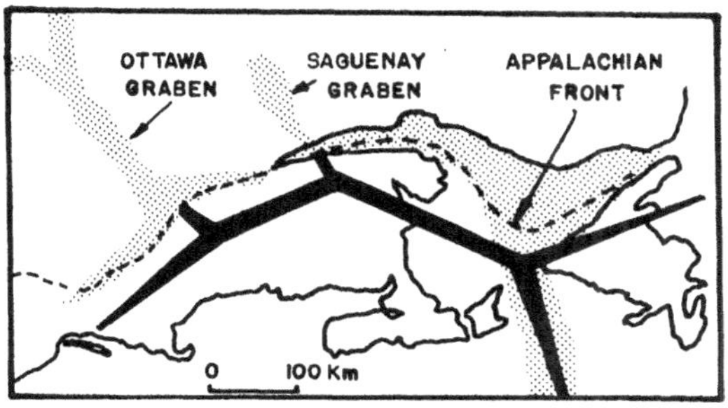


Fig. 6. Sketch illustrating probable relationships between the eo-Appalachian rift system (black) and the St. Lawrence rift system (stippled).

Salient and thence across the southern part of the St. Lawrence Valley and along the Ottawa graben [42]. It later became a failed arm, and the other two arms became parts of the geosyncline. This explains the transverse trend of the Ottawa graben to the fold-belt, and the location of the former at a salient of the latter. The second triple junction lay under the present site of the Gulf of St. Lawrence. One arm of this three-pronged rift extended along the site of the outer part of the Laurential Channel, the other two arms became site of the geosyncline. Although the former lay across the site of the geosyncline, it survived the mountain building processes as a line of structural weakness. The Saguenay graben may have originated as a minor branch of the eo-Appalachian rift system. In the late Hadrynian-early Cambrian times, the Shield area traversed by the Ancestral Structures became eroded to sea-level, to allow the deposition of the Paleozoic platformal cover. Later, in the mid-Mesozoic, possibly as a result of tensional stresses related to the opening of the Atlantic ocean, the unhealed faults along the Marginal Segment and the various failed arms became reactivated and integrated to form the St. Lawrence rift system. The above scheme takes into account all the five requirements mentioned earlier.

## REFERENCES

1. J.A. Jacobs, R.D. Russel and J. Tuzo Wilson, Physics and Geology, McGraw-Hill, 1959.
2. P.S. Kumarapeli and V.A. Saull, Can. Earth Sci., 3, 639, 1966.
3. P.S. Kumarapeli, Geol. Assoc. Can. Spec. Pap., 14, 301, 1976.
4. B. Voight, Am. Assoc. Pet. Geol. Mem., 12, 955, 1969.
5. R.E. Sheridan and C.L. Drake, Can. Earth Sci., 5, 337, 1968.
6. L.H. King and B. MacLean, Can. Earth Sci., 7, 1470, 1970.
7. L.E. Stephens, A.K. Goodacre and R.V. Cooper, Earth Physics Br. Ottawa, Gravity Map Series no. 123, 1971.

8. R.T. Haworth, Can. Geol. Surv. Pap., 74-30, 1975.
9. P.S. Kumarapeli, Can. Min., 10, 421, 1970.
10. H.R. Wynne-Edwards, Geol. Assoc. Can. Spec. Pap., 11, 264, 1972.
11. J.W. Ambrose, Am. J. Sci., 262, 817, 1964.
12. H. Cloos, Geol. Rundsch., 30, 405, 1939.
13. V.V. Beloussov, Am. Geophys. Un. Monog., 13, 539, 1969.
14. W.M. Cady, Geol. Soc. Am., Mem., 120, 1969.
15. A.E. Wilson, Can. Geol. Surv., Mem., 241, 1946.
16. M. Kay, Geol. Soc. Am., Bull., 53, 585, 1942.
17. A.W. Quinn, J. Geol., 41, 113, 1934.
18. P.M. Clifford, Am. Assoc. Pet. Geol., Mem., 12, 647, 1969.
19. D.W. Lewis, Can. J. Earth Sci., 8, 853, 1971.
20. D.F. Strong and H. Williams, Geol. Assoc. Can. Proc. 24, 43, 1972.
21. N.A. Florensov, V.P. Solonenko and N.A. Logachev, 23rd Int. Geol. Cong. Prague, 1968.
22. I.G. Gass, Tectonic and magmatic evolution of the Afro-Arabian Dome, in: African Magmatism and Tectonics, ed. by T.N. Clifford and I.G. Gass, Oliver and Boyd, 1970.
23. W.F. Farhig, Can. Geol. Surv., Econ. Geol. Rept., 1, 131, 1926.
24. V. Stukas and P.H. Reynolds, Earth Planet Sci. Lett., 22, 256, 1974.
25. G.S. Murthy, Can. J. Earth Sci., 8, 802, 1971.
26. K.L. Currie, Can. Min., 10, 411, 1969.
27. A.R. Philpotts, The Monteregian Province, in: Alkaline Rocks, ed. by H. Sørensen, Wiley Interscience, 1974.
28. M.D. Higgins and R. Doig, Nature, 267, 40, 1977.
29. R. Doig and J. Barton Jr., Can. J. Earth Sci., 5, 1401, 1968.
30. R. Doig, Can. J. Earth Sci., 7, 22, 1970.
31. B.C. King, Vulcanicity and rift tectonics in East Africa, in: African Magmatism and Tectonics, ed. by T.N. Clifford and I.G. Gass, Oliver and Boyd, 1970.
32. Canada Dept. Energy, Mines, Resourc., Ottawa, Map GMS74-1.
33. P.M. Duncan and G.D. Garland, Can. J. Earth Sci., 14, 145, 1977.
34. I.B. Ramberg, Nor. geol. Unders., 325, 1, 1976.
35. M.L. Sbar and L.R. Sykes, Geol. Soc. Am. Bull., 84, 1861, 1973.
36. D.S. Rankin, R. Rawindra and D. Zwicker, Can. J. Earth Sci., 6, 771, 1969.
37. D.J. Hearty, R.F. Mereu and C. Wright, Can. J. Earth Sci., 10, 2284, 1973.
38. R.C. Bailey, R.N. Edwards, G.D. Garland, R. Kurtz and D. Pitcher, J. Geomag. Geoelec., 26, 125, 1974.
39. J.M. Bird and J.F. Dewey, Geol. Soc. Am. Bull., 84, 1031, 1973.
40. A.H. Chidester and W.M. Cady, Nature Phys. Sci., 240, 27, 1972.
41. W.J. Morgan, Am. Assoc. Pet. Geol., Bull., 56, 203, 1972.
42. K. Burke and J. Dewey, J. Geol., 81, 406, 1973.

# SOME PROBLEMS OF RIFTING DEVELOPMENT IN THE EARTH'S HISTORY

E.E. Milanovsky

Moscow State University, Moscow, USSR

## 1. INTRODUCTION

During the last 15-20 years our knowledge of the Earth's rift zones has widened and become more profound. It has become evident that rifting is a global geotectonic process which, at least in the Cenozoic, has equalled the geosynclinal process in importance. The theory of rifting is developing rapidly. Some general regularities of the structure and development of rift zones have been revealed, and a structural-genetic classification has been proposed [1,2]. However, a number of important rifting problems remain and is the subject of considerable debate. Some of these problems, in particular those of rifting development in the Earth's history and their correlations with volcanic history and geosynclinal processes are briefly considered in this communication. It must be emphasized that this review is based upon the study of continental rift zones as the structure and geological history of these are much better known and understood than that of oceanic rift zones. Before we turn to the main subject a few words must be said about the rift zone classification accepted by the author and terminology used by him.

According to the character of the deep structure of the Earth's crust the rift zones of the Earth may be divided into four basic categories:

1) continental (or intracontinental) rift zones with crust of continental type, although somewhat thinned, fragmented, anomalously heated and underlain by a lens of the upper mantle material with lower density;

*I. B. Ramberg and E.-R. Neumann (eds.), Tectonics and Geophysics of Continental Rifts, 385-399.*

2) intercontinental where the axial part of the rift has a crust similar to that of oceanic rift zones, the peripheral parts have somewhat thinned and reworked continental crust, and the shoulders have typical continental crust (rift zones of Red Sea, Gulfs of Aden and California);

3) oceanic (or intraoceanic) where both the axial rift valley and the flanks of the rift zone have crust similar to oceanic crust, but are underlain by a "pillow" of anomalously heated mantle material with lower density;

4) pericontinental (or perioceanic) rift zones differing from all the preceding types in having sharply asymmetric structures: such rift zones are bounded by very different areas as they have continental type crust on one side and oceanic type on the other. At present such rift zones exist mainly in a passive buried state (along the periphery of continents bordering the Atlantic and Indian oceans), they were active during the Mesozoic and early Cenozoic.

Intercontinental rift zones are evidently the result of the development of continental rifting to a point where total severance of the continental crust has occurred. Only some oceanic rift zones (e.g. the Mid-Atlantic rift belt) may be assumed to result from continued development of an intercontinental rift zone. Others (e.g. the East-Pacific rift belt) are younger than the oceanic space containing them and were undoubtedly initiated in the oceanic crust.

Pericontinental rift zones occur in the process of horizontal extension at an early stage of oceanic trough formation. However, the wide-spread view that these zones are "halves" of continental rifts separated far from each other by the "opening" of the ocean seems unlikely. It is more probable that such zones developed on opposite flanks of intercontinental (subsequently oceanic) rift belts on the margins of continental blocks framing them with fragmented thinned and partially reworked crust [3].

The following basic criteria must be taken into consideration when classifying modern continental rift zones; a) The peculiarities of tectonic position, basement structure and preceding geological history specific to the area which became the scene of rifting; b) the character of the tectonic structures generated in the process of rifting and the regularity of their formation and c) the role, scope and character of magmatic processes accompanying, or sometimes preceding rifting.

On the basis of (a), continental and intercontinental rift belts and zones may be divided into two groups; 1) rift belts and zones of platforms (or epiplatform) where rifting started after a fairly prolonged (200-500 m.y. or more) stage of platform development and 2) rift belts and zones of young folded structures (or epiorogenic) where rifting immediately succeeded the completion of the geosynclinal development (i.e. the orogenic stage) or was even combined with phenomena peculiar to epigeosynclinal orogenesis. The epiplatform rift belts are characterized by rift zones with large single axial grabens and subalkaline or alkaline products of accompanying volcanism. Sometimes alkaline ultrabasites and carbonatites are present. In contrast, epiorogenic rift belts and zones are characterized by combinations of numerous narrow grabens, horsts and unilateral blocks (e.g. Basin-and-Range Province in Cordillera, USA). Volcanic formations in them are of calc-alkali type.

Continental epiplatform rift zones occur mainly on outcrops of folded platform basement which have been subjected to continuous stable uplift before rifting. These zones are seldom found in areas of platform cover development. They are mostly associated with areas of Late-Proterozoic folding or tectono-magmatic regeneration, being superimposed on, or "adjusted" to, previous structural inhomogeneities of the substratum such as longitudinal faults or strike direction of fold zones. As a rule they evade areas of Archean consolidation with development of granite-gneiss domes and other isometric structures. The latter serve as "the frame" of rift belts or form median massifs within them (as for example the Victoria massif in the southern part of the African-Arabian rift belt). During the period of platform development "weakened" zones of the fundament may be repeatedly activated. They serve either as zones of increased permeability facilitating the rise of magmatic melts and emplacement of intrusions, or as zones of fault and graben development. Two extreme types may be distinguished among epiplatform rift zones. They differ significantly in the character of their structures, the relative role of volcanism and formation history. Several transitional types between the two extremes have been recognised.

a) Rift zones of the arch-volcanic type (e.g. Kenyan and Ethiopian rift zones in Africa) are characterized by intense and long sub-aerial volcanic activity. Volcanism appearance is preceded by steady growth of an oval arch-uplift. Subsequently a rather shallow axial graben develops in the "weakened" axial zone.

b) Rift zones of the crevice type are characterized by grabens of greater depth, that is 3-4 km (the Upper Rhine rift) and even 5-7 km (the Baikal rift). The zones of marginal uplifts are here much narrower, and may be discontinuous or totally absent. Within the rift, however, inner narrow zones of uplift sometimes occur (Ruwenzori in the Tanganyika rift zone). Volcanism in crevice zones is either relatively weak, local, and of episodic nature or totally absent. The formation of crevice zones begins with the formation of long, narrow grabens. Marginal and inner uplifts appear later, creating the illusion of arch collapse. Considerable heat leakage seems to be associated with the grabens of crevice rift zones.

Table I. Main categories and types of rift zones (R.Z.) and rift belts (R.b.)

| categories R.Z. and R.b. according to their deep structure | epiplatform R.Z. | | | | epiorogenic R.Z. | | |
|---|---|---|---|---|---|---|---|
| | types of R.Z. | predominant structures and movements | volcanism: intensity | volcanism: composition | types of R.Z. | predominant structures and movements | volcanism |
| continental (intracontinental) R.Z | arch-volcanic R.Z. | oval arch uplifts complicated by relatively shallow axial, sometimes also transverse grabens. Very weak extension of the crust. | very intense preceding and accompanying all stages of rifting | volcanic and subvolcanic rocks of subalkaline and alkaline series (from ultrabasic to acid, predominantly basic) and carbonatites. | continental epiorogenic R.Z. (in the inner or marginal parts of continents). | very long linear grabens and horsts, "keyboards" of blocks, systems of one-side tilted blocks. General considerable "diffuse" extension and attenuation of the crus . | usually powerful wide-spread subaerial volcanism mainly on the early stages of rifting. Volcanic rocks of calc-alkali series (acid and later basic). |
| | R.Z. of crevice type: in the areas of general uplift | deep grabens often bordered from one or two sides by marginal uplifts and sometimes complicated by inner uplifts. Relatively greater extension of the crust. | weak, sporadic, episodic volcanism, on the different stages of rifting (little volcanic R.Z.). | volcanism is absent (non-volcanic R.Z.) | | | |
| | R.Z. of crevice type: in the areas of general subsidence | deep grabens without the marginal uplifts | | | | | |
| inter-continental R.Z. | intercontinental R.Z. with the platform "frame" (interplatform R.Z.). | very deep and wide "step-like grabens" with the separation of the blocks of continental crust in the axial zone. Very considerable extension. | on the flanks of R.Z.-relatively weak and sporadic, in the axial zone--more intense(?) volcanism, hydrotherms and high heat flow. | basic rocks of calc-alkali series in the axial zone, subalkaline and alkaline on its ends and flanks. | intercontinental R.Z. with the frame of young folded structures (interorogenic R.Z and R.b.). | the same as in the intercontinental R.Z. with the platform "frame". | calc-acid volcanism. |
| pericontinental R.Z. and R.b. | | subdivision is not given | | | | | |
| oceanic (intraoceanic R.Z. and R.b.) | | | | | | | |

## 2. DEVELOPMENT OF RIFTING PROCESSES WITH GEOLOGICAL TIME

The earliest "roots" of rifting extend into the distant past; structures similar to recent continental rift zones have existed since the Late Proterozoic, while related but not identical structures appeared in the Early-Middle Proterozoic period. However, in the Mesozoic and particularly in the Cenozoic, the rifting activity increased markedly, and rifting phenomena were much more widespread, whereas there was a corresponding decrease in the total area of geosynclinal belts and formation of "secondary" "secondary" oceanic depressions. The relative importance of the various categories and types of rift zones and the tectonic environments in which they originated and developed have also changed considerably in the course of geological history (Fig. 1).

In Late Proterozoic the most ancient structures morphologically similar to the Cenozoic continental epiplatform crevice rift zones, the so-called aulacogenes, began to appear. But unlike the Cenozoic continental rift zones many Riphean aulacogenes were somehow tectonically related to the Baikal geosynclinal belts surrounding these platforms. Most of them were their offsets on, or parallel extensions to their margins. Some of the most "mature" aulacogenes lay between platforms and miogeosynclinal troughs and were subjected to folding of moderate intensity. Aulacogenes are not recorded from the Early and Middle Proterozoic periods, but at this time within the consolidated cores of some future platforms the so-called paleo-aulacogenes started forming. These were narrow superimposed graben-like troughs whose development was accompanied by magmatic activity and crowned by folding. Many Phanerozoic rifting areas notable for their peculiarities of development have been subjected to processes of tectonomagmatic activization and repeated magmatism since the Early of Late Proterozoic.

In the Paleozoic the tectonic conditions under which rift zones were formed became more varied. In the Middle Paleozoic (Devonian) some aulacogenes on the ancient platforms were reactivated.

In Early and especially in Middle Paleozoic (Cambrian and Devonian) the first epiorogenic rift zones overlying the young folded epigeosynclinal areas, came into being (for examples in the Baikal folded belt of East Siberia and in the Caledonian folded belt in the North Atlantic). On the platforms of some southern continents the first rift zones appeared in Early-Middle Paleozoic (Australia) and Late Paleozoic (India, Madagascar, S.E. Africa). Their formation seems to be connected with the early stages of Gondwana supercontinent destruction.

In the Triassic-Jurassic the rifting areas continued to

widen and rifting intensity increased. Simultaneously a long epoch of strong trappean magmatism began on the continents, and from the Middle Jurassic an even greater activity of basaltic volcanism may be traced in the oceans. Continental rift zones "opening" towards the Indian Ocean trough and pericontinental rift zones parallel to its margins began to develop on the Hindustanian, Australian and African platforms.

In the Late Triassic the epiorogenic rift system on the western shore of the Atlantic came into being. It is superimposed on the Hercynian folded area of the Appalachians. In the Jurassic powerful basaltic eruptions took place in the marginal zones of the Central Atlantic Ocean. In the Early Triassic a grandiose trappean volcanism took place on the ancient Siberian platform, in the Taimyr and in the Paleozoic folded area of West Siberia. The epiorogenic West Siberian rift system possibly extending as far north as the Arctic Ocean was formed in the Triassic-Lias. During the Mesozoic a system of grabens confined to zones of long repeated tectono-magmatic activity developed in Zabaikalye. Both these rift systems were subjected to some compressional deformations at the end of their development.

In the Cretaceous-Paleogene the rifting and trappean magmatism spread over many areas adjacent to the South and North-Atlantic and continued to spread along the periphery of the Indian Ocean.

Trappean eruptions occurred on the Hindustanian platform during the Cretaceous period, being especially intensive in the Early Paleogene. Later the Cambay-Bombay rift zone was formed in its western margin. An early stage of the Aden rift zone formation is assigned to the Jurassic-Cretaceous, while the origin, or rather regeneration of the Red Sea rift zone may be dated to the Late Cretaceous-Paleogene period. In the Late Cretaceous-Paleogene powerful eruptions of alkaline basalts occurred on Ethiopian territory, which was subjected to epiplatform rifting in the Neogene. During the Early Cretaceous numerous continental and pericontinental rift zones were formed on the margins of the South American and African platforms adjoining the South Atlantic. Basaltic eruption took place on the bottom of the marginal zones of the Southern and Central Atlantic.

The intercontinental rift system of Baffin Bay was formed (or rather regenerated), and the intracontinental rift system of St. Laurence in East Canada became active in the Cretaceous. In the Paleogene the northernmost part of the Atlantic Ocean became an area of voluminous basalt eruptions. Simultaneously the Rhine-Liberyan rift zones appeared. These crossed both the platforms and the Mediterranean geosynclinal belt and continued to develop in the Neogene-Anthropogene.

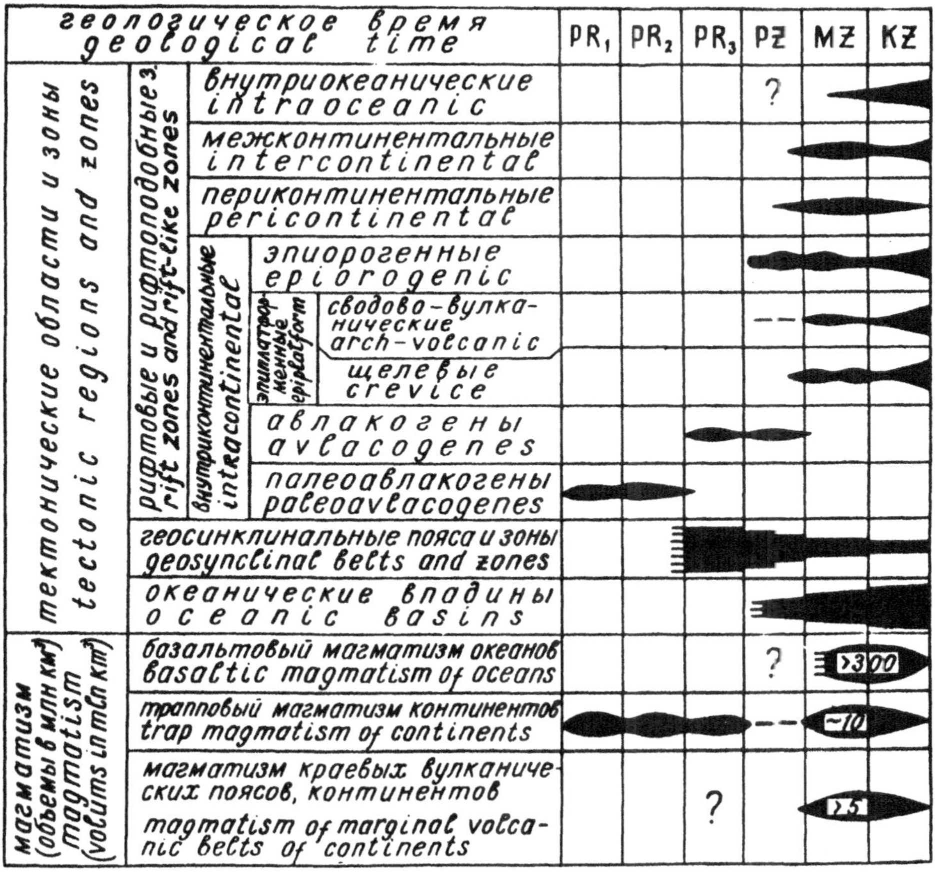


Fig. 1. Distribution of some types of tectonic regions and zones, and magmatic processes in the Earth's history.

In the Late Paleogene-Neogene-Anthropogene the African Arabian rift belt was formed. This included a number of intracontinental epiplatform as well as the two intercontinental rift zones. At the same time the isolated Baikal epiplatform rift belt appeared in Siberia and an epiorogenic rift belt appeared in western North America, superimposed on the Mesozoic folded area of the Cordillera. Like the Afro-Arabian rift belt these represent offsets of intraoceanic rift ranges. The northern segment of the Mid-Atlantic rift belt crossing Iceland was not formed until the Neogene-Anthropogene. It started developing in the North Atlantic trappean basin.

The greatest activation of continental and oceanic (?) rifting in the Earth's history possibly occurred in the Late Cenozoic (Pliocene-Anthropogene) following the main paroxysm of alpine folding (Oligocene-Miocene). At the same time the importance of pericontinental rift zones as well as that of basaltic volcanism in the oceans decreased.

Trappean volcanism on the continents is absent from the end of the Eocene.

In comparison with the Paleozoic the Mesozoic-Cenozoic was in general a period of more powerful manifestations of rifting, with the formation of intracontinental (epiplatform and epiorogenic) and numerous pericontinental, intercontinental and intraoceanic rift zones and belts. Their development was closely connected with the formation of the depressions of the Atlantic, Indian and Arctic oceans and with the "rejuvenation" of the Pacific Ocean floor, and was accompanied by intense volcanic eruptions both in the oceans and on the continents.

## 3. CONNECTION OF RIFTING WITH NON-GEOSYNCLINAL VOLCANISM

Figs. 1 and 2 show that variations with time in the intensity of rifting can be directly correlated with variations in both continental and oceanic non-geosynclinal volcanism. Such a correlation may be naturally assumed a priori, since both phenomena take place under similar geodynamic conditions - horizontal extension of the Earth's crust. The most important types of non-geosynclinal volcanism are as follows: 1) trappean volcanism of platforms, 2) alkaline-basic volcanism of tectonomagmatic activation areas, 3) essentially intermediate-acid sialic volcanism of marginal volcanic belts and, finally, 4) basaltic volcanism of oceans which in volume exceeds significantly all the other types of volcanism on the Earth.

If we confine ourselves to consideration of the Phanerozoic volcanism we shall see that the first three types appear on the Earth almost exclusively in the Meso-Cenozoic. They are almost unknown in the Paleozoic, being recorded only in the Devonian which was also characterized by a noticeable upsurge of epiplatform and epiorogenic continental rifting. However, the scale of continental volcanism in the Devonian was much lower than that in the Mesozoic or Cenozoic.

The total volume of products of trappean Meso-Cenozoic volcanism of platforms comprises $10\text{-}15 \cdot 10^6$ km$^3$, in marginal volcanic belts it is not less than $10 \cdot 10^6$ km$^3$. The volume of products of alkaline-basic volcanism of activated areas, including that in epiplatform rift zones, is several millions of km$^3$. The total volume of all products of non-geosynclinal Meso-Cenozoic volcanism on continents is therefore less than $25\text{-}30 \cdot 10^6$ km$^3$.

All the basaltic volcanism in oceans known so far occurred in the Mesozoic (Jurassic, Cretaceous) and Cenozoic. The total volume of the products may be estimated to be minimum $300 \cdot 10^6$ km$^3$, the average thickness of the second seismic layer in the oceans being about 1.5 km. This means that this volume exceeds that of synchronous non-geosynclinal volcanism of the continents by at least one order of magnitude. The scale of Meso-Cenozoic rifting in the oceans is thus many times greater than that on the continents.

If, during the Phanerozoic, there has been direct correlation between continental rifting intensity and continental non-geosynclinal volcanism, it is natural to assume a similar correlation to have existed in the oceans as well. Therefore we may assume that the intensity of rifting and volcanism within the oceans also increased sharply in the Mesozoic and Cenozoic (as compared to the Paleozoic) when the formation of the Atlantic, the Indian and the Arctic Ocean basins with their rift belts and the "rejuvenation" of the Pacific ocean floor took place. The sharp increase in non-geosynclinal volcanism on the continents and, probably also in the oceans, and the formation of the latter in the Mesozoic and Cenozoic undoubtedly indicate a radical redistribution of energy resources over the Earth's surface. It probably also signifies an essential change in the total energy regime of our planet in the Mesozoic and Cenozoic in comparison with the Paleozoic.

## 4. CORRELATIONS BETWEEN RIFTING AND GEOSYNCLINAL PROCESS

During the Cenozoic rifting and geosynclinal processes are represented virtually directly opposing mechanism. Rifts and geosynclinal belts are zones of maximum tectonic and magmatic activity. In rift belts deep material rises to the surface and horizontal extension of the Earth's crust takes place. In geosynclinal belts there is submersion of material and horizontal compression,

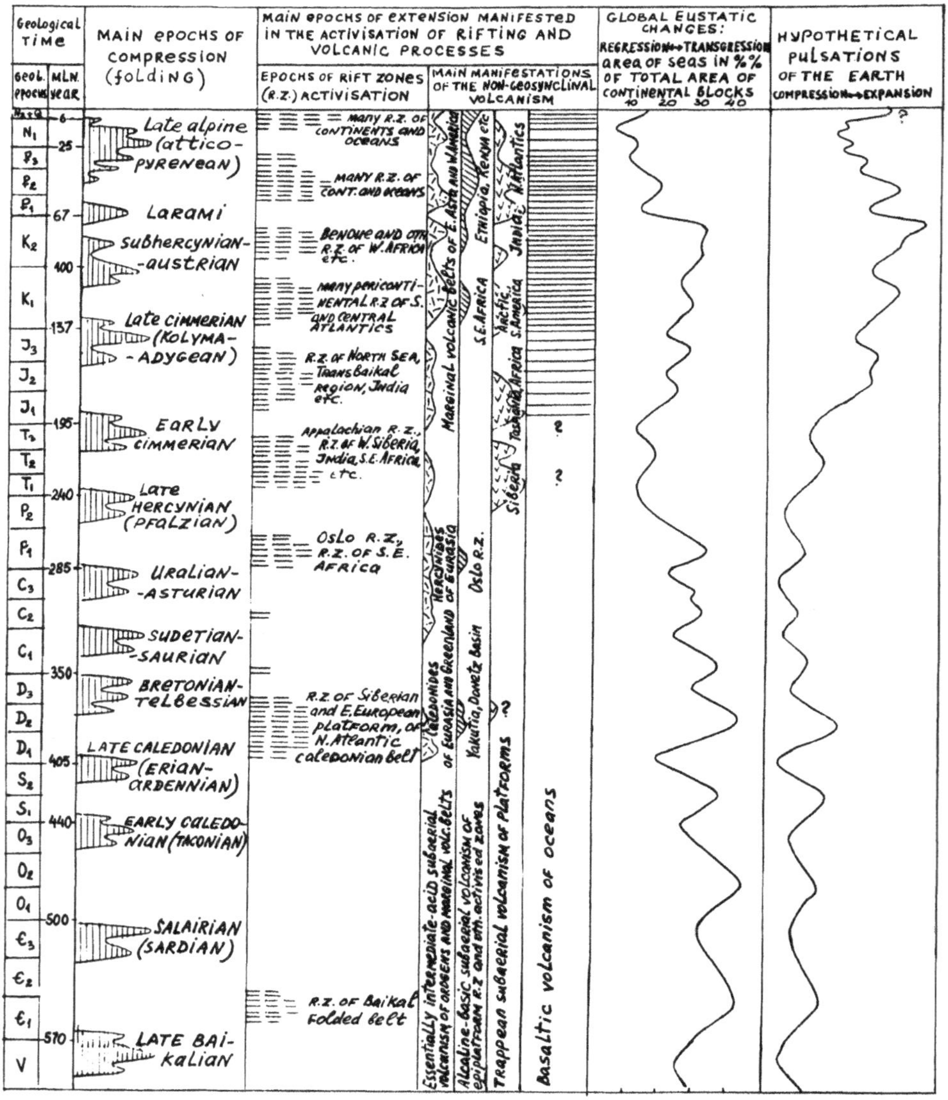


Fig. 2. Correlation of the global epochs of tectonic deformations and volcanism, global eustatic cycles and hypothetical pulsations of the Earth during Phanerozoic. Compiled by E.E. Milanovsky, 1977.

contraction of the Earth's crust prevails. On a global scale these processes approximately compensate each other. In the past, however, such a compensation cannot have taken place since the relative importance of the geosynclinal process has decreased successively in the course of the Earth's history, whereas that of rifting has increased. If we trace the geological history backwards we also see that the radical differences between rifts and geosynclinal zones grow gradually weaker and finally disappear. It must be noted that horizontal extension phenomena are not confined to rift zones alone, but are also observed in geosynclinal zones. The converse is also true, since compression phenomena have been identified in rifts as well as in geosynclinal zones. In the development of all geosynclinal belts, and in eugeosynclines in particular, phases of essential extension with generation of oceaníc type crust have taken place. Such phases are recorded especially in the so-called ophiolite zones of eugeosynclines. Later, but long before the end of the geosynclinal "cycle", the latter were subjected to the strongest compression and crushing with formation of tectonic mélange. Essential extension may also occur during the late orogenic phases of development of geosynclinal belts (e.g. in the Mediterranen belt during the Pliocene-Anthropogene). On the other hand, many pre-Cenozoic rift zones underwent compression phases with the formation of linear folds and overthrust structures. Examples of such deformations can be found in the Cretaceous rift of Benoue in Africa, the Mesozoic (Jurassic, Cretaceous) grabens of Transbaikal area, the Triassic grabens of Western Siberia, the Middle Paleozoic Donetsk rift (or aulacogene), subjected to folding late in the Paleozoic, the Devonian graben of Spitsbergen, the Cambrian grabens of the Baikal folded area, etc. There was also a compression phase in the history of the Oslo graben at the very beginning of the Devonian. Compression phases are also thought to have taken place in the history of development of oceanic rift belts, in the Mid-Atlantic, in particular [4]. Upper Proterozoic and, partially, even Paleozoic continental rift zones (the so-called aulacogenes) were structures intermediate in nature between the platform and miogeosynclinal depressions. The largest and most mature among them branched off from geosynclinal belts and intruded into the body of adjacent platforms. These inherited the position and strike of the Proterozoic folded zones in their basement. Still closer (related) to geosynclinal zones are the so-called paleoaulacogenes - narrow linear volcanically active depressions of Early and Middle Proterozoic age - were revealed in the structure of the Siberian and East European platform. Thus, the most ancient - Proterozoic analogs of continental rift zones are structures close to underdeveloped epicraton geosynclinal depressions. This indicates common historic "roots" for geosynclinal and rift zones. The situation and orientation of most Proterozoic and Phanerozoic rift zones has been determined by the structural pattern of their consolidated epigeosynclinal substratum. This also confirms the relationship.

In an earlier time - in the Archean - there were evidently no rift zones differing in principle from geosynclinal zones. Numerous but relatively narrow protogeosynclinal zones in this epoch of the Earth's history were separated not by vast platforms but by relatively small consolidated massifs similar to the median massifs in the geosynclinal belts of the Phanerozoic. Therefore those tectonic tendencies which were more "polarized" later on into geosynclinal and rift areas were also typical of the Archean and Early Proterozoic mobile protogeosynclinal belts.

## 5. PERIODICITY OF RIFT PROCESS DEVELOPMENT AND ITS CORRELATIONS WITH PERIODICITY OF GEOSYNCLINAL DEVELOPMENT

As the study of rift zones and belts has progressed, it has become more and more evident that their development is a complicated and long process. The duration may be comparable with the development of geosynclinal belts. At the same time it is now clearer that this development is not regular in time. There are relatively short phases of strong rifting activation alternating with periods of "weakened" activity. Such an irregularity of rift zones development in time was demonstrated recently by Pilger and Rösler [5]. They used the Afro-Arabian rift system as an example while Kazmin [6] described such a periodicity of rifting as a more general phenomenon. This irregularity in development of rift zones in time may be observed clearly in the Rhine rift system, the rift zones of Iceland, etc. It must be emphasized that the variations are not only in intensity of horizontal extension (which at certain moments may be even replaced by some compression), but also in orientation of the dynamic field. The latter type of variation was well described by Illies and Greiner [7] for the Rhine rift system.

The author's research shows that the phases of extension are not restricted to particular rift zones but common for whole rift belts and evidently for the whole rift system. Thus periodicity in rift belt development is a phenomenon similar (though opposite in its main kinetic tendency) to that of geosynclinal belt development. The latter is indicated by the existence of global phases and epochs of folding as stated by Stille [8]. In this case the so-called folding (compression) phases in geosynclinal areas, and extension (taphrogenesis) phases in rift zones represent different examples of the same general global irregularity and periodicity of geodynamic activity in the course of the Earth's development. If this is true a question arises - are the phases of folding (horizontal compression) in geosynclinal belts simultaneous with those of taphrogenesis (horizontal extension) in rift belts or, do they, on the contrary, alternate with each other. Many researchers consider these phenomena to be simultaneous (e.g. [9]). This view agrees well with the "plate tectonics" concept assuming mutual relation and kinematic "equilibrium" between spreading in some zones of the Earth and collision of lithospheric plates (subduction, obduction, etc.) in others. The assumption that the Earth's radius is constant is a natural consequence of the "plate tectonics" concept [10]. There are,however, a number of facts which do not agree with the idea of temporal coincidence between compression and extension paroxysms in the different tectonically active zones of the Earth. It has for example already been stated that, with the exception of recent ones, rift zones of different age demonstrate that periods of extension (prevailing on the whole) alternate with periods of compression. Periods of slight compression or no extension in rift zones correlate in time with folding phases in geosynclinal belts. It has also been established that quite significant crustal extensions have occurred in geosynclinal areas. In particular, these are phases of regeneration of the oceanic type crust in ophiolite zones like that in the Alpine Mediterranean belt in the middle of the Mesozoic. Another example is the Pliocene-Quaternary epoch of extension in the Southern Mediterranean part of the Alpine geosynclinal belt which followed the epoch of intensive fold-overthrust deformations in this belt in the Miocene. At the same time the Pliocene-Quaternary epoch was characterized by a sharp intensification of rifting process in numerous rift zones, continental (Baikal and others), intercontinental (the Red Sea rift) and oceanic. Thus, according to Steiner [11] the extension rate in Median rift zones of the Atlantic, Indian and Pacific oceans in the Pliocene increased considerably as compared to the Miocene. Epiorogenic rift zones occur within the young folded epigeosynclinal areas as a result of extension. They were subjected to this extension after considerable compression which created their folded structure. Extension phases in rift zones coincide in time with world wide phases of volcanism activity. The latter took place in the Devonian, Mesozoic and Cenozoic. The whole Meso-Cenozoic epoch of rifting intensification is an epoch of marked reactivation of non-geosynclinal continental and evidently oceanic volcanism.

An attempt to correlate directly in time the main global epochs of folding (compression) increase in geosynclinal belts and extension in rift zones is made in Table 2 which illustrates the lack of coincidence and marked tendency to alteration of these phenomena. On the whole extension weakening in rift zones (which may even invert to slight compression) corresponds to phases of intensification of compression (folding) in geosynclinal belts. Conversely, strong extension activity in rift zones is accompanied by compression decrease in geosynclinal belts and even by the appearance of extension structures in them, an example being many orogenic belts in the Pliocene-Anthropogene. Global epochs of extension intensification are synchronized with the largest positive phases of sea level eustatic fluctuations ("world transgressions"), while the epochs of marked compression correlate with the negative phases of the latter ("world regressions").

The concept briefly stated above [12] naturally requires the assumption of the alternation of general compression and extension periods and epochs in the course of the Earth's development. This brings us back to the idea of the existence of global pulsations of the Earth. Already 40 years ago that was expressed by an American geologist,Bucher [13], and later by the Soviet geologists Obrutchev [13] and Ussov [15]. Periods of general compression on the Earth are global epochs of folding intensification and rifting when volcanism decrease. General extension periods are periods of folding attenuation, rifting activation and massive volcanism. The rate of pulsation of the Earth during the Phanerozoic is approximately 40-50 m.y. Together with these rhytmic oscillations it is possible to detect a general trend of expansion of the Earth in the Mesozoic and Cenozoic compared with the Paleozoic. It is indicated by the sharp increase in rifting and volcanism that involved a large part of the planet's surface at that time, by the formation of modern oceans and by the successively diminishing role played by geosynclinal belts in the general structure of the Earth.

The idea of some expansion of the Earth in the Phanerozoic put forward originally by Hilgenberg and Holm in the thirties looks rather likely from a geological point of view. It is therefore not accidental that it has acquired ever growing support of present day geologists (B. Heezen, S. Carey, B. Cirič, J. Steiner, N.P. Kropotkin, M.V. Muratov, Yu. V. Chudinov and others). According to the data on volcanism and rifting development the main epoch of Earth expansion covers the Mesozoic and Paleogene. Apparently it was suspended or even stopped in the Oligocene and, especially, in the Miocene but renewed again in the Pliocene-Anthropogene.

REFERENCES

1. E.E. Milanovsky, Tectonophysics, **15**, 65, 1972.
2. E.E. Milanovsky, Rift zones of continents, Nedra, Moscow, 1975 (in Russian).
3. W.L. Van der Linden, Mar. geophys. Res., **3**, 209, 1977.
4. A.V. Peive, The tectonics of the Middle Atlantic ridge, Geotectonica, **5**, 3, Moscow, 1975 (in Russian).
5. A. Pilger and A. Rösler, Temporal relationships in the tectonic evolution of the Afar depression (Ethiopia) and the adjacent Afro-Arabian rift system, in: Afar between continental and oceanic rifting, ed. by A. Pilger and A. Rösler, E. Schweizerbart'sche, Stuttgart, 1976.
6. V.Q. Kazmin, Epochs of riftogenesis and some problems of origin of rift structures, Geol. Geophiz., **9**, 3, Novosibirsk, 1975 (in Russian).

7. H. Illies and G. Greiner, Oberrhein. Geol. Abh., 25, 1, 1976.
8. H. Stille, Abh. Preuss. Geol. Landesanstalt, N.F.H. 95, 110, 1925.
9. W. Schwan, Clausthales geol. Abh., 17, 1, 1974.
10. X. Le Pichon, I. Francheteau and I. Bonnin, Plate tectonics, Elsevier Sci. Publ. Comp., Amsterdam, 1973.
11. I. Steiner, Geology, 5, 312, 1977.
12. E.E. Milanovsky, Some regularities in tectonic development and volcanism of the Earth during phanerozoic (The problems of pulsations and expansion of the Earth), Geotectonica, 5, Moscow, 1978 (in Russian).
13. W.H. Bucher, The deformation of the Earth crust, Princeton Univ. Press, 1938.
14. V.A. Obructhev, The pulsation theory of geotectonics, Izvestia Akad. Nauk USSR ser. geol., 1, 12, 1940, (in Russian).
15. M.A. Ussov, Structural geology, Gosgeolizdat. Moscow, 1940, (in Russian).

# RIFT VALLEYS ON EARTH, MARS AND VENUS

C.A. Wood and J.W. Head

Department of Geological Sciences, Brown University, Providence, Rhode Island 02912, U.S.A.

ABSTRACT. Rift-like structures observed on Mars and Venus are grossly similar to their terrestrial counterparts. Rifts on both Mars and Venus cut regional uplifts and the martian rift is associated with extensive volcanism. The major differences between terrestrial and extra-terrestrial rifts are scale and abundance: the martian rift is ~2200 km long and 90-120 km wide and the venusian trough has a length of ~1400 km and a width of ~150 km. These large dimensions suggest that the crusts of the two planets may be significantly thicker than the Earth's crust. Only one major rift is seen on Mars, and only one has been detected to date on Venus. A lack of fold mountains, strike-slip faults, and subduction zone style volcanoes on Mars, and preservation of ancient cratered terrain, imply that unlike the present-day Earth, the martian rift was not formed as a result of plate tectonics. The possibility of additional rifts on Venus, and the presence of ancient cratered terrain, suggest that Venus may be analogous to the earliest stages of terrestrial rifting and plate tectonics.

## 1. INTRODUCTION

Mariner 9 photography of Mars and radar imagery of Venus have revealed giant troughs with morphologies grossly similar to terrestrial rift valleys. The existence of rift-like structures on three planets provides a fascinating opportunity to compare morphologically similar structures formed in totally different environments and possibly by different processes.

*I. B. Ramberg and E.-R. Neumann (eds.), Tectonics and Geophysics of Continental Rifts,* 401-408.

## 2. CHARACTERISTICS OF RIFTS

Earth. Milanovsky [1] classified terrestrial rifts into 3 fundamental categories: (1) oceanic - formed entirely of oceanic crust, e.g., Mid-Atlantic Ridge; (2) intercontinental - the rift is floored by oceanic crust but its shoulders are continental crust, e.g., Red Sea; (3) continental - both rift floor and shoulders are of continental crust, e.g., Kenya Rift. It is widely accepted that continental rifts are precessors of intercontinental rifts (see for example, Burke and Dewey [2]), while oceanic rifts can be considered simply as long active intercontinental rifts. For reasons discussed below it appears that neither oceanic crust nor extensive plate tectonic processes have acted on Mars or Venus, thus only continental rifts will be considered here.

Graben-like continental rifts were further subdivided by Milanovsky into two types. Arch-volcanic type rifts, exemplified by the Kenya rift (Figs. 1 and 5), develop along the axial part of crustal swells and have extensive associations of basic and intermediate alkaline lavas that occur both as plateau deposits and as cones on rift floors and on the swells [3, 4]. Crevice type rifts are characterized by deep grabens and small amounts of uplift and volcanism (example: Baikal Rift, Figs. 2 and 5).

Rift-transected crustal swells [5] average ~750 km in diameter and 1 to 1.5 km in height. Individual rifts are commonly 1 to 3 km deep (not counting sediment/lava accumulations), and range in length from 300 to 1500 km, being related to swell diameter. Rift widths are typically 35-55 km wide, approximately equalling crustal thickness, as proposed by Cloos [6].

Mars. The Mariner 9 spacecraft photographed a vast system of troughs, fractures and depressions collectively known as Valles Marineris (Fig. 3). At the center of the assemblage is a ~2200 km long canyon called Coprates Chasma. Where not enlarged by subcircular zones of subsidence the canyon is 90-120 km wide and 3-4 km deep. To the east the rift-like central canyon opens into an area of irregular depressions with jumbled rubble-strewn floors (the Chaotic Terrain of Sharp [7]), while to the west the canyon devolves into a zone of reticulate fractures similar to those formed in the center of uplifted areas [8]. And indeed the intricately fractured zone is at the summit of a broad (~2500 km wide) dome - the Tharsis plateau [9] - that rises 6 km above the ancient cratered crust of Mars. Unlike most terrestrial rifts, however, Valles Marineris does not transect the Tharsis "swell" but rather is radial to it. The smooth surface of the entire Tharsis plateau is interpreted as a vast volcanic plain [10], similar to the Trap basalts which cap the uplifted Ethiopian swell [11]. There are no volcanic mountains on the rift floor or on its shoulders, but 3 large shield volcanoes (Vs of Fig. 3) and 6 smaller ones occur on

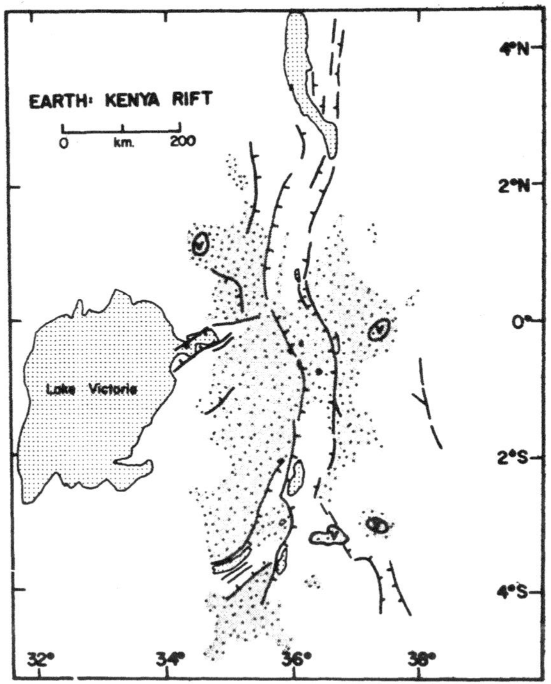


Fig. 1. Kenya Rift, uplift (elevations above 1500 m stippled), and volcanic cones (V). Data from [3] and others.

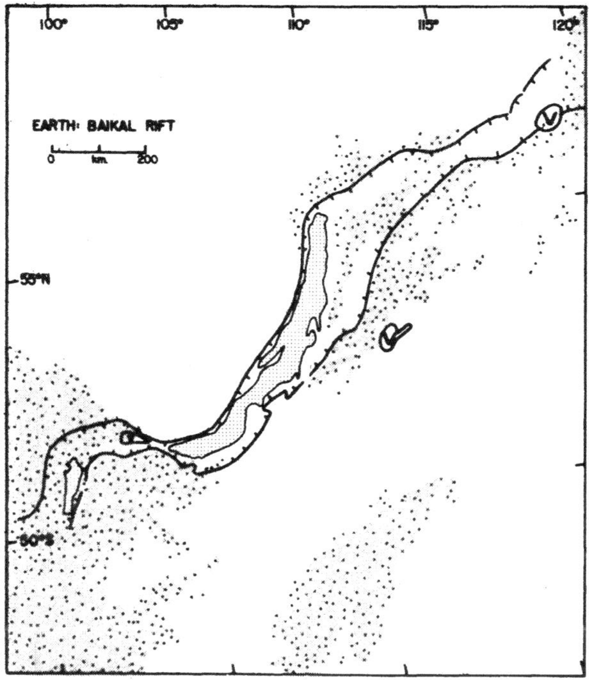


Fig. 2. Baikal Rift, uplift (elevations above 1500 m stippled), and volcanic fields (V). Data from [23].

the north and west flanks of the Tharsis swell, and the giant shield Olympus Mons is NW of Tharsis.

Venus. Radar imagery now available for about 2% of Venus provides tantalizing impressions of diverse surface features [12]. Only tentative interpretations can be proposed presently because of (1) ambiguities due to limited observations, and (2) horizontal resolutions of ~30 km, however, a number of different landforms have been reasonably convincingly identified. Large craters or basins (of presumed impact origin) appear to be nearly as abundant as those on the Moon [13], and shield volcanoes comparable to Olympus Mons (diameter about 500 km) have also been interpreted from imagery [12]. The most intriguing image, however, is of a long trough bisecting a dome (Fig. 4). Malin and Saunders [12] report that this feature is approzimately 1400 km long, 150 km wide, and 3 km deep. To the north the topographic low of the trough appears to be continued as a series of three craters, each ~100 km wide. The southern portion of the trough bifurcates (at an angle of ~60°), just as do the Kenya Rift, the Rhine Graben, and many other terrestrial rifts [2]. A median ridge runs through part of the trough, reminiscent of ridges on the floors of Valles Marineris and the Baikal Rift. Apparently the elevation of the floor of the venusian rift mimics the topography of the swell, being high in the middle and low at the edges; the Kenya Rift does exactly the same.

## 3. DISCUSSION

The descriptions above document many similarities in gross morphology for rift-like structures on Earth, Mars and Venus. Rifts on all three planets are generally related to domed regions and volcanism (except on Venus where the resolution of the imagery is inadequate for identification of small-scale volcanic features). On Earth there are perhaps a dozen well defined rifts, as well as another 3 dozen possible failed rifts [2], all of which tie together - with oceanic ridges - into a global rift system [14]. In contrast, only a single rift-like structure has been found on Mars. The detection of a single rift on the 2% of Venus currently imaged with moderate resolution radar suggests that the known rift may not be unique.

The troughs on Venus and Mars are considerably larger than most terrestrial rifts: typical rifts on Earth are 500-700 km long, whereas Coprates Chasma is 2200 km long and the Venus trough is 1400 km in length. These dimensions are comparable to the diameters of associated swells, which, for the Earth, are commonly believed to be crustal reactions to mantle convection plumes [2, 15]. Mathematical modelling by Withjack [16] has shown that crustal uplifts similar to those observed for the Kenya and Ethiopia rifts ( ~600 km wide, 3 km high) can be produced by mantle plumes

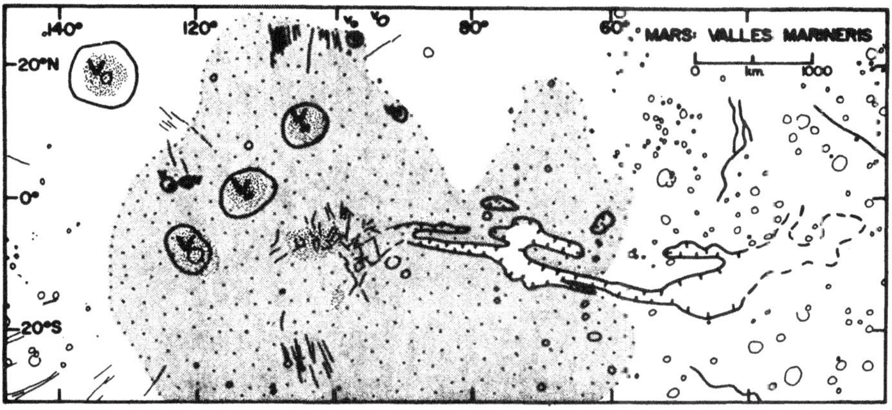


Fig. 3. Valles Marineris, topography (elevations between 5 & 10 km, fine stipple; 10 km coarse stipple), volcanic cones (V), and impact craters (circles) [24].

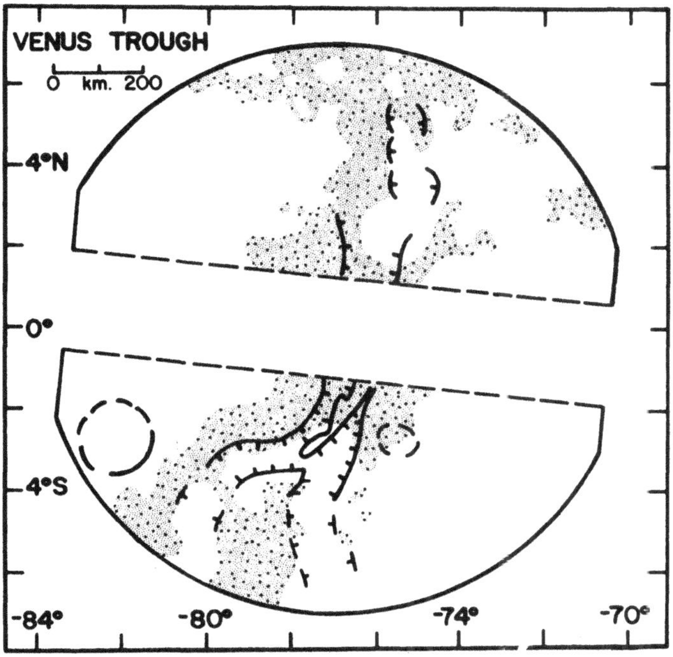


Fig. 4. Venusian trough, topography (elevations >800 m above lowest contour stippled), and suspected craters (circles) [12].

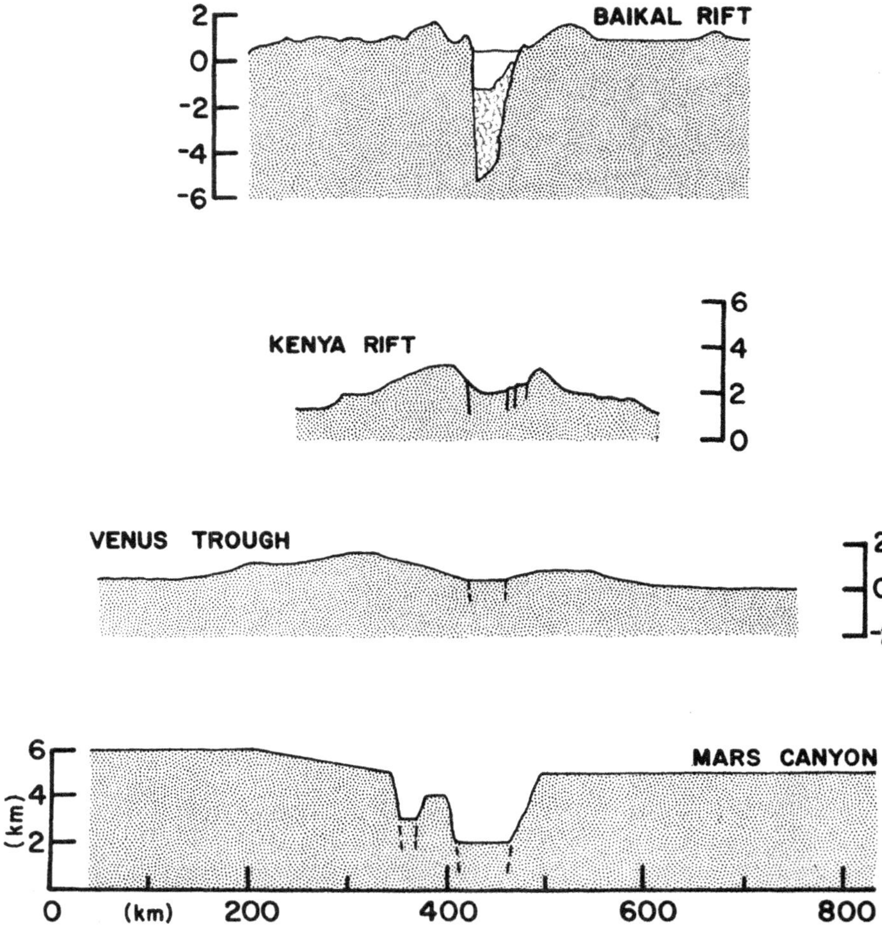


Fig. 5. Comparison of rift structures on three planets.

200 km wide and reaching up to the bottom of the crust. According to this model, wider and higher uplifts can result from broader plumes and/or greater crustal thicknesses. Thus, the great size of the Tharsis plateau of Mars may reflect a thick crust, an inference that is consistent with the 100-120 km wide width of Coprates Chasma; in terrestrial cases rift width approximately equals crustal thickness. Additional support for a thick martian crust comes from gravity studies [17]. The similarly great width of the Venusian rift implies that that planet may also have a much thicker crust than the Earth.

On Earth rift valleys result from crustal tension preceding plate tectonic rupture of continents [18]. Presently observed continental rifts are either young rifts initiating future continental drifting, or they are abortive attempts to do so. The large number of failed rift-arms and longitudinally split rifts on continental margins [18] evidences the intimate relation of rifting and plate tectonics on the Earth. For Mars, however, there is no compelling evidence of a plate tectonic history - no fold mountains, strike-slip faults, nor island arc/subduction zone volcanoes. Additionally, both Mars and Venus contain large tracts of ancient cratered terrain whose preservation implies that plate tectonic activity has not rejuvenated their surfaces. If plate tectonics has not been significant on Mars or Venus their rifts may simply have resulted from tensions produced by crustal arching as originally proposed for terrestrial rifts by Cloos [6]. If crustal swells formed during the early histories of Mars and Venus, then it is reasonable to infer they may have also been formed and ruptured by rifts on Earth. Thus evidence for rifting in ancient terrestrial rocks is not necessarily evidence for early plate tectonics.

An alternative explanation for the existence of tensional structures on Mars and Venus is suggested by modelling of planetary thermal evalutionary development by Solomon and Chaiken [19]. Assuming that Valles Marineris and other fractures are a consequence of thermal expansion of Mars, they found that the planet may have increased in radius by 19 km over the first 3.6 b.y. of its history. Various lines of evidence have been presented to suggest that the Earth may also have increased slightly in volume throughout time, especially since the breakup (and attendent rifting) of Pangaea in the late Jurassic [20]. Such an expansion would have precipitated rifting. Although little thermal modelling has been done for Venus, its close similarity in diameter, mass, and density to the Earth implies that it may have had an analogous thermal history, although plate tectonics may not have been widespread. Nonetheless, the presence of a rift valley indicates that horizontal tensions have been generated on Venus.

It is intriguing that Earth, Venus, and Mars - the three largest terrestrial planets - have crustal swells, rift valleys, and large shield volcanoes. Neither the Moon nor the 40% of Mercury photographed by the Mariner 10 spacecraft have these major tectonic features. Indeed, Mercury is shrivelled by a system of scarps [21] that apparently resulted from thermal contraction. The consistency of occurrence of features indicative of crustal tensions in large planets, and their lack in smaller planets, strongly implies that rifting may be a fundamental process in evolution of planets larger than a critical size of mass, and that plate tectonics on the Earth stems from an abundance of rifting.

Future radar investigations of Venus will hopefully reveal details of the tectonic history of the planet. With the present sparse data, Venus appears more Mars-like than Earth-like [22]. Closer investigation may reveal Venus to be more analogous to the earlier history of the Earth, near the inception of global plate tectonics. If additional rifts are discovered it may be possible - by analyzing many samples rather than a single structure - to investigate the significance of a planet's composition, volatile content, mass, and gravity on tectonic development.

REFERENCES

1. E.E. Milanovsky, Tectonophysics, 15, 65, 1972.
2. K. Burke and J.F. Dewey, J. Geol., 81, 406, 1973.
3. B.H. Baker, P.A. Mohr and L.A.J. Williams, Geol. Soc. Am. Spec. Pap., 136, 1972.
4. P.A. Mohr and C.A. Wood, Earth Planet. Sci. Lett., 33, 126, 1976.
5. M.J. Le Bas, Nature Phys. Sci., 230, 85, 1971.
6. H. Cloos, Geol. Rundschau, 30, 401, 1939.
7. R.P. Sharp, J. Geophys. Res., 78, 4073, 1973.
8. T.A. Mutch, R.E. Arvidson, J.W. Head, K.L. Jones and R.S. Saunders, The Geology of Mars, Princeton, 1976.
9. K.R. Blasius and J.A. Cutts, Proc. Lunar Sci. Conf. 7th, 3561, 1976
10. M.H. Carr, H. Masursky and R.S. Saunders, J. Geophys. Res., 78, 4031, 1973.
11. P.A. Mohr, Tectonics of Africa, UNESCO, 447, 1971.
12. M.C. Malin and R.S. Saunders, Science, 196, 987, 1977.
13. G.G. Schaber and J.M. Boyce, Proc. Symp. on Cratering Mechanics (Pergamon), in press.
14. J.H. Illies, Tectonophysics, 8, 5, 1969.
15. W.J. Morgan, Nature, 230, 42, 1971.
16. M. Withjack, Lithospheric thinning and crustal uplift produced by mantle convection (in prep.).
17. R.J. Phillips, R.S. Saunders and J.F. Conel, J. Geophys. Res., 78, 4815, 1973.
18. K. Burke, Tectonophysics, 36, 93, 1976.
19. S.C. Solomon and J. Chaiken, Proc. Lunar Sci. Conf. 7th, 3329, 1976.
20. J. Steiner, Geology, 5, 313, 1977.
21. R.G. Strom, N.J. Trask and J.E. Guest, J. Geophys. Res., 80, 2478, 1975.
22. J.W. Head, C.A. Wood and T.A. Mutch, Am. Sci., 65, 21, 1977.
23. N.A. Florensov, Tectonophysics, 8, 443, 1969.
24. U.S.G.S., Topographic Map of Mars, Map I-961, 1976.

# PALEORIFTS - CONCLUDING REMARKS

E.-R. Neumann [1] and I.B. Ramberg [2]

[1] Mineralogical-Geological Museum
University of Oslo, Oslo 5, Norway
[2] Department of Geology, University of Oslo,
Blindern, Oslo 3, Norway

ABSTRACT. It is not yet clear if continental rifting occurs in response to one or more causal mechanisms. However, evidence exists that at least some of the rifts in areas affected by plate-to-plate interactions have formed over sublithospheric thermal anomalies. All present hot spots identified on the North American and Eurasian plates (not falling near or on spreading ridges) lie in areas of high seismic activity on the continental side cf collision type or transform plate boundaries. It is suggested that when a plate passes over a thermal anomaly of sufficient magnitude, the chances of this thermal anomaly causing penetrative magmatism and rifting, are greatest in those parts of the plate which is already under tension.

Geochemical variations among rift magmatic rocks may be divided into two main groups: a) systematic variations related to the melting conditions (like depth and degree of partial melting), and b) unsystematic variations caused by second-order mantle heterogeneities and complex combinations of processes affecting the melts during and after melting. It is possible that the general pattern of systematic geochemical variations observed among magmatic rocks in the transition area between a continental rift and an oceanic spreading ridge may represent mixing between "enriched" lower lithosphere and "depleted" low velocity layer material.

## 1. INTRODUCTION

One of the problems discussed at the NATO Advanced Study Institute at Sundvollen, 1977, was the definition of the term "rift". Perhaps surprisingly, this seemingly rather simple problem has been

*I. B. Ramberg and E.-R. Neumann (eds.), Tectonics and Geophysics of Continental Rifts,* 409-424.

repeatedly subjected to debate in recent years. However, at the Meeting, the variety in geology and tectonic environment, of those graben features termed "rifts" by different geoscientists was well illuminated through the lectures. Moreover, the problem of the definition of the term is entangled with some of the most important, but still unanswered questions concerning the process of rifting: Are rifts the lithospheric responses to more than one causal mechanism? Are these causal mechanisms to be sought entirely below the lithospheric plates, or within, or as a result of plate-to-plate interaction? What are the necessary and sufficient requirements for rifting to occur, and for a continental rift to develop into an oceanic spreading segment? It was agreed at the Meeting that as long as these questions remain unanswered, the term "rift" should be used in a broad sense. Hence, it is suggested that Burke's [1] definition of rifts as "places where the entire thickness of the lithosphere has ruptured under tension" is used.

In the following sections differences and similarities among geologic features termed rifts, as presented at the Meeting, are briefly summarized, together with a discussion of necessary and sufficient conditions for rifting to occur. Compositional variations among rift magmatic rocks are discussed on the basis of present knowledge about heterogeneities in the mantle.

For more detailed discussions of various rifts and rifting processes the reader is referred to all the articles in this volume and in Proceedings Vol. I ("Petrology and Geochemistry of Continental Rifts" ed. by E.-R. Neumann and I.B. Ramberg). In this paper references to other papers in Proceedings Vol. I and II has been kept to a rigid minimum.

## 2. CHARACTERISTICS OF RIFT SYSTEMS

The characteristics of continental rift systems as presented directly and indirectly during the Meeting at Sundvollen, are summarized below:

a) Continental rifts may vary between ones with strong volcanicity and close interrelation between magmatic activity and extensional tectonics, and others with weak magmatic activity where the volcanism seems to be essentially independent of the faulting [2, 3].

b) Rift magmatic products are generally mildly to strongly alkaline; but tholeiitic and calc-alkaline rock-types may also occur [e.g. 3-8].

c) Continental rifts frequently are spatially associated with domal uplifts (for instance the Rhine Graben, and the Baikal, Rio Grande and East African rift systems). A genetic progression has been proposed [9, 10] from "unrifted" domes through rifted domes to triple spreading ridge junctions. The early stages of development may or may not be accompanied by the occurrence of "mid-plate" volcanism.

d) Some continental rift systems have an approximately uniaxial extension (like the East African rift system), whereas others (like the largely Triassic North Sea rift system) are geographically complex and lack an obvious main axis [3]. In the southwest part of the USA the Rio Grande rift, regarded as an example of the uniaxial type, merges southward with the wide Basin and Range Province [5, 6, 11].

e) Some continental rifts are, or have been, linked to the network of accreting plate boundaries; the classical example is the East African rift system [e.g. 12]. Other rifts seem completely isolated [e.g. 1, 4, 13], although this impression may in some instances merely result from inadequate mapping of the surroundings.

f) There is a tendency for rifts to be concentrically distributed about cratons and to show a radial outward decrease in age of initial activity; i.e. a migration of rift systems from areas of earlier to those of later cratonization [14].

g) Rift zones commonly bear evidence of several separate periods of rifting and volcanism [2, 3, 6, 15-18].

h) Two periods of rifting within the same region may develop quite differently, implying that the composition and structure of the lithospheric plate itself is not the overall controlling factor. (However, the stress field may be quite different). For instance, the northerly extension of the Oslo rift, the socalled Sparagmite region or Østerdal aulacogene [19], formed well-defined fault-basins in the Late Precambrian (Eocambrian) [18, 20]. 1500-3000 m of sedimentary rocks are still preserved in these depressions [20]. This period of rifting was, however, accompanied by only very minor magmatic activity as evidenced by a few flows of alkali basalt [21], in clear contrast to the intense magmatic activity associated with the Permian event in the Oslo Graben.

i) The crust under rifts is generally abnormally thin [e.g. 1, 3, 6, 13, 22-26]. In paleorifts, the depth to the top of the "paleorift cushion" roughly represents the thickness of the original rigid layer which is a measure of the degree of development that was reached. The detailed tectonic pattern of the individual rift will depend on the thickness and constitution of the rigid layer.

j) The upper mantle under active rifts is invaded by low-velocity, low-density mantle material [e.g. 6, 13, 23, 25]. A corresponding axial body of high-density material is found at the crust-mantle boundary and as part of the crust under paleorifts [17, 18]. No difference has been reported between the character of the upper mantle under rifts with much and rifts with little volcanism.

The tectonic environments of rifts vary considerably, but have some common factors [e.g. 1, 3]:

a) Rifting occurs in response to tension.

b) Rifts form at various stages of the Wilson cycle [ 1].

c) Rifting generally takes place along preexisting zones of weakness the same zone in a given area will therefore tend to give way during each of several episodes.

It has been suggested that there exists a correlation between rift characteristics and tectonic regime. Rift systems closely connected in time and space with orogeny tend to be isolated and to exhibit complex horst and graben provinces, poorly defined triple junctions, and calc-alkaline volcanism; on the other hand, rifts believed to represent embryonic accreting plate boundaries have an obvious main axis, well-defined triple junctions, and alkaline to tholeiitic magmatism [e.g. 3, 5, 15, 27, 28].

It is not known if these differences are related to the causal mechanisms of rifting, or if they result from differences in the structure of the lithosphere. The question of what are necessary and sufficient conditions for the initiation of rifting and mid-plate volcanism is still being debated. Gass et al. [10] have proposed that mid-plate volcanism almost entirely can be explained in terms of plate thickness, plate velocity and the magnitude of the causal thermal anomaly. Considerable data are available which testify to the dependence of rifting and mid-plate volcanism on the first two of these factors. It is, however, possible that lithospheric properties other than thickness also influence the distribution of rifts and hot spots. Both recently and in the past, some continental rifting seems to have occurred in association with orogeny (including compressional plate boundaries) [e.g. 1, 3, 5, 6, 14, 15, 27, 29 ]. On the North American and Eurasian plates all hot spots away from spreading ridges, as identified by Burke and Wilson [30], lie in areas of high seismic activity on the continental side of collision type or transform plate boundaries (Fig. 1). No hot spots have been identified on the Atlantic Ocean part of the North American plate. Although the true global distribution of sublithospheric thermal anomalies is not known, the strong correlation between hot spot distribution and crustal tension can hardly be fortuitous. It is possible that rifts may form as the passive biproducts of inter-plate motion alone (without any sublithospheric thermal anomaly), for example impactogens

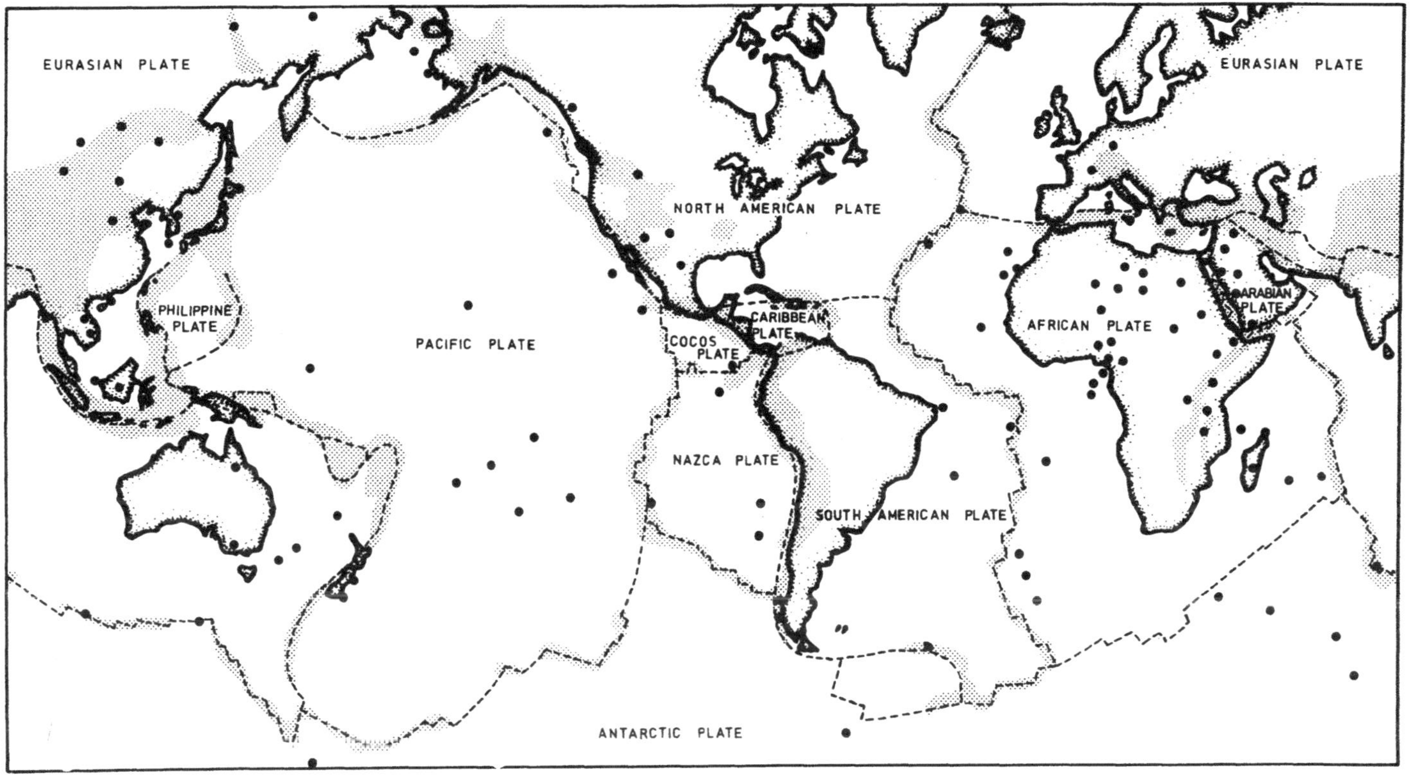


Fig. 1. Distribution of hot spots active in the last 10 million years (after Burke and Wilson [30]), together with areas of recent high seismic activity [31, 32].

(formed by plate collisions [1, 27, 29], propagating fractures [33], or wrench faulting [5, 6]).

An alternative possibility is that preexisting regional tension increases the chances for penetrative magmatism and rifting when the plate passes over a thermal anomaly. The Rio Grande rift may be an example of this. It has been suggested that both the Basin and Range Province and the Rio Grande rift represent parts of a regional shear zone accommodating oblique divergence of the North American and Pacific plates. A detailed study of the structural pattern of the southern Rio Grande rift and the adjacent part of the Basin and Range Province revealed, however, that in contrast to the shear tectonics of the Basin and Range Province, the southern Rio Grande rift is characterized by extension tectonics [11]. Ramberg et al. [11] has concluded that the stress field of the southern Rio Grande rift results from the combined effects of the regional shear stress, and extension over a major shallow basaltic intrusion or upwarp of the upper mantle.

If all rifting is related to sublithospheric thermal anomalies, then rifts in any tectonic regime may develop into spreading ridges, if not, rifts may be divided into two groups, "passive" rifts and rifts caused by sublithospheric thermal anomalies, of which only the latter may be expected occasionally to develop into spreading ridges. In any case rift characteristics may be important clues to the assignment of specific rifts to causal mechanisms.

The identification of the necessary and sufficient conditions for rifting to occur and to develop into spreading ridges can only be answered on the basis of detailed investigations of the characteristics, and reconstruction of the tectonic settings of ancient rifts that are known to have failed, and such that caused continental rupture and separation.

## 3. COMPOSITIONAL VARIATIONS IN THE MANTLE

Most continental rifting is accompanied by magmatic activity. The geochemistry of the magmatic products typical of the various phases of rifting is a very important clue to our understanding of the rifting processes. The significance of different compositional characteristics can, however, only be evaluated on the basis of knowledge about the geochemistry of the mantle source region for these magmas.

In the following summary a comparison of the chemistry of intra-plate and spreading ridge magmatism has been used as a basis for a discussion of compositional variations in the mantle.

Continental rift magmatic products generally have a highly

alkaline nature unparalleled in other tectonic regimes. Mafic alkaline rocks (alkali basalt, nephelinite etc.) are markedly enriched in alkalis, incompatible elements (including rare earth elements (REE)) and volatiles (particularly $CO_2$) as compared to tholeiitic basalts, and are enriched in light relative to heavy REE. Tholeiitic continental rift and plateau basalts are enriched in alkalies and incompatible elements relative to abyssal tholeiites, and are enriched (or at least undepleted) in the light relative to the heavy REE. Furthermore, continental mafic rocks have a wide range in initial Sr and Pb isotope ratios and significantly lower $^{143}Nd/^{146}Nd$ (and $^{143}Nd/^{144}Nd$) ratios than mid-ocean ridge basalts (MORB) [e.g. 34-45].

Alkaline magmatic rocks and tholeiitic basalts of ocean islands, both within plates and along spreading axes, are also generally enriched in incompatible elements compared to ocean ridge tholeiites with similar Mg/(Mg+Fe) ratios. However, exceptions have been reported. Tholeiitic and nepheline-normative basalts from seamounts on the Nazca Plate have light-REE depleted patterns similar to those of the bordering East Pacific Rise spreading axis. Ocean island basalts have a range in initial $^{87}Sr/^{86}Sr$ ratios intermediate between those exhibited by continental and MOR basalts. $^{143}Nd/^{146}Nd$ ratios are significantly lower than those in MOR basalts. Significant differences in incompatible element concentrations and radiogenic Sr, Pb and Nd isotope ratios have been found both within single groups of ocean islands, and between different groups [e.g. 34, 37, 42, 43, 45-55, 57-61, 63].

Ocean floor basalts are generally depleted in alkalis and incompatible elements relative to continental and ocean island mafic rocks, and have low uniform Sr and high Nd isotope ratios. Significant uncorrelated variations among incompatible element, Sr and Pb isotope ratios, have, however, been observed. Such variations are best documented among basalts restricted to the "zero-age" ocean floor of the mid-ocean ridge rift valleys [e.g. 34, 35, 42, 43, 46, 48, 50, 52-54, 56, 61-75].

The possibility that all basalts (and highly alkaline mafic rocks) may have originated from a homogenous mantle has been examined very carefully in the light of the observed geochemical differences. Variations in the processes operating during and after melting are clearly capable of producing major chemical differences. Numerous such processes have been discussed (equilibrium and disequilibrium partial melting, dynamic open-system melting, fractional crystallization, zone refining, magma mixing, multiple stage melting or enrichment, and crustal contamination) and the role they play has been analyzed in a large number of papers (e. g. by Sun and Hanson [45, 76], Hanson [49], Cox et al. [36], O'Nions and Pankhurst [55], Lambert and Holland [72], Kay and Gast [39], Gast [48], Church and Tatsumoto [67], Sun and Jahn [60],

Langmuir et al. [73], Cumming [68], DePaolo and Wasserburg [46], Richard et al. [42], O'Nions et al. [54], Schilling [44, 57, 58, 74], White et al. [63]). These authors are, however, unanimous in the conclusion that not all observed variations can be explained by such processes, and that the mantle must therefore be heterogeneous. What is more, the heterogeneities must have existed for several hundred million years or more. Nephelinites, alkali basalts and tholeiites in continents and ocean islands are believed to be derived from a mantle source region relatively enriched in incompatible elements (e.g. K, Ba, Rb, light REE) and radiogenic Sr and Pb isotopes. Nephelinites and alkali basalts reflect very limited partial melting - no more than a few percent [38, 45]. Ocean ridge basalts are derived from a relatively depleted source.

Most authors also conclude that the large uncorrelated variations in isotope and trace element ratios found among continental and ocean island basalts are due to second-order heterogeneities within the "enriched" mantle source region. The magnitude of these variations is, however, uncertain as it is strongly dependent on the behaviour of isotopes and trace elements during and after partial melting. Second-order heterogeneities have also been postulated in the "depleted" source region. O'Nions and Pankhurst [56] have suggested that some isotopic variations may be due to disequilibrium partial melting. O'Hara [77] has shown that significant variations in incompatible element concentrations and ratios may result from a complex process involving fractional crystallization of a periodically refilled magma chamber, with or without remelting of previously crystallized gabbros.

The depths and geometries of the mantle source regions are still not established with any certainty. The "depleted" source region giving rise to ocean ridge basalts is generally believed to be the asthenosphere as suggested by Morgan's mantle plume model [78, 79].

Opinion is divided as to whether the "enriched" source region lies above or below the asthenosphere. Morgan [78, 79] considered that ocean islands and continental rift and plateau magmatism could be the surface expressions of plumes or blobs rising adiabatically from an "undepleted" mantle region below the asthenosphere, possibly as deep as the core-mantle boundary. Although, the mantle plume hypothesis has gained much support, it has as yet neither been proven nor disproven. An additional or alternative possibility is the existence of an "enriched" lower lithosphere. Evidence of metasomatic enrichment of LIL-elements and volatiles in the upper mantle is provided by numerous ultramafic xenoliths [e.g. 7, 41, 80, 81], and is even suggested by geochemical variations in lavas extruded on the flanks of the inner rift valley of the Mid-Atlantic Ridge [66, 70]. Such "high level" enrichment is considered to have arisen either by trapping of enriched residual asthenospheric material at the base of the lithosphere [82], thermal accretion by freezing of "enriched" asthenosphere material before it was depleted, lithospheric underplating by "undepleted" mantle diapirs and blobs [34], or diffusion of LIL-element rich volatiles from either the asthenosphere [66, 70] or the deeper mantle [7, 41] into the lithosphere. Measured diffusivities of divalent cations in basaltic melts indicate that diffusion may be important in small-scale processes, such as local enrichment below the continental and oceanic lithosphere [83]. Although many enriched magmatic rocks may well be derived from an "enriched" lower lithosphere, this can hardly be true for all such rocks. It is, for example, difficult to explain the enriched character of MOR basalts extruding within the general areas of hot spots (e.g. the Azores and the Galapagos hot spot [58, 63, 93]) in terms of either partial melting of "enriched" lower lithosphere material, or mixing between such and "depleted" low velocity layer material. Nd isotope data furthermore suggest that the alkali basalts on some ocean islands have a <u>less</u> complex previous history than depleted MOR tholeiites [42].

## 4. CONTINENTAL RIFT MAGMATISM

### 4.1 Unsystematic geochemical variations

The above discussion clearly indicates that there is substantial evidence to indicate that the source region of continental rift magmas is generally "enriched" mantle exhibiting marked heterogeneities, and that the processes affecting the melt composition during and after melting may lead to additional "unsystematic" geochemical variations.

Experimental investigation of divalent cation diffusion rates in basaltic systems [83] indicate that local heterogeneities may be preserved even during periods of partial melting. In a completely crystalline rock the range of equilibration will not exceed a few centimeters per $10^8$ - $10^9$ years. In a partially molten mantle with a stagnant interstitial melt phase, local equilibration may occur over distances of the order of 10 m in $10^6$ years (or 100 m in $10^8$ years) [83]. The partially molten mantle material underneath an active continental rift will rise adiabatically, and the system thus cannot be regarded as completely stagnant. Separate periods of intensive magmatic activity in continental rifts seem, however, only to last a few tens of millions of years (about $5 \cdot 10^7$ years in the Oslo Graben [84]). It is thus highly unlikely that equilibration (caused by lateral magma mixing and diffusion) in the <u>melting</u> region underneath an active continental rift will have time to affect distances of a higher order of magnitude than 1 km. Near the crust/mantle boundary where magma chambers form, equilibration will, of course, be heavily dependent on the lateral extent of these chambers.

Mantle heterogeneities will not only cause variations in the trace element and isotope contents of the melts. $CO_2$ and $H_2O$ are among the compounds believed to be heterogeneously distributed in the source region prior to melting [e.g. 7, 41, 80, 81]. As the alkalinity of a melt is strongly dependent on the activity of $CO_2$ relative to $H_2O$ in the melting system [85], local variations of these components in the region of partial melting may cause considerable variations in the alkalinity of the melts. The unsystematic compositional variations caused by mantle heterogeneities and complex processes may therefore be both regional and local. They will be superimposed on the systematic ones related to simple process parameters such as depth and degree of primary melting and will cause an increase in the total range of compositional variations which may completely obscure the systematic ones.

Many of the compositional variations reported from continental rifts may more easily be explained in terms of unsystematic variations than by the abrupt shifts in depth or degree of melting suggested by some geologists. Examples of such unsystematic variations are uncorrelated variations among incompatible element concentrations (the Oslo rift [86]), and between incompatible element concentrations and Sr isotope ratios (Afar [4]), pluss local or short term alternations between magmatic products belonging to different main rock series (Gulinskaya, East Sibir [2]; the Kenya rift[8]; the Rio Grande rift [87]).

The geochemical data from continental rifts thus confirm that unsystematic variations arise and are preserved in the magmatic products even within single rift segments. Geophysical studies of continental rifts indicate that such areas provide excellent opportunities for geochemical evolution of a variety of melts to go on for long periods of time in more or less separate magma chambers before extrusion or intrusion into the upper crust. The upper mantle under active rifts consists of anomalously hot, low-density, low-viscosity masses [e.g. 13, 23, 25] which correspond to the masses of dense material at the crust/mantle boundary found along the entire length of a number of paleorifts [17, 88]. These dense masses are interpreted as solidified mafic intrusives, representing the residues of fractionation processes [e.g. 88]. Mafic melts from deeper parts of the mantle are supposed to stabilize near the crust/mantle boundary because they lose their buoyancy when encountering the light crust [89]. The masses of partially molten or crystallized mafic intrusives and accumulates found near the crust/mantle boundary in active continental rift areas are commonly referred to as a single continuous body. Evidence exists, however, that this is a misinterpretation. Deep seismic soundings and micro-earthquake surveys indicate separate intrusive bodies and a magma chamber of restricted size in the deep crust under the Rio Grande rift [25]. The lower crust beneath the Rhine Graben has a lamellar structure probably due to alternation between high- and low-velocity material [23]. More or less continuous inflow of new melts into the deeper parts of magma chambers, where fractionation is taking place at the same time may lead to considerable variations - or greatly enhance pre-existing differences - in trace and minor element compositions without necessarily giving the melt a more "evolved" character [77].

## 4.2 Systematic geochemical variations

In spite of the heterogeneous nature of the mantle source region for continental rift magmatism, systematic overall variations of the type postulated by Gast [48] are also recognized. Such variations include decreasing alkalinity with time (e.g. the Kenya rift [8, 90] and the Oslo rift [91]), and increasing alkalinity with distance from the rift axis (e.g. the East African rift system [4, 8, 90]).

The Afar Region provides a very good example of two types of "systematic" chemical variations in a hot spot area. The region is particularly interesting as it lies on the junction between a continental rift and two oceanic spreading axes. Barberi and Varet [4] have, in the Ethiopean rift, found a gradual decrease in the magma production rate, combined with an increase in the alkalinity of the magmatic products with increasing distance from Afar. Similarly, within the Afar Region they found that the $K_2O/(K_2O + Na_2O)$ ratios of basalts increased away from the axial ranges. Along the Red Sea and Gulf of Aden mid-ocean ridges, however, there is a negative correlation between the La/Sm ratios of tholeiitic basalts and distance from Afar [44] and the $K_2O/(K_2O + Na_2O)$ ratios are low [4].

This geographic distribution of magmatic products with different chemical characteristics corresponds closely to what may be expected in an area with magmatic activity from two source regions. On the continent all magmas are derived from an "enriched" source region, and the overall correlation between alkalinity and degree of partial melting postulated by Gast [48], Kay and Gast [38], Sun and Hanson [45], and others, may be observed. Passing from Afar to the Red Sea or the Gulf of Aden, we primarily observe the effect of mixing of "enriched" magmas with gradually increasing relative amounts of "depleted magmas", as suggested by Schilling [44].

The relations described from the Afar Region are not unique. On Iceland the magma production rate increases, and the alkalinity of the extruded basalts decreases towards the active spreading axis [92]. However, another type of systematic variation is exposed by sampling only rocks from the active spreading axis. Tholeiitic basalts erupted along the Reykjanes ridge are relatively enriched in LIL-elements and radiogenic Sr and Pb isotopes close to Iceland (with considerable scatter perhaps indicating a heterogenous source), but becomes progressively more depleted with increasing distance from Iceland [50, 57, 74].

It is an important fact that the general pattern of magma mixing as described by Schilling from the Reykjanes Ridge, Afar, and the Azores [44, 50, 57-59, 74] is not restricted to mixing between "depleted" low velocity layer (DLVL) material and primordial deep mantle material. The same type of pattern may be produced by, for example, mixing of DLVL material and "enriched" lower lithosphere material.

## 5. SUMMARY

The main features of the tectonomagmatic pattern and evolution, lithospheric structure, and geochemistry of continental rift areas and continental rift - spreading ridge transition zones are now fairly well known, as are some of the lithospheric conditions favouring rifting and penetrative magmatism. A number of basic problems, however, remain unsolved: Do rift formation and magmatism represent simply the responses of pre-existing anisotropies in the lithosphere to stresses acting on the plate? Does continental rifting on the contrary require the coincidence of a zone of lithopsheric weakness with an underlying disturbance? What type of processes cause magmatic activity to continue over tens of millions of years? If sublithospheric (thermal) anomalies exist, what causes them and where do they form? How large are they? When and why does a continental rift develop into a spreading segment?

It is quite clear that partial or complete answers to some of the above questions may be obtained through compilation and synthesis of the large amounts of data already available on continental rifts and other intra-plate magmatic events. Further data are, however, also needed. Some lines of study which may prove valuable to our understanding of the rifting processes are outlined below.

Whether rifting is caused by processes below or within the lithospheric plates may be analyzed by further study of surface features such as volcano spacing and frequency, trends and spacing of faults, and age gradients along rifts.

Comparisons of the tectonic settings of ancient rifts known to have failed and those which developed into continental break-up, and detailed correlations of tectonic and magmatic events over long periods prior to continental rupture and separation, may yield further information about the necessary requirements for a continental rift to develop into a spreading ridge.

The various hypotheses for the cause of sublithospheric thermal anomalies are interlocked with ideas about the relative positions of the "enriched" and "depleted" mantle source regions. Tedious testing of existing and additional trace element and radiogenic isotope data from numerous locations against the various models aimed at explaining the geochemical heterogeneities assumed to exist in the mantle is likely to shed further light on this problem. Nd isotope studies will become an important tool for such tests, but so far only scattered data exist.

Finally, complete understanding of the processes leading to continental rifting and magmatism can only be obtained by comparing data from continental with those from oceanic intra-plate magmatic-tectonic events.

## ACKNOWLEDGEMENT

P. Vogt, W.L. Griffin and B.B. Jensen are gratefully acknowledged for critically reviewing the manuscript.

## REFERENCES

x Petrology and Geochemistry of Continental Rifts, ed. by E.-R. Neumann and I.B. Ramberg, 1978.

1. K. Burke, this volume.
2. N.A. Logatchev, this volume.
3. E.E. Milanovsky, this volume.
4. F. Barberi and J. Varet, Proceedings I, 1978 x.
5. W.E. Elston, Proceedings I, 1978x.
6. J.H. Stewart, this volume.
7. D.K. Bailey, Proceedings I, 1978 x.
8. B.H. Baker, R. Crossley and G.G. Goles, Proceedings I, 1978 x.
9. K. Burke and J.F. Dewey, J. Geol., 81, 406, 1973.
10. I.G. Gass, D.S. Chapman, H.N. Pollack and R.S. Thorpe, Proceedings I, 1978 x.
11. I.B. Ramberg, F.A. Cook and S.M. Smithson, Geol. Soc. Am., Bull., 89, 107, 1978.
12. D.P. McKenzie, D. Davies and P. Molnar, Nature, 226, 243, 1970.
13. N.A. Logatchev, V.A. Rogozhina, V.P. Solonenko and Yu.A. Zorin, this volume.
14. I.B. Ramberg and E.-R. Neumann, Proceedings I, 1978 x.
15. P.A. Ziegler, this volume.
16. A.R. Philpotts, Proceedings I, 1978 x.
17. B.G.J. Upton and D.J. Blundell, Proceedings I, 1978 x.

18. I.B. Ramberg and N. Spjeldnæs, this volume.
19. D. Roberts and G.H. Gale, The Caledonian-Appalachian IAPETUS Ocean. In: The Evolution of the Earth's Crust, ed. by D.H. Tarling. Academic Press, London-New York, 1977 (in press)
20. K. Bjørlykke, A. Elvsborg and T. Høy, Nor. Geol. Tidsskr., 56, 233, 1976.
21. K. Bjørlykke, Nor. Geol. Tidsskr., 49, 313, 1969 (in Norwegian)
22. R.W. Girdler, this volume.
23. S. Mueller, this volume.
24. J.H. Illies, this volume.
25. R.J. Bridwell, this volume.
26. M.F. Sheridan, this volume.
27. P. Tapponier and P. Molnar, J. Geophys. Res., 82, 2905, 1977.
28. E.H. Francis, this volume.
29. P. Molnar and P. Tapponier, Geology, 5, 212, 1977.
30. K.C. Burke and J.T. Wilson, Sci. Am., 235, 46, 1976.
31. M. Barazenga and J. Dorman, Seismol. Soc. Am., Bull., 59, 369, 1969.
32. P.J. Grim, Terrestrial Heat Flow Data, World Data Center A, for solid Earth Geophys., Boulder, Colorado (USA), 1976.
33. G.P. Eaton, R.L. Christiansen, H.M. Iyer, A.M. Pitt, D.R. Mabey, H.R. Blank, I. Zietz and M.E. Gettings, Science, 188, 787, 1975.
34. C. Brooks, D.E. James and S.R. Hart, Science, 193, 1086, 1976.
35. W.B. Bryan, F.A. Frey and G. Thompson, Contrib. Min. Petrol., 64, 223, 1977.
36. K.G. Cox, C.J. Hawkesworth, R.K. O'Nions and J.D. Appleton, Contrib. Min. Petrol., 56, 173, 1976.
37. D.H. Green, Phil. Trans. Roy. Soc. Lond. A, 268, 707, 1971.
38. R.W. Kay and P.W. Gast, J. Geol., 81, 653, 1973.
39. S.E. Kesson, Contrib. Min. Petrol., 42, 93, 1973.
40. J.D. Kramers, Earth Planet. Sci. Lett., 34, 419, 1977.
41. F.E. Lloyd and D.K. Bailey, Phys. Chem. Earth, 9, 389, 1975.
42. P. Richard, N. Shimizu and C.J. Allègre, Earth Planet. Sci. Lett., 31, 269, 1976.
43. J.-G. Schilling, Phil. Trans. Roy. Soc. Lond. A, 268, 663, 1971.
44. J.-G. Schilling, Nature Phys. Sci., 242, 2, 1973.
45. S.-S. Sun and G.N. Hanson, Contrib. Min. Petrol., 52, 77, 1975.
46. D.J. DePaolo and G.J. Wasserburg, Geophys. Res. Lett., 3, 249, 1976.
47. E. Bonatti, C.G.A. Harrison, D.E. Fischer, J. Honnorez, J.-G. Schilling, J.J. Stipp and M. Zentilli, J. Geophys. Res., 82, 2457, 1977.
48. P.W. Gast, Geochim. Cosmochim. Acta, 32, 1057, 1968.
49. G.N. Hanson, J. Geol. Soc. Lond., 134, 235, 1977.
50. S.R. Hart, J.-G. Schilling and J.L. Powell, Nature Phys. Sci., 246, 104, 1973.
51. R. Hekinian, Contrib. Min. Petrol., 43, 125, 1974.
52. R. Hekinian and G. Thompson, Contrib. Min. Petrol., 57, 145, 1976.
53. D.R.C. Kempe and J.-G. Schilling, Contrib. Min. Petrol., 44, 101, 1974.

54. R.K. O'Nions, P.J. Hamilton and N.M. Evensen, Earth Planet. Sci. Lett., 34, 13, 1977.
55. R.K. O'Nions and R.J. Pankhurst, Earth Planet. Sci. Lett., 21 13, 1973.
56. R.K. O'Nions and R.J. Pankhurst, J. Petrol., 15, 603, 1974.
57. J.-G. Schilling, Nature, 242, 565, 1973.
58. J.-G. Schilling, Earth Planet. Sci. Lett., 25, 103, 1975.
59. J.-G. Schilling and A. Noe-Nygaard, Earth Planet. Sci. Lett., 24, 1, 1974.
60. S.-S. Sun and B.-m. Jahn, Nature, 255, 527, 1975.
61. S.-S. Sun, M. Tatsumoto and J.-G. Schilling, Science, 190, 143, 1975.
62. W.M. White and W.B. Bryan, Geol. Soc. Am., Bull., 88, 571, 1977.
63. W.M. White, S.R. Hart and J.-G. Schilling, Carnegie Inst. Wash., Year Book, 74, 224, 1975.
64. H. Bougault and R. Hekinian, Earth Planet. Sci. Lett., 24, 249, 1974.
65. W.B. Bryan and J.G. Moore, Geol. Soc. Am., Bull., 88, 556, 1977.
66. W.B. Bryan and G. Thompson, Can. J. Earth Sci., 14, 875, 1977.
67. S.E. Church and M. Tatsumoto, Contrib. Min. Petrol., 53, 253, 1975.
68. G.L. Cumming, Earth Planet. Sci. Lett., 31, 179, 1976.
69. F.A. Frey, W.B. Bryan and G. Thompson, J. Geophys. Res., 79, 5507, 1974.
70. R.H. Hekinian, J.G. Moore and W.B. Bryan, Contrib. Min. Petrol. 58, 83, 1976.
71. R. Kay, N.J. Hubbard and P.W. Gast, J. Geophys. Res., 75, 1585, 1970.
72. R.St.J. Lambert and J.C. Holland, Can. J. Earth Sci., 14, 809, 1977.
73. C.H. Langmuir, J.F. Bender, A.E. Bence, G.N. Hanson and S.R. Taylor, Earth Planet. Sci. Lett., 36, 133, 1977.
74. J.-G. Schilling, J. Geophys. Res., 80, 1459, 1975.
75. J.-G. Schilling, Earth Planet. Sci. Lett., 25, 93, 1975.
76. S.-S. Sun and G.N. Hanson, Geology, 3, 297, 1975.
77. M.J. O'Hara, Nature, 266, 503, 1977.
78. W.J. Morgan, Nature, 230, 42, 1971.
79. W.J. Morgan, Am. Assoc. Pet. Geol., Bull., 56, 203, 1972.
80. A.L. Boettcher, J.R. O'Neil, K.E. Windom, D.C. Stewart and H.G. Wilshire, Sec. Int. Kimberlite Conf. Abstr., 1977 (Abstr.).
81. A.J. Erlank and R.S. Rickard, Sec. Int. Kimberlite Conf. Abstr., 1977 (Abstr.).
82. R. Kay, Eos, 56, 1077, 1975 (Abstr.).
83. A.W. Hofmann and M. Magaritz, J. Geophys. Res., 82, 5432, 1977.
84. B. Sundvoll, Proceedings I, 1978 x.
85. B.O. Mysen and A.L. Boettcher, J. Petrol., 16, 549, 1975.
86. E.-R. Neumann, Proceedings I, 1978 x.
87. W.S. Baldridge, Proceedings I, 1978 x.
88. I.B. Ramberg, Norg. Geol. Unders., 325, 1976.

89. R.C.O. Gill, Nature Phys. Sci., 242, 41, 1973.
90. S.J. Lippard and P.H. Truckle, Proceedings I, 1978 [x].
91. T.V. Segalstad, Proceedings I, 1978 [x].
92. S.P. Jakobson, Lithos, 5, 365, 1972.
93. J.-G. Schilling, R.N. Anderson and P. Vogt, Nature, 261, 108, 1976.

SUBJECT INDEX

GPSR Compliance

*The European Union's (EU) General Product Safety Regulation (GPSR) is a set of rules that requires consumer products to be safe and our obligations to ensure this.*

*If you have any concerns about our products, you can contact us on ProductSafety@springernature.com*

In case Publisher is established outside the EU, the EU authorized representative is:

Springer Nature Customer Service Center GmbH
Europaplatz 3
69115 Heidelberg, Germany

**Batch number: 10761599**

Printed by Printforce, the Netherlands